DRUG TESTING *in* HAIR

DRUG TESTING
in HAIR

Edited by
PASCAL KINTZ

CRC Press
Boca Raton   New York   London   Tokyo

Publisher:	Robert B. Stern
Acquiring Editor:	Joel Claypool
Project Editor:	Joan Moscrop
Marketing Manager:	Susie Carlisle
Direct Marketing Manager:	William J. Boone, Jr.
Cover design:	Dawn Boyd
PrePress:	Carlos A. Esser
Manufacturing:	Sheri Schwartz

Library of Congress Cataloging-in-Publication Data

Drug testing in hair / edited by Pascal Kintz.
     p. cm.
     Includes bibliographical references and index.
     ISBN 0-8493-8112-6 (alk. paper)
     1. Hair--Analysis. 2. Drugs--Analysis. 3. Chemistry, Forensic.
I. Kintz, Pascal.
RB47.5.D78 1996
614'.1--dc20

               95-47255
               CIP

© 1996 by CRC Press, Inc.

No claim to original U.S. Government works
International Standard Book Number 0-8493-8112-6
Library of Congress Card Number 95-47255
Printed in the United States of America 1 2 3 4 5 6 7 8 9 0
Printed on acid-free paper

PREFACE

Morphological, serological, and chemical examination of human hair for forensic and medical purposes was initiated several years ago. A single human hair is sometimes the only remnant at the scene of a crime. In many cases, it serves to confirm or exclude a possible suspect.

In the 1960s and 1970s, hair analysis was used to evaluate exposure to toxic heavy metals, such as arsenic, lead, or mercury. Simultaneously in the U.S. and West Germany, it was possible to denominate various organic drugs by means of radioimmunoassay (RIA) at the beginning of the 1980s. Generally, the courts of justice only recognize the results of chemical-toxicological analyses when they are confirmed by a second independent method. For this reason, gas chromatography coupled with mass spectrometry (GC/MS) is now the method of choice for hair analysis.

Technically, testing of hair for drugs is no more difficult or challenging than testing in many other "alternative" matrices (for example, liver, bone, etc.). In fact, the application of analytical methods and instrumental approaches are in most cases quite similar, regardless of the initial matrix. At present, hair analysis is routinely used as a tool for detection of drug use in forensic science, traffic medicine, occupational medicine, and clinical toxicology.

Special thanks must go to all the international authors that have agreed to give a chapter to what, I hope, is a worthwhile contribution. It was my intention to cover both theoretical and practical aspects of hair analysis in order to provide the state of the art in this rapidly growing area. As it will be seen by the readers, various opinions, sometimes controversial or contradictory, have emerged from the different authors. I find it helpful to define what the areas of agreement are among the active investigators and what the issues are that require further efforts to get a consensus.

Pascal Kintz, Pharm.D., Ph.D.

THE EDITOR

Pascal Kintz, born in Strasbourg, France, received his doctorate in pharmacy in 1985 from the University of Strasbourg. In 1989 he was awarded his Ph.D. in toxicology from the University Louis Pasteur, Strasbourg.

Since 1990 he has been an associate professor of toxicology and head of the toxicology laboratory at the Institut de Médecine Légale et de Médecine Sociale in Strasbourg. He is chairman of the hair testing committee of the French Society of Toxicology, and in 1995, he was the guest editor of a special issue on hair analysis in *Forensic Science International*. He has published numerous articles in worldwide journals.

Dr. Kintz' interests include:

- diagnostic application of hair, sweat, and saliva analyses to drug abuse,
- determination of the cause of death,
- poisoning with plants (mushrooms, colchicine, digoxin, atropine, etc.),
- development of new analytical methods (SFE, LC/MS, etc.), and
- drug metabolism.

CONTRIBUTORS

Werner A. Baumgartner, Ph.D.
Psychemedics Corporation,
Culver City, California and
Radioimmunoassay Laboratory
Nuclear Medicine Service
West Los Angeles Veterans Administration
  Medical Center
Los Angeles, California

David L. Blank, Ph.D.
Bureau of Naval Personnel
Navy Drug and Alcohol Program
Washington, D.C.

Vincent Cirimele, M.S.
Institut de Médecine Légale
Strasbourg, France

Edward J. Cone, Ph.D.
National Institute on Drug Abuse
Baltimore, Maryland

Patrick Edder, Ph.D.
Analytical Pharmaceutical Chemistry
University of Geneva
Geneva, Switzerland

Hans P. Eser, Dipl.-Chem.
Institute of Legal Medicine
University of Saarland
Homburg/Saar, Germany

Diana Garside, Ph.D.
University of Florida
College of Medicine
Gainesville, Florida

Bruce A. Goldberger, Ph.D.
University of Florida
College of Medicine
Gainesville, Florida

Virginia A. Hill, B.S.
Radioimmunoassay Laboratory
Nuclear Medicine Service
West Los Angeles Veterans Administration
  Medical Center
Los Angeles, California

Marilyn A. Huestis, Ph.D.
National Institute on Drug Abuse
Baltimore, Maryland

Robert E. Joseph, Jr., B.S.
National Institute on Drug Abuse
Baltimore, Maryland

David A. Kidwell, Ph.D.
Chemistry Division
Naval Research Laboratory
Washington, D.C.

Pascal Kintz, Pharm.D., Ph.D.
Institut de Médecine Légale
Strasbourg, France

Patrice Mangin, M.D., Ph.D.
Institut de Médecine Légale
Strasbourg, France

Manfred R. Moeller, Ph.D.
Institute of Legal Medicine
University of Saarland
Homburg/Saar, Germany

Hans Sachs, Ph.D.
Institute of Legal Medicine
Munich, Germany

Christian Staub, Ph.D.
Institute of Forensic Medicine
University of Geneva
Geneva, Switzerland

Irving Sunshine, Ph.D.
Institute of Pathology
Case Western Reserve University
Cleveland, Ohio

Antoine Tracqui, M.D., Ph.D.
Institut de Médecine Légale
Strasbourg, France

Jean-Luc Veuthey, Prof.
Analytical Pharmaceutical Chemistry
University of Geneva
Geneva, Switzerland

TABLE OF CONTENTS

INTRODUCTION

Irving Sunshine

Is drug analysis alive and well? Yes? No? Well, it depends — for some specimens yes, for some no, and for still others, maybe.

In the 1970s analytical toxicology matured. Many novel applications of the rapidly growing, innovative analytical techniques were applied to biological specimens. Chromatographic and immunological procedures utilizing these instrumental advances permitted detection and quantitation of many analytes which were previously considered esoteric and too difficult for analysis by many toxicologists. The classical Stas Otto procedure, with its demand for 500 g of tissue or 50 mL of blood, became antiquated. As the requirements for sample size decreased, analysis of most biological specimens became viable. Armed with new powerful tools, toxicologists explored new analytical worlds with awe and achieved remarkable results. Among those specimens that previously had been disregarded was hair. The potential of this specimen could now be evaluated because the required sensitivity and specificity were available. Although only one casual mention of its potential had been reported previously,[1] many curious investigators began to explore the potential of hair analysis.

One group of researchers recognized the ability of atomic absorption to detect trace metals. They hypothesized that hair was a preferable specimen to blood and urine for evaluating environmental hazards.[2] They proposed that hair could store substances for an extended time period, hence analyses of whole or segmented hairs might provide objective data on the extent of the individual's exposure and its potential for harm. Lead, arsenic, mercury, and cadmium were among the first to be studied, with varying degrees of success.

The potential of hair analysis also interested some physicians who postulated that there was a direct relationship between the trace metals concentration in people's hair and their intelligence and learning ability. Others expanded this concept to encompass mental illnesses and mental disorders.[3,4] By the end of the 1970s, sufficient experience was obtained that warranted discarding this practice despite the very extensive, but unsuccessful efforts to justify it.

Probably the most popular and, as it proved to be, the most contentious use of hair analysis was its application to the evaluation of nutritional deficiencies.[5,6] Many food faddists, pharmacists, nutritionists, and some physicians were convinced that a person's well being was influenced by the concentration of trace metals in his hair. The demands for hair analysis became very large. Many laboratories, some with dubious competence, were established to satisfy what seemed to be a limitless, ever growing service demand. With the passage of time, however, proponents of this

0-8493-8112-6/96/$0.00+$.50
© 1996 by CRC Press, Inc.

technology became disenchanted. This change of attitude is best illustrated by Hambridge's experience.[5-7] The title of his 1982 article,[7] "Hair Analysis: Worthless for Vitamins, Limited for Minerals," reflects his change of thought. The coup de grâce to this practice of analyzing hair for trace metals was administered by Barett's article, "Commercial Hair Analysis: Science or Scam?"[8] Despite some minor flaws, most of his objections to hair analysis were valid. They sounded the death knell of this application.

As the tableau just described was disappearing, another scene was being staged in the wings. Immunoassays for the detection of drugs appeared on the scene. Applications were developed for the analysis of blood and urine for the presence of drugs. Dissatisfied with their experience with trace metal analysis of hair, Baumgartner et al.[9] decided to apply immunoassays to the detection of morphine in hair. Their 1979 seminal paper was supplemented by their further efforts and those of other pioneers in this field, Smith and Pomposini,[10] Arnold and Puschel,[11] Valente et al.,[12] and Suzuki et al.,[13] which confirmed the value of analyzing hair to ascertain whether or not it contained any drug of concern. These pioneering efforts led to worldwide interest and activity by many competent and creative scientists who, to date, have generated a bibliography of over 300 peer-reviewed articles on the analysis of hair for its drug content, the interpretation of these results, and the related pharmacokinetics and pharmacodynamics.[14] Surprisingly, this flood of articles has not led to a consensus. Many unanswered questions remain. It is hoped that renewed efforts will be made to resolve the existing dilemmas.

At this writing, there is reasonable agreement that the qualitative results are valid, but the interpretation of both these results and the quantitative data is still debatable. Some of the reasons for this discordance are concerns about a suitable preparation of the specimens for analysis, the lack of acceptable standards for reference material, the ability to distinguish endogenous from exogenous drug content, the purported bias due to hair color or ethnic origin of specimens, the value and validity of segmental analysis, the dose vs. concentration relationship, acceptable cutoff values, and the lack of accreditation of laboratories. These concerns are generic to any biochemical assay and are not peculiar to hair analysis. Time and continued studies will resolve them. However, despite these concerns, the prevailing opinion held by experienced scientists in this field indicates that they think hair analysis has sufficient scientific validity to warrant its clinical use, but that those who use hair analysis for other purposes should be aware of its limitations. Those readers who desire a more detailed historical perspective should consider reading the excellent review article by Mieczkowski.[15] A current assessment of the present status of hair analysis is presented in the material that follows.

REFERENCES

1. R.W. Goldblum, L.R. Goldbaum, and W.N. Piper. Barbiturate concentrations in the skin and hair of guinea pigs. *J. Invest. Dem.* 22:121–128, 1954.
2. A. Chatt et al. Scalp hair as a monitor of community exposure to environmental pollutants. *Hair, Trace Elements & Human Illness.* A.C. Brown and R. Crouse, Editors. Praeger Publishers, New York, 1980, pp. 46–73.
3. R.O. Pehl and M. Parkes. Hair analysis on learning and behaviour problems. *Hair, Trace Elements & Human Illness.* A.C. Brown and R. Crouse, Editors. Praeger Publishers, New York, 1980, pp. 128–143.

4. P.J. Barlow and M. Kepel. Metal and sulfur content of hair in relation to certain mental status. *Hair, Trace Elements & Human Illness.* A.C. Brown and R. Crouse, Editors. Praeger Publishers, New York, 1980, pp. 105–127.

5. K.M. Hambridge, D.O. Rodgerson, and D. O'Brien. Concentration of chromium in hair of normal children and children with juvenile diabetes mellitus. *Diabetes* 15:517–519, 1968.

6. K.M. Hambridge, P.A. Walrarens, and R.M. Brown. Zinc nutrition of preschool children to the Denver Head Start Program. *Am. J. Clin. Nutr.* 29:734–738, 1976.

7. K.M. Hambridge. Hair analysis: worthless for vitamins, limited for minerals. *Am. J. Clin. Nutr.* 36:943–948, 1982.

8. S. Barrett. Commercial hair analysis: science or scam? *JAMA* 254:1041–1045, 1985.

9. A.M. Baumgartner, P.F. Jones, W.A. Baumgartner, and C.T. Black. Radioimmunoassay of hair for determining opiate-abuse histories. *J. Nucl. Med.* 20:748–752, 1979.

10. F.P. Smith and D.A. Pomposini. Detection of phenobarbital in bloodstains, semen, seminal stains, saliva, saliva stains, perspiration stains, and hair. *J. Forensic Sci.* 26:582–586, 1981.

11. W. Arnold and K. Puschel. Experimental studies on hair as an indicator of past or present drug use. *J. Forensic Sci. Soc.* 21:83, 1981.

12. D. Valente, M. Cassini, M. Pigliapochi, and G. Vansetti. Hair as the sample in assessing morphine and cocaine addiction. *Clin. Chem.* 27:1952–1953, 1981.

13. O. Suzuki, H. Hattori, and M. Asano. Detection of methamphetamine and amphetamine in a single human hair by gas chromatography/chemical ionization mass spectrometry. *J. Forensic Sci.* 29:611–617, 1984.

14. C. Walls. Drug testing in hair: a selective review in SOFT conference on hair, 1994. Published by Society of Forensic Toxicologists, Mesa, AZ.

15. T. Mieczkowski. New approaches in drug testing: a review of drug analysis. *Ann. Am. Acad. Polit. Soc. Sci.* 521:132–150, 1992.

Chapter **1**

TECHNICAL AND LEGAL ASPECTS OF DRUGS OF ABUSE TESTING IN HAIR

Marilyn A. Huestis

CONTENTS

I. INTRODUCTION

Analysis of human hair is purported to provide evidence of the use or lack of use of drugs of abuse. According to some investigators, the extent and timing of drug exposure may also be estimated. Hair test results for drugs of abuse have recently been entered more frequently as evidence in the U.S. courts and have generated much interest and controversy in the forensic toxicology community. Toxicologists agree that no single technique or specimen can provide answers to all toxicological questions. More information is obtained with a variety of analytical approaches and different biological specimens. Hair analysis offers a unique perspective on human drug use by providing a wider window of drug detection, and may offer advantages over other drug testing methods in differentiating the source of some drug exposures, i.e., heroin vs. poppy seed exposure in positive opiate cases. It also may be more difficult to evade drug detection in a hair test, as compared to a urine test, due to an inability to dilute the sample by ingesting large quantities of liquids.

In forensic toxicology, the strengths and weaknesses of each method must be characterized to determine the scientific and legal defensibility of test results. Although hair test methods utilize well-defined analytical techniques, this fact alone does not imply forensic acceptability. Method validation requires an objective demonstration of the accuracy, precision, sensitivity, and specificity of an assay. Toxicologists continue to disagree on whether or not some of these fundamental factors have been adequately evaluated. Another factor is the lack of proficiency testing and inspection programs to objectively appraise assay and laboratory performance. An even more difficult and challenging task may be the resolution of questions that impact the interpretation of hair test results, including minimum detectable concentrations for drug exposure, the time course of appearance of drugs in hair, the presence or absence of dose-response relationships, the stability of drugs in hair and effects of hair treatments, the appropriate analytical cutoff concentrations, the possible racial or sex biases in drug detection, and the recognition of contamination of hair from external sources of exposure. The results of tests for drugs in hair have been submitted as evidence in recent cases in the American courts. These cases illustrate the potential value of hair drug test results, and also some of the objections to the use of hair test programs. Forensic toxicologists must be prepared to critically examine and employ this new source of information regarding human drug use.

II. COMMON LAW SYSTEM AND ADMISSIBILITY OF EVIDENCE

The common law system of the U.S. is based mainly on court decisions and legal precedents, rather than on legislative acts. Judges are given the power to make law through their decisions and opinions. In addition, the U.S. Constitution serves to protect the rights and liberties of American citizens and serves as the final arbitrator on this type of issue. An individual is presumed to be innocent until proven guilty beyond a reasonable doubt in matters of public law and innocent unless otherwise proven guilty by the preponderance of evidence in matters of civil law. But just as we confront disparate interpretations of scientific data, conflicting interpretations of the laws and of the meaning of the Constitution are apparent.

Until recently, the standard of admissibility of scientific evidence into American courts was based on a decision put forth in 1923 in *U.S. v Frye* which required that a scientific principle or discovery, the technique used for applying the scientific principle, and the specific application on which the expert testimony is to be based to be sufficiently established to have gained general acceptance.[1] This decision of the Court of Appeals for the District of Columbia ruled on the admissibility of evidence from a systolic blood pressure deception test, a precursor to the polygraph machine or lie detector test. The Frye opinion stated that it was difficult to determine when a scientific principle or discovery crossed the line between the experimental and demonstrable stages. "Somewhere in this twilight zone the evidential force of the principle must be recognized, and while courts will go a long way in admitting expert testimony deduced from a well-recognized scientific principle or discovery, the thing from which the deduction is made must be sufficiently established to have gained general acceptance in the particular field in which it belongs."[1] The Federal Rules of Evidence were adopted in 1975 and aimed to balance the relevance, reliability, and helpfulness of evidence against the likelihood of waste of time, confusion, and prejudice.[2] Rule 702 Testimony by Experts in the Federal Rules of Evidence is more liberal than the Frye standard and states that "if scientific, technical, or other specialized knowledge will assist the trier of fact to understand the evidence or to determine a fact in issue, a witness qualified as an expert by knowledge, skill, experience, training or education, may testify thereto in the form of an opinion or otherwise." Despite the enactment of the new rules, the Frye standard remained the predominant determining factor on the admissibility of evidence until the U.S. Supreme Court's decision in *Daubert v Merrell Dow* in 1993.[3] In this case, pregnant women who had been prescribed Bendectin®, an anti-nausea medication, contended that the drug may have produced birth defects in their children. None of the company's scientific studies demonstrated a cause and effect relationship between birth defects and drug exposure. The plaintiff's scientific data were not admitted under the Frye rule due to the criteria for general acceptance. The U.S. Supreme Court ruled that the Federal Rules of Evidence supersede the Frye standard in determining the validity of scientific evidence. "General acceptance" is not required for admissibility. The evidence must be grounded in the method and procedures of science and must be relevant to the case at issue. Furthermore, test results from any method submitted as evidence in court will be subject to review of the specimen collection, handling, and testing procedures to evaluate data reliability. It is clear that in the future judges will have more latitude in deciding what evidence to hear. The relaxation of the requirements for admissibility of evidence will most likely result in an increase in the submission of hair test results in the courts and an increased debate among expert toxicology witnesses in the interpretation of these data.

III. DRUG HAIR TEST RESULTS IN THE AMERICAN COURTS

A. Criminal Cases

The first time that hair drug test data were submitted to an American court proceeding was in May 1982 in the *State of Alaska v Richard Gene Majdic*.[4] The court admitted into evidence radioimmunoassay (RIA) test results demonstrating the presence of cocaine in the hair, perspiration, and menstrual stain of an alleged sexual assault victim. These findings were used to implicate cocaine use by the victim, a claim she denied in testimony, but later admitted, according to Dr. F. P. Smith, an expert witness. No confirmatory analyses were performed. Challenges to the testimony included questions as to the specificity of the RIA for cocaine metabolites and the possibility of external drug contamination. The calibrators used to determine the concentration of drug in the sample were not prepared in hair, which may have introduced a matrix effect in the test results.

In 1985 in *People v Miel*, Martin Miel was charged and convicted of first-degree murder.[5] The defense introduced hair test results to indicate that a key prosecution witness had lied when he said he was not using drugs at the time of the crime. Also in 1985, Robert Korner was convicted of robbery in *People v Korner*; however, the defense had submitted hair test results from the accused to prove that he was under the influence of drugs at the time of the crime and therefore could not have formed criminal intent.[6] In these three cases, positive hair test results were used to demonstrate a prior history of drug use.

In 1987 in *State of New Jersey v Samuel L. Davis*, the defense was able to obtain a court order forbidding a sexual assault victim from cutting her hair until hair testing could be performed.[7] The law clerk claimed that she had been raped by an attorney after he spiked her drink with cocaine. In this case, the finding of no cocaine in the hair would be used to refute the woman's testimony. Minimum detectable drug concentrations in hair have not been clearly established to date; a single drug exposure may not be sufficient to produce a positive drug hair test.

In 1990 in *U.S. v Riley*,[8] the district court suppressed evidence seized pursuant to a search warrant. The U.S. Appeal Court reversed the lower court decision and permitted collection and testing of hair for drugs of abuse. Also in 1990, in *U.S. v Foote*,[9] the appeal court affirmed the judgment of a district court that denied a motion to require an undercover detective to submit to a hair drug test. The intrusive nature of the "experimental RIA analysis" and the lack of evidence of drug use were the basis for the denial. Hair testing was also denied by the court in *People v Thurman* in 1990.[10] The defense council had requested the collection and testing of fingernail and hair samples from a police informant to document cocaine use. In 1992 in *Maull v Warren*,[11] a Delaware State Trooper appealed his suspension on drug use charges. He had consented to blood and hair tests for drugs, both of which were found to be negative. These test results were admitted into evidence at the hearing; however, it was noted that during a search of the officer's home, hypodermic needles and syringes and anabolic steroids were found. In 1993 in *Colorado v Allen Thomas*,[12] hair testing data were not admitted. The defense counsel wanted to document diminished capacity of the homicide defendant with positive hair test results.

B. Military Court-Martials

In 1987 a marine stationed at Camp Pendleton, CA was acquitted of the wrongful use of cocaine in a military court-martial, *U.S. v LCpl Steven M. Piccolo*.[13] A positive urinalysis for cocaine metabolites had been obtained; a negative hair analysis was submitted to refute this information. The military judge refused to admit into evidence the negative hair test results.

In two additional court-martial cases in 1989, *U.S. v RM1 Brian J. Dunn*[14] and *U.S. v TM1 Gary D. Noble*,[15] the defense attempted to submit into evidence negative hair test results from head and pubic hair specimens to refute positive urinalysis results for cocaine. In both cases, the defense attempted to claim inadvertent drug exposure via passive inhalation, subterfuge, or contamination of the urine specimen, rather than wrongful use of cocaine. The positive urinalysis results were claimed to be "evidentiary false positives," that is, the drug may have been present in the specimen; however, the individual did not knowingly use the drug. The final disposition of the charges is unknown, as is the admission or nonadmission of the evidence.

In a 1993 U.S. Air Force case, *U.S. v Jenkins*,[16] a military judge admitted mass spectrometry/mass spectrometry (MS/MS) hair test data documenting the presence of cocaine in pubic hair. A positive urine cocaine test was appealed in *U.S. v Nimmer*[17] in 1994. The court refused to admit negative hair test data due to the inability of a hair test to detect single drug use. In general, hair test data have been judged to be variable and unreliable in military court-martials and data have usually not been admitted when urine results are positive and hair results are negative. Hair test data have been admitted more frequently when used to substantiate other positive test results.

C. Child Custody and Adoption Cases

In 1988 in *Burgle v Burgle*, NY Superior Court Appellate Division, the court ordered that hair be collected and tested for cocaine to determine the fitness of the custodial parent in a child custody dispute.[18] Furthermore, the mother was ordered not to cut or chemically treat her hair. This decision was appealed and affirmed. The father claimed in the pending divorce action that the mother had engaged in a pattern of habitual and uncontrollable cocaine abuse. The mother admitted to occasional social drug use with the father, an admission considered to be relevant to the court's decision due to her placing her physical and mental health at issue. The court determined that the novelty of the hair drug testing procedure was relevant to result admissibility not to the discovery process and that a preliminary hearing on the reliability and validity of hair analysis was unnecessary.

In 1990 in *Garvin v Garvin*, the court ordered the mother and a third party to submit hair samples for testing for marijuana use.[19] The higher court reversed the decision due to the unreliability of the hair test in detection of marijuana use and the fact that only marijuana use was suspected.

In 1993 in *Adoption of Baby Boy L.*, the biological mother revoked her adoption consent.[20] The court ordered hair testing by RIA and confirmation by gas chromatography/mass spectrometry (GC/MS), to evaluate the biological mother's and father's cocaine use in order to determine the best interests of the child.

D. Probation Revocation Proceedings

In 1990 in an important probation revocation case, Judge Weinstein supported the admissibility of RIA hair testing results in *U.S. v Anthony Medina*.[21] Judge Weinstein, considered to be an expert on court evidence, determined that RIA results were sufficiently reliable and had attained sufficient acceptance in the scientific community to be admissible as novel scientific evidence, if performed in a careful and accurate manner on an appropriate sample. The defendant had pleaded guilty in 1988 to violation of the narcotic laws; his 10-year sentence was suspended and he was given five years probation. A year later, a positive urinalysis test for cocaine was obtained and the defendant admitted drug use. The court later ordered a hair test for drugs which resulted in a positive finding for cocaine, and hence, a violation of the probation order.

In 1994 in *U.S. v Neely*, positive hair tests provided by the FBI were not admitted in a probationary hearing.[22] The court-directed urinalysis program had detected no drug use for 60 days. However, the probation period was extended and the number of urine tests performed per week were increased by the court.

E. Unemployment Compensation Cases

In a recent unemployment compensation case, *In the Matter of Lloyd A. England II [L. & L. Fittings]*, an Administrative law judge in Indiana determined that an individual had been discharged with just cause following a positive cocaine hair test.[23] In 1993, the claimant was temporarily hired and required to take a hair analysis drug test. The initial test was positive for cocaine; however, he was permitted to continue working as a material handler with the requirement that future hair tests would indicate no further drug use. The claimant did not attend two scheduled hair tests over the next four months. He was informed that if he missed the next test he would be terminated. A month later the next hair test was positive for cocaine and he was discharged. He was found to have knowingly violated a reasonable and uniformly enforced employer rule.

Also in 1993, in *Holmes v Hotel San Remo*, the Appeals Tribunal State of Nevada Employment Security Department found that the claimant was ineligible for benefits due to discharge for misconduct violation of a known and reasonable rule.[24] Employees were notified that the employer would comply with the Drug Free Workplace Act and random hair analysis testing would occur following a 90-day grace period. Holmes' hair test indicated recent cocaine use. She was found to be ineligible for benefits.

F. Preemployment Testing

Employees of Harrah's Club filed a complaint for the issuance of a preliminary and permanent injunction against the employer's substance abuse policy. They claimed that the policy was unlawful and constituted an invasion of their rights of privacy. In 1990 the 9th District Court of Nevada determined in *Koch v Harrah's* that the employer's substance abuse policy statement was valid, reasonable, fair, and lawful and did not violate any constitutionally protected right.[25] The policy required employees' hair to be tested by RIA and GC/MS for the presence of drugs. If the employee tested positive and did not admit drug use, the employee would be required to participate in an unannounced urine testing program for 60 days. Any positive urine test would result in termination for violation of company policy. The court also determined that an RIA screening test was insufficient when used alone to terminate employees. In addition, Harrah's stipulated that employees henceforth would be given the option of providing a hair specimen or participating in an unannounced urine drug testing program over a 60-day period to fulfill the drug testing requirement.

G. Exclusion of Evidence

In 1993 in *Bass v Florida Department of Law Enforcement*, it was determined that a hearing officer had erred in not permitting the appellant's expert witness to testify about the results of a negative hair test for the accused.[26] The appellant had been hired as a corrections officer in 1982 and maintained a good employment record for more than 10 years. During a biannual physical she provided a urine sample that tested positive for cocaine, and was terminated from employment. Subsequently, she provided a hair sample for drug analysis that was negative for all drugs. She attempted to introduce the hair analysis evidence and testimony from an expert to explain how a false positive reading could have been obtained on the urinalysis result. The court ruled that this testimony should have been admissible to explain and rebut the urinalysis scientific evidence. The decision was reversed and remanded for further proceedings.

IV. ANALYTICAL METHODS

All forensic data, including hair test data, are subject to a critical peer review evaluation of their acceptability. Some important factors for evaluation include carefully designed and characterized analytical methods, complete chain of custody documentation, comprehensive records, experienced and knowledgeable personnel, evaluation of the adequacy of test reagents, appropriate quality control/quality assurance protocols, consistent reporting criteria for positive/negative results, and well-maintained instrumentation. The sensitivity, specificity, accuracy, precision, linearity, and potential for carryover should be established for each assay. All aspects of the method should be evaluated for possible cross contamination between samples or specimens. Calibrators and controls should be prepared in the same matrix as the tested sample, e.g., hair specimens or extracts. The preparation of homogenous and stable quality control materials has also been a challenge in the development of hair assays. Procedures should include quality control samples designed to monitor each step of the procedure, including washing procedures, immunoassay, digestion, extraction, and chromatography. The assay should be monitored for both within- and between-batch performance. The criteria for a positive hair test result should include positive results in two fundamentally distinct analytical techniques, e.g., positive results in a highly sensitive initial immunoassay test such as RIA, fluorescent polarization immunoassay (FPIA), enzyme immunoassay (EIA), and confirmation of findings in a highly specific chromatographic test, e.g., GC/MS, GC/MS/MS, MS/MS, or high-performance liquid chromatography (HPLC). Preferably the initial and confirmation tests should be performed on two separate aliquots of the specimen. Validation of forensic toxicology procedures requires adherence to these fundamental requirements. Guidelines for forensic testing have been published by the Department of Health and Human Services and the College of American Pathologists for forensic urine drug testing, and by the Society of Forensic Toxicologists and the Toxicology Section of the American Academy of Forensic Sciences Laboratory Guidelines Committee for forensic toxicology testing.

V. INTERPRETATION OF HAIR TEST DATA

A more difficult task is the interpretation of hair test data. The mechanisms of incorporation of drug into hair continue to be debated. Some toxicologists argue legitimately that this information may not be available for other types of biological samples; however, the potential for incorporation of drug from external sources has been demonstrated. It is apparent that drug may be deposited in hair by multiple routes including from the bloodstream, sweat and sebaceous glands, and from external exposure. Some investigators indicate that exposure from exogenous and endogenous sources can be distinguished through specific washing procedures and the use of mathematical algorithms. What is the magnitude of interindividual differences in drug deposition in hair? Is there a potential bias in the identification of drug users due to differences in drug incorporation rates based on racial characteristics, sex, or hair type? Is this a critical issue, in light of the wide interindividual differences in drug metabolism and excretion, and variability in observed drug effects in controlled clinical studies of drugs in humans?

Another difficult interpretation issue is the use of negative hair test results to refute positive drug test results in other biological fluids. The minimum detectable concentrations of drugs in hair have not been well established. This also is true for other biological fluid testing methods; however, the different windows of drug detection and wide interindividual variability of metabolism and excretion are utilized to interpret, rather than refute other test data. A positive urinalysis test and a negative hair test may indicate that the individual did not use drugs chronically, that the amount of drug exposure was inadequate to produce a positive hair test result, that the hair specimen was taken at the same time as the urine specimen and was not within the window of detection for the hair specimen, or that there was contamination of the urine specimen. Other scenarios are possible. In the same manner, there are many ways to interpret a positive hair test and a negative urine drug test. With our current state of knowledge it is dangerous to use negative hair test results to unilaterally state that drug exposure did not occur, as was done in the cited case of the law clerk whose drink was allegedly spiked with cocaine, or that the individual was unaware of drug exposure, as was attempted in the cited court-martial cases.

The time course of appearance and disappearance of drugs in hair and determination of the dose-response relationship between drug use and deposition in hair are important factors in the interpretation of data when hair test results attempt to describe patterns of drug use over time. It is also unclear whether or not drug may migrate within the hair shaft after initial deposition of the drug analyte. Do drug concentrations in hair decrease over time with normal hygienic practices? Does drug deposit from sweat or the environment down the length of the hair shaft? Are the answers to these questions dependent upon the drug class or the sex or race of the individual? How does the type of hair, whether coarse, fine, pigmented, or chemically treated, affect these questions? Can appropriate cutoff concentrations for drugs in hair be defined to circumvent some of this variability?

One of the major areas of disagreement that significantly impacts the forensic acceptability of hair testing results relates to the potential for false positive test results due to external contamination of hair with drug from the environment. Similar troubling scenarios have been developed for the possibility of positive urine tests following passive inhalation of marijuana smoke. Following heroin use, heroin and 6-acetylmorphine are the predominant analytes detected in hair, while morphine is detected in the urine. Hair analysis therefore appears to be a superior method for the detection of heroin use, although the possibility of externally deposited heroin and the stability of heroin in the hair remain unanswered questions. Systems have been developed to deal with some of these urine drug test interpretation issues, including use of assay cutoff concentrations, review of an individual's clinical history, presence of needle track marks, measurement of 6-acetylmorphine, etc. Can similar systems be developed to identify the possibility of external drug contamination in hair?

The external drug contamination issue has been addressed through the addition of washing regimes, measurement of drug concentrations in the wash solutions, derivation of mathematical algorithms to identify possible contamination, measurement of drug metabolites, and parent drug to metabolite ratios. To date, research in this area has produced much conflicting data. Resolution of the disparate drug contamination data may be accomplished by the opportunity for peer review of analytical methods and positive drug identification criteria, reproduction of decontamination test results in other laboratories, and laboratory identification of blind quality control samples that have been externally contaminated with drug. In addition, the degree of external drug exposure should be realistic. Extreme conditions of marijuana smoke exposure may lead to the production of positive urine drug test results due to passive inhalation of drug. Resolution of the external drug contamination issue is critical to the forensic acceptability of hair test results, especially when these results would stand alone as a single piece of evidence, e.g., for preemployment testing.

VI. SOCIETY OF FORENSIC TOXICOLOGISTS (SOFT) CONSENSUS STATEMENTS

In 1990 the Society of Forensic Toxicologists issued a consensus opinion summarizing the current applicability of hair analysis for testing for drugs of abuse. It was determined that the use of hair analysis for workplace drug testing was premature and interpretation of test results was insupportable with current information. Hair was judged to be a useful specimen in forensic investigations when supported by other evidence of drug use. No generally accepted procedures for hair analysis were available and the accuracy, precision, sensitivity, specificity, and cutoff levels defining positive and negative test results were not established. The reporting of a positive result based upon a single or replicate immunoassay was judged to be unacceptable. Reference materials were unavailable to standardize analytical methods, and peer review of data had not always been permitted. Many important questions on the mechanisms of drug incorporation into hair, minimum detection levels, drug stability, and other questions requiring extensive research efforts were enumerated.

A revised consensus opinion on the applicability of hair analysis for drugs of abuse was issued in 1992. Due to the lack of information on key issues necessary to interpret hair test results the use of hair analysis for employee and preemployment drug testing was determined to be premature. Hair test results were stated to be useful when data are supported by other competent evidence of drug use and when assays are performed under generally accepted guidelines for forensic drug testing. A thorough method validation must be accomplished and available for peer review. The use of two fundamentally different analytical techniques is necessary prior to the reporting of a positive test result. Important unanswered questions on the incorporation, retention, stability, and analytical detection of drugs in hair were listed.

VII. SUMMARY

Hair test results have been submitted as evidence in child custody cases, military court-martials, adoption and probation revocation proceedings, and unemployment compensation cases in the U.S. courts. Preemployment hair testing has been upheld as a legitimate component of a drug-free work environment plan.

The constitutionality of hair testing has been challenged and will continue to be debated. Does hair testing represent an unlawful invasion of privacy? Certainly the collection of hair specimens is less invasive in some senses than the collection of urine specimens; however, the wider window of detection of drug use may constitute a greater privacy invasion. Drug use may be detected over a much longer time span, and may provide more information on the pattern, frequency, and amount of drug use by an individual. Past court decisions indicate that hair testing may be permissible and may not be judged to be an unlawful invasion of privacy, although it is apparent that future challenges will better define this issue.

There may also be challenges to the admissibility of the scientific evidence. The greatest challenge will be based upon the possibility of external contamination of hair with drug from the environment and the subsequent failure of the test to reliably identify true drug use. However, the use of common drug testing methodologies, the ability of scientists to critically evaluate testing data, and the new rules of evidence admissibility make it is less clear that these challenges will prohibit evidence introduction.

Data interpretation will be difficult. Hair testing has the potential of providing a great deal of information on human drug use. Useful applications include, but are not limited to, monitoring of parole populations, validation of drug self-report data, identification of *in utero* drug exposure, providing information on the pattern and magnitude of drug use, and as a diagnostic aid in drug treatment programs. Controlled clinical studies and additional research techniques, e.g., infrared microscopy, are needed to resolve many of the unanswered questions in order to effectively interpret hair test results. Establishment of minimum detectable concentrations for drug exposure may be necessary prior to hair test results being utilized to refute other evidence of drug use. Determination of the time course of appearance and disappearance of drugs in hair, the stability of drugs in hair, and the effects of hair treatments may be necessary prior to accurately describing the use or lack of use of drugs at specific time points. Resolution of the presence or absence of drug dose-response relationships may be necessary prior to describing the frequency and amount of drug use. The potential bias in identification of drug users due to differences in drug incorporation rates that may be linked to racial or sex characteristics must be evaluated for fairness and equal treatment of all individuals. The major technical area of disagreement that impacts forensic acceptability relates to the potential for false positives due to external contamination with drug from the environment. It is evident that these important issues will have a greater impact on the weight supplied to the hair test evidence, rather than to its admissibility, in future court proceedings.

REFERENCES

1. *U.S. v Frye*, 54 Appellate District Court 46, 293 F 1013, 1923.
2. Federal Rules of Evidence, 403, 28 United States Code, 1975.
3. *Daubert v Merrell Dow*, United States 113 Supreme Court 2786, 125 L.Ed. 2d 469, 1993.
4. *State of Alaska v Richard Gene Majdic*, No. K081-367CR, 3rd Judicial District, Kodiak, 1982.
5. *People v Martin Miel*, No. A 804003, Los Angeles Superior Court, 1985.
6. *People v Robert Korner*, No. 154558, Santa Barbara Superior Court, 1985.
7. *State of New Jersey v Samuel L. Davis*, No. 87050901, New Jersey Superior Court, 1987.
8. *U.S. v Riley*, 906 F 2nd 241 (2nd Cir), 1990.
9. *U.S. v Foote*, 898 F 2nd 659, 669 (8th Cir), 1990.
10. *People v Thurman*, 787 P2nd 646, CO, 1990.
11. *Maull v Warren*, WL 114111 (Del. Super.) 1992.
12. *State of Colorado v Allen Thomas, Jr.*, Adams Co. 91CR190, Denver, CO, 1993.
13. *U.S. v LCpl Steven M. Piccolo*, USMC, Camp Pendeleton, CA, 1987.
14. *U.S. v RM1 Brian J. Dunn*, USN, Agana, Guam, 1990.
15. *U.S. v TM1 Gary D. Noble*, USN, Orlando, FL, 1990.
16. *U.S. v Airman Jenkins*, USAF, Valdosta, GA, 1993.
17. *U.S. v Todd A. Nimmer*, USN, Washington, D.C., 1994.
18. *Burgle v Burgle*, 141 AD 2d 215, 533 New York State 2d 735 2d Dept., 1988, or No. 1651E, 1651 AE, New York Superior Court Appellate Division, 1988.

19. *Garvin v Garvin*, 162 AD2d 487, NY, 1990.
20. *Adoption of Baby Boy L.*, 157 Misc. 2d. 353, 596 New York State 2d 997 New York Family Court, 1993.
21. *U.S. v Anthony Medina*, 749 F. Supp. 59, Federal Court U.S. East District Court of New York, 1990 or No. 87 CR-824-3, U.S. District Court ED, 1990.
22. *U.S. v Neely*, Northern District, Roanoke, VA, 1994.
23. *In the Matter of Lloyd A. England II* [L & L Fittings], Indiana Workforce Development Unemployment Insurance Appeals, Case No. 93-IBA-1108 (January 27, 1994).
24. *Holmes v Hotel San Remo*, Nevada Employment Security Department Office of Appeals, Decision No. V3-1403 (June 10, 1993), affirmed on appeal *In the Matter of Cynthia Holmes* [Hotel San Remo], Nevada Employment Security Department Board of Review, Decision No. BV3-0625 (V3-1403) (August 20, 1993).
25. *Koch v Harrah's*, No. 23740, 9th District Court of Nevada, 1990.
26. *Bass v Florida Department of Law Enforcement*, Case No. 92-2669, Third District Court of Appeal of Florida, 1993 or Lexis 12321; 18 Fla. Law W. D2639.

Chapter 2

ENVIRONMENTAL EXPOSURE — THE STUMBLING BLOCK OF HAIR TESTING

David A. Kidwell and David L. Blank

CONTENTS

I. INTRODUCTION

The Devil is in the details, and there are many details concerning the analysis of hair for drugs of abuse and the interpretation of the analytical result. Several controversies have arisen over the past few years surrounding many of these details. The controversies have focused on two broad areas: (1) the mechanisms for appearance in hair and binding to hair of drugs of abuse and (2) removal of external contamination. The conclusions reached in these two areas and the weight given to the scientific facts greatly color the interpretation of any hair analysis result. This review outlines the historical evidence accumulated in these two areas of research and discusses how the interpretation of these data has led to two disparate models of incorporation and removal of drugs from hair. Also, new data are presented to clarify areas of controversy. An understanding of the data and its interpretation is critical to the proper application of hair analysis results.*

A. Historical Concerns for Passive Exposure in Urinalysis

For more than a decade, the Department of Defense (DoD) has placed increased emphasis on detection and deterrence of drug use, with urinalysis being the cornerstone in this deterrence program.[1] The U.S. Navy has been a strong proponent of urinalysis and has become one of the largest drug screening organizations in the

---

* In this chapter, all concentrations of drugs have been converted to ng of drug/mg of hair as was recommended in the 2nd Interational Conference on Hair Testing held in Genova, Italy, June 1994. Attention must be given to the units of measure of the results presented here when a comparison is made to the previous results of other authors.

world. As the use of urinalysis increased and military personnel became more aware of the disciplinary measures, the self-reported use rate (past 30 days) in the Navy dropped from 33% in 1980 to under 4% in 1995.[2]

Often, the only evidence of drug use is a positive laboratory result. Falsely accusing an individual of drug use could be devastating to the individual and have serious repercussions to the laboratory.[3] Therefore, laboratories must follow reliable, valid, and generally accepted procedures, and the procedures and results must be subjected to internal and external review. Procedures must be employed that attach ultimate importance to avoiding false accusation of drug use. Many of the procedures developed in the DoD laboratories have been adopted for use in the civilian sector. Although these procedures evolved after consideration of the best science available at the time, numerous problems still arose that had to be addressed. Unfortunately, these problems had already affected the lives of some people.

The lessons learned in conducting over 100 million urine tests in the past decade have brought several changes to military urinalysis protocols. Some of the protocol changes addressed procedural issues, such as chain of custody, whereas others addressed passive or accidental exposure to drugs that resulted in falsely accusing individuals of drug use. Because each military drug positive result is thoroughly investigated, occasionally it was found that occurrence of events considered to be improbable were observed.[4] This was due to the testing of a large number of samples associated with several low probability events. Perhaps the best way to illustrate the simultaneous occurrence of low probability events is to describe some examples of actual instances where they occurred.

One example involved the misidentification by a private laboratory in 1990 that resulted in accusing three individuals of methamphetamine use.[4] Both then and now, the accepted protocol was to have each urine sample tested by an immunoassay and positive results confirmed by a gas chromatography/mass spectrometry (GC/MS) analysis. In this case, some individuals metabolized over-the-counter nasal decongestants (ephedrine) to a form detected by the immunoassay (low probability event number one). Fortunately, three people were found positive within a few days of each other and all denied use of methamphetamine (low probability event number two). Investigation of the procedures in these cases uncovered that the GC/MS confirmation was modified from the published protocol in an attempt to shorten the confirmation time (low probability event number three). The changes made to the published GC/MS procedures were so minor that most chemists would agree that no misidentification should have occurred. However, further testing of these three specimens by different methods showed that all were negative for methamphetamine. The procedures have since been modified to avoid this identified problem with methamphetamine analysis.

Example number two concerns false positive urinalysis as a result of poppy seed ingestion.[5] Early in the urinalysis program it was thought that the morphine contained in poppy seeds was present in too small an amount to generate an opiate positive. In fact, testing the urine of several individuals who consumed poppy seeds failed to detect opiates. For reasons still unknown, an increase in the number of opiate positives was observed in 1986. Not only did the individuals deny use, but their stories were credible (low probability event number one). Later it was learned that some types of poppy seeds contain large amounts of opiates, depending upon the country of origin and the method of preparation. For an individual to be positive, these poppy seeds must be eaten in sufficient quantities (low probability event number two) by an individual with a high absorption rate for opiates (low probability event number three) and shortly before a urine test is administered (low probability event number four). As a result of these false identifications, opiate cutoff levels were raised and procedures put in place for a medical review officer to evaluate each positive opiate urinalysis for supporting evidence of opiate use.

A third example was the presence of cocaine in an individual's urine due to accidental ingestion. Inca tea, a tea made from purportedly decocainized cocaine leaves purchased in a health food store, was brewed and ingested by an individual who was later found to be positive for cocaine. If the cocaine had been removed from the leaves as stated, the product would have been perfectly legal and could not have contributed to a positive urinalysis (low probability event number one). Likewise, if the individual had not been tested within a few hours of consuming the tea, the results probably would have been negative[6] (low probability event number two). In this case, a sample of the tea bag was produced, analyzed, and found to contain cocaine in sufficient quantities to cause a positive urinalysis (low probability event number three). Further investigation revealed that Inca tea leaves were ground and dried cocaine leaves and not decocainized. The product was immediately removed from the market.[7,8]

The most frequently cited low probability event, that became apparent as a defense argument for urinalysis, was passive exposure to marijuana smoke. An explanation for a marijuana positive urinalysis was that the service member was exposed to significant quantities of marijuana sidestream smoke in close proximity to the service member.* Because this defense may have had validity, the Navy commissioned subsequent scientific studies to resolve this issue. These studies showed conclusively that environmental exposure would result in a marijuana positive, providing the exposure was severe enough.[9,10] However, in the experiment the exposure needed to be repeated for many hours and at a level such that the subjects needed to wear eye goggles to protect themselves from the marijuana smoke.[11,12] Exposures at these levels could hardly be considered to be inadvertent or unwilling.[12] As a result of these studies, cutoff levels could be defended and the credibility of the passive exposure defense was diminished.

All of these examples share several common characteristics. One or more low probability events have transpired to produce a urine positive. However, sufficient belief in the testimony of the accused led to an investigation of the claims of innocence. Except for the first methamphetamine example (see above), the accuracy of the analysis itself was not called into question — only the way in which results were interpreted. Despite the good intentions of chemists, false positives occur. It is the ethical responsibility of the chemist in these cases to resolve these injustices. Individuals using hair analysis or urinalysis have responsibilities for accurate analyses and interpretation of data. Furthermore, the survival of any testing program depends upon the perceived credibility of the program by the public at large. False positives, by their very nature, contribute to a poor perception by the general public and the ultimate demise of any testing program.

B. Historical Concerns for Passive Exposure in Hair Analysis

The human body has the ability to cleanse itself of drugs by metabolism and excretion. Thus, any exposure to a drug must be recent for the drug to be detected in blood, saliva, sweat, or urine. On the other hand, hair is a unique matrix because

---

* The phenomena of passive exposure is familiar to most nonsmoking individuals. Like many analytical instruments, the nose is an exquisite detector for some compounds, tobacco smoke being one. After exposure to tobacco smoke in a smoke-filled room and upon returning to his/her spouse that evening, the spouse may readily be able to tell that he/she had been in the presence of a smoker due to smoke clinging to the hair or clothing.

no active metabolism/excretion is present to remove drugs once deposited. There are only two removal mechanisms for drugs in hair: replacement/cutting of hair, with a time frame of months to years, and the slow hygienic removal of bound substances. For hair analysis, this is both good news and bad news. The good news is that, if drugs enter the hair exclusively as a result of ingestion, they can be detected for long periods of time. The bad news is that, if drugs become bound to hair, even partially by other mechanisms such as passive (accidental) exposure, they may be difficult to remove and distinguish from actual use. Passive exposure has historically been a problem for hair analysis for trace metals.

Lenihan[13] has pointed out that hair "is a mirror of the environment." However, many investigators have noted that while in some cases blood/urinary concentrations of trace metals are associated with elevated hair concentrations, this is not always true. Numerous other authors have discussed the problems of trace metal analysis with respect to passive exposure from the environment.[14-17] A detailed discussion of the literature on hair analysis for heavy metals is beyond the scope of this review, but one may be found in Chatt and Katz[18] and a brief review may be found in Manson and Zlotkin.[19] Chatt and Katz[18] state three factors that prevent the use of hair analysis for the diagnosis of disease and the assessment of nutritional status. These are (1) a difficulty in differentiating external deposition of trace metals from ingestion; (2) an inability to define normal ranges of trace metal concentrations in hair; and (3) a dearth of information on mechanisms of the incorporation of trace elements into hair. Likewise, Harkey and Henderson[20] have contrasted hair testing for drugs of abuse with hair testing for nutritional status and noted many similarities.[20] The latter procedures have been considered by some to be pure quackery.[21,22]

The potential for environmental exposure and the difficulty in distinguishing ingestion from external sources has greatly diminished the value of hair analysis for heavy metals. Are drugs uniquely different from heavy metals in terms of their binding mechanisms, removal mechanisms, or presence in the environment, or are molecules just molecules? We believe that drugs are similar to heavy metals and the literature on heavy metal analysis must be weighed when evaluating hair analysis data for drugs of abuse.

In 1986, several studies were started to validate hair analysis as a useful forensic tool. The results of these initial studies were surprising[23] and led to alternative explanations of the incorporation of drugs into hair.[24,25] These studies hypothesized that drugs were incorporated into hair via sweat, and if so, external contamination had to be considered in data interpretation. Although a report of deposition of heavy metals in hair from sweat was reported prior to this time,[17] to our knowledge, sweat as a mechanism of *transfer of drugs of abuse into hair* was not considered prior to these reports. This concept has since polarized the hair testing community. More recent findings from this laboratory, and those of other investigators, have given support to this early contention.[26,27]

II. MECHANISMS FOR INCORPORATION OF DRUGS INTO HAIR

A. Why Consider Mechanisms of Drug Incorporation?

Hair analysis is only useful if the drugs that are measured in hair arise from ingestion rather than from other sources. Therefore, it is imperative that drugs arising from the external environment be removed prior to analysis. If this cannot be accomplished, the analysis must take the external drug into consideration either by accounting for these sources of drugs in terms of absolute level or by the presence of unique compounds derived from *in vivo* metabolism. To evaluate whether or not passive exposure can contribute to drugs in hair, the mechanisms of appearance and binding of drugs to hair both from external (passive) sources and ingestion must be understood. Furthermore, an understanding of these mechanisms may give clues for differentiation between the different modes of drug deposition.

B. Models for Drugs Incorporation

The earliest theoretical position explaining the incorporation of drugs into hair has been given much attention.[28] In this model, drugs present in the bloodstream are entrapped in inaccessible regions of the hair during the hair growth process (Figure 1). After the hair emerges from the scalp, these drugs form bands that are in direct proportion to the concentrations present when the hair was formed. The entrapped drugs are protected by the hair matrix from removal or change by the external environment. Because hair grows at a relatively constant rate, this model predicts that hair analysis would provide a history of drug consumption in both time and amount. We call this model the "entrapment model." As will be demonstrated, there is little or no direct evidence in support of the hypothetical construct of inaccessible regions in hair. Therefore this model is to be considered purely theoretical at this time.

We questioned this early hypothesis on incorporation of drugs and proposed an alternative for drug incorporation.[29] In our model (Figure 2), some drugs are incorporated during hair growth from the compounds present in the bloodstream. However, water-soluble drugs are also excreted into sweat/sebum, which bathe the hair, and may be incorporated after the hair emerges from the skin. In this model, drugs can come from three sources. The first source is the blood, as described above for Figure 1. The second source is excretion of the drug and/or metabolites into sweat and subsequent incorporation into the hair. The third source is from passive exposure of the hair to the drug, either from vapor phase (e.g., smoke) or solid phase contact (e.g., drugs on furniture or clothing) followed by dissolution of the drug into once drug-free sweat or other aqueous media. Because both of the latter two sources of drugs are in aqueous solution, they are indistinguishable after they are incorporated into the hair.

We call the model depicted in Figure 2 the "sweat model" to emphasize the contribution of drugs from external aqueous media of moderate ionic strength. The sweat model predicts few or no regions in the hair inaccessible to the external environment. Several lines of evidence will be offered in support of this model for drug incorporation. A similar model to the sweat model for drugs of abuse had been proposed for heavy metal ions where a substantial fraction of the heavy metal ions present in the hair come from an external source such as sweat.[17]

A schematic model for sources of drugs in hair is shown in Figure 3. As Figure 3 further illustrates, there is usually some passage of time between ingestion and hair analysis. During that time, drugs loosely bound to the surface of the hair could be washed away by normal hygienic hair care. The removal of drugs will depend upon several variables, not the least of which are the characteristics of the solutions used to wash or treat the hair. In fact, one might visualize hygienic practices as an *in vivo* extraction of drugs of abuse and contrast them with laboratory decontamination/extraction procedures. The variable cleansing of hair by an individual before the sample is taken for hair analysis greatly complicates the classification of external contamination. How personal hygiene affects hair analysis will be discussed in detail in the theoretical framework section of this chapter.

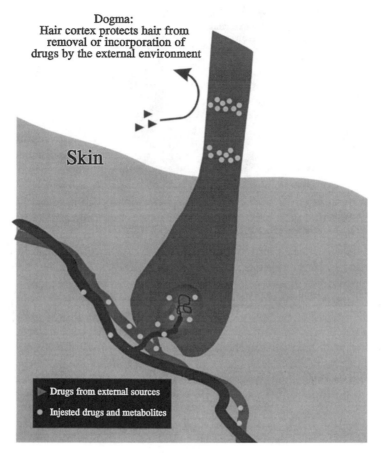

Figure 1
Entrapment model of drug incorporation. Drugs are incorporated into hair from the bloodstream during the growth phase. Their concentration in the hair reflects that present in the bloodstream. A central dogma of the entrapment model is that hair is resistant to incorporation of drugs from outside contamination.

C. Can Exogenously Applied Drugs be Incorporated into the Hair Matrix?

Yes. Some authors have claimed that drugs deposited as a solid phase, such as crack cocaine smoke, can be successfully removed by certain decontamination procedures.[30,31] In contrast to these two papers, a larger number of authors have failed to remove all the external contamination applied.* For example: Welch et al.[32] evaluated several decontamination procedures to remove cocaine placed on the hair in the solid form and found that none were successful in removing all external contamination. Also, when Kidwell and Blank[33] exposed hair to vapors of phencyclidine (PCP), large amounts of PCP could be removed by washing, but large amounts of PCP still remained in the hair when analyzed. In fact, they observed a greater concentration of PCP in these passive exposure experiments than was found in analyses of user's hair.[34,35] Similarly, Cone et al.[36] found that substantial quantities of cocaine (but not all) externally applied in an aqueous media could be removed if careful attention was paid to the experimental details of decontamination. Henderson and colleagues[37]

---

* Even some of the wash procedures of Baumgartner and Hill[31] may be questioned as to their completeness. See the section entitled "Can Externally Applied Drugs be Removed?"

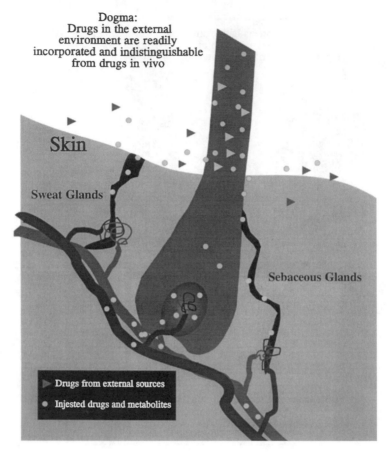

Figure 2

Sweat model of drug incorporation. Drug incorporation occurs from drugs excreted into sweat and/or from the external environment with minimal incorporation during hair formation. A central dogma of the sweat model is that hair is NOT resistant to penetration by outside contamination and that this contamination may be permanently incorporated into the hair matrix. External contamination may come from endogenous drugs present in the sweat or exogenous drugs that have been dissolved in sweat or other aqueous media.

reported that a single exposure of hair to cocaine in vapor form ("crack smoke") resulted in 15 to 53 ng of cocaine per milligram of hair, even after extensive washing. These quantities compared with the upper range of concentrations for cocaine users.[38-42] Henderson and co-workers[37] performed their experiments in two ways. After exposure, they analyzed the hair without decontamination and after decontamination with various washing procedures. In the unwashed hair, the amounts of cocaine were relatively constant between hair types. However, in the washed hair, substantial variations in amounts of cocaine occurred, suggesting that the effectiveness of incorporating drugs from the external environment may depend on hair type.

Drugs placed on the surface of hair must have some mechanism for entry into the hair matrix. If a solid is placed on the hair surface, most of it can be readily removed. However, after the deposition of the solid drugs onto the hair surface, hair may be bathed at some point in an aqueous media, be that sweat, sebum, normal hygienic solutions, or during the hair analysis procedure. Alternatively, an individual might come in contact with a solution of a drug by transfer of sweat from another individual. It is for these reasons that we and others studied solution phase transfer as the vector for admission and incorporation of drugs into hair.

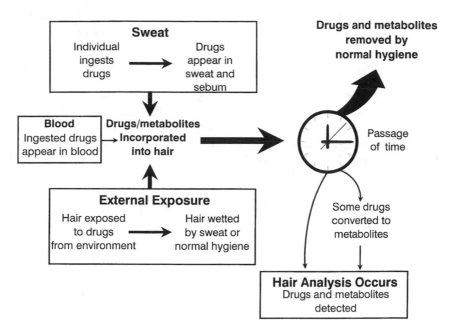

Figure 3
Theoretical framework for the incorporation of drugs into and removal from hair.

Kidwell[43,44] and Kidwell and Blank[45] have shown that hair readily adsorbs drugs from an aqueous solution. When drugs are applied in the aqueous phase, it appears they are incorporated into the hair matrix and are difficult to remove by either normal hygienic washing or laboratory procedures. This work had been extended with fluorescently labeled drugs[33] and more recently by use of radioactive tracers.[46,47] The concentration of drug in the applied aqueous media can be very small. Henderson and co-workers[37] have shown that hair soaked in a solution of cocaine at the 15-ng/mL level and repeatedly washed can produce a measurable amount of cocaine in the hair. Cone et al.[36] suggest that much of the cocaine applied externally can be removed. However, their work exposed hair to a solution of cocaine at 1 mg/mL for 24 h. Based on experience from this laboratory, a solution of this concentration vastly exceeds that necessary to incorporate substantial quantities of cocaine into hair. It is possible that Cone et al.[36] may have reached a saturation point of the hair matrix for cocaine and thereby shifted the amount removed in the decontamination step to much higher values. Sellers[48] has demonstrated that, depending upon hair type, as much as 31% of the available cocaine in a contaminating solution could be incorporated into the hair sample over an 8-h period.

Blank and Kidwell[47] developed a simple method to follow incorporation and removal of cocaine from hair using radioactive tracers, thus permitting them to perform a large number of experiments under a variety of conditions and with high precision. Using this technique, the quantity of cocaine in various wash steps could be carefully monitored. They found that, regardless of the decontamination technique, a substantial amount of drug could still be found in the final hair digest.* The

---

* Some possible criticisms of the use of radiotracers for monitoring the uptake and release of cocaine are that the tracer chemically reacts with the hair matrix or contains radiolabeled impurities which show preferential binding. Both criticisms are unlikely because (1) the radiotracer never exceeds 1% of the unlabeled drug, and (2) hair exposed to drugs without the radiotracer and analyzed by GC/MS show the parent drug present in amounts indicated by the radiotracer analysis.

binding capacity of hair for cocaine is not known. Theoretically, one would expect that the binding capacity of hair would be finite making it easier to remove excess drugs.

Table 1 reviews data from several studies of external contamination showing both the percentage of external contamination removed and the concentration of drug remaining in the hair sample. Depending upon the particular study, between 77.9 and 99.986% of externally applied cocaine can be removed. Nevertheless, the cocaine remaining in the hair ranges from 1.7 to 400 ng/mg of hair. These levels are compared to the published data on the distribution of cocaine concentrations in the hair of cocaine users. It is clear that these externally contaminated *and extensively decontaminated* specimens would all be considered positive. In fact, they would be in the middle to upper range of cocaine positive specimens. The percentage removed and remaining would not necessarily be the same if the hair samples had been contaminated to a smaller degree because hair is limited in binding capacity. High removal percentages can give the analyst a false sense of security about the robustness of the analytical procedure unless the amount of drug remaining is carefully measured and compared to normal user levels.

TABLE 1.

Concentration of Cocaine Possible in Hair with External Contamination

Wash solvent	Percent removed	Percent remaining	Remaining cocaine ng/mg hair	Approximate% of users below this level[a]
From Baumgartner et al.[28]				
20 × Ethanol	99.985	0.015	2.7	47%
20 × Phosphate	99.979	0.021	3.5	53%
20 × Prell® shampoo	99.986	0.014	1.7	42%
From Welch et al.[32] (black hair)				
3 × Methylene chloride	?	?	120	100%
3 × Phosphate buffer	?	?	260	100%
3 × Ethanol	?	?	310	100%
6 × Methanol	?	?	400	100%
From Cone et al.[36]				
2 × Methanol #1	95.6	4.4	5.1	42%
2 × Methanol #2	96.8	3.2	10.2	74%
2 × Methanol #3	97.2	2.8	34.3	100%
1 × Ethanol, 3 × phosphate				
Sample #1	84.8	15.2	18.5	84%
Sample #2	80.3	19.7	94.9	100%
Sample #3	77.9	22.1	62.9	100%

[a] User levels from Baer et al.[116]

D. Can Exogenously Applied Drugs Mimic Drug Usage?

Yes. In the hope of distinguishing the source of the drugs by their extraction pattern, the removal of drugs from external sources was compared to the removal of endogenous drugs by exposing the hair from drug users to various drug analogs.[33] Samples of hair from two cocaine users, a crack smoker, and an intravenous cocaine user were exposed to an aqueous solution of *p*-bromobenzoylcocaine (a derivative of cocaine) at 10 μg/mL for 1 h. The hair specimens were then rinsed, air dried, and extracted according to literature procedures for hair analysis available at that time.[34,49] Cocaine and *p*-bromobenzoylcocaine were quantitated in all solutions by GC/MS. The results for the crack cocaine smoker are shown in Figure 4. We were surprised

that the wash-out kinetics for the cocaine and the externally introduced
p-bromobenzoylcocaine were very similar. This implied that external contamination
could mimic drug use, even with a contaminating solution as low as 10 µg/mL.

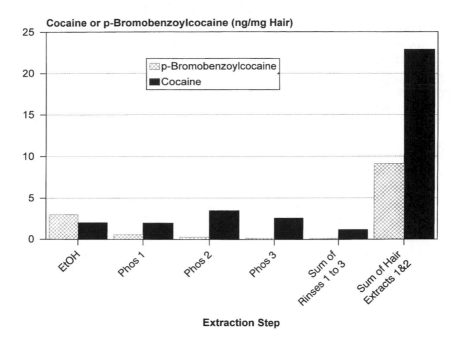

Figure 4
Extraction profile of a cocaine smoker's hair exposed to a cocaine surrogate (*p*-bromobenzoylcocaine).
After exposure to 10 µg/mL of *p*-bromobenzoylcocaine, the hair was washed once with ethanol, three
times with phosphate buffer (pH 7), rinsed with water three times, and then the cocaine remaining in the
hair was extracted two times with 0.6 *N* HCl. The cocaine and *p*-bromobenzoylcocaine were quantitated
in each solution by GC/MS with selected ion monitoring. (From Kidwell, D. A. and Blank, D. L., in *Hair
Testing for Drugs of Abuse: International Workshop on Standards and Technology*, E. J. Cone, M. J. Welch, and
M. B. Grigson Babecki, Eds., National Institutes of Health Publication #95-3727, Superintendent of Documents, U.S. Government Printing Office, Washington, D.C., 1995, pp. 19-90.)

The one criterion reported to evaluate the possibility of passive exposure was the
ratio of drug in the hair to the final phosphate wash. A ratio greater than 10 had been
suggested (employing a different extraction procedure) to distinguish between active
and passive exposure.[50] In our study, although the hair was not dissolved in the drug
extraction step (suggesting more may be present due to inadequate extraction), significant amounts of both cocaine and *p*-bromococaine were found in the final hair extract.
The ratio of the drug surrogate (*p*-bromococaine) in the hair to that present in the third
phosphate wash was 54:1 and the ratio of cocaine from the individual's use to the third
phosphate wash was 8.9:1. Using the proposed factor of 10,[50] this cocaine user would
have been considered a user of *p*-bromococaine rather than cocaine.

Analogous experiments were performed with PCP.[33] In these experiments, hair
from a PCP drug user was exposed to a solution of *p*-methylPCP at a concentration
of 1 µg/mL for 1 h. The final extract contained 2.3 ng of PCP/mg of hair, whereas
p-methylPCP was found at a concentration of 12 ng/mg of hair. More of the drug
surrogate was found in the hair than the PCP incorporated *in vivo*. Neither the
extraction kinetics nor the ratio of PCP in hair to that of the last phosphate wash
clearly distinguished this hair specimen as contaminated by *p*-methylPCP.

E. Are Drugs Present in Sweat and Can They be Incorporated into Hair?

It appears so. Drugs in sweat have been found by several authors.[51,52] For example, Balabanova and Schneider[53] found that as much as 9.2 µg of cocaine/cm^2/24 h could be deposited in the axillary region of a drug user. In a controlled dosage study using deuterated cocaine as tracers, Henderson and colleagues[37] measured concentrations of the deuterated cocaine at levels as high as 50 µg/mL in sweat 1 h after intranasal administration of cocaine at a dose of 0.6 mg/kg. Other authors[54] found varying amounts of amphetamine and dimethylamphetamine in sweat after low dosages of the drugs were administered.

The hypothesis that sweat was a vehicle for the transfer of drugs into hair became more apparent after examining hair from a male Korean who had used medicinally prescribed dihydrocodeine.[33] This individual had taken a total of 135 mg of dihydrocodeine over a one-week period after a surgical procedure. One week after the dihydrocodeine ingestion ceased, the hair was cut 2 to 4 cm distal from the scalp to avoid collection of hair that could have grown during administration of the dihydrocodeine. The subject did not undertake strenuous exercise/work during this two-week period, and he washed his hair daily. Analysis by GC/MS showed dihydrocodeine at a level of approximately 0.04 ng/mg of hair. Considering the dose, the amount detected was high, particularly when compared with most opiate users (0.1 to 2 ng of opiates/mg hair).[55-60]* Clearly, some mechanism other than deposition during hair growth had to contribute to drug incorporation in this hair sample.

A more convincing experiment was performed by Henderson and co-workers.[37] Five volunteers were administered deuterated cocaine. After 2 h these individuals held drug-free hair in their hands for 30 min. All the hair became contaminated with deuterated cocaine. The concentration of deuterated cocaine in these samples was as high as 48 ng/mg of hair before decontamination of the hair samples and 11 ng/mg of hair after decontamination. Henderson and colleagues[37] concluded, "Even after washing, the amount of deuterated cocaine in this externally contaminated hair sample was higher than that found in his hair when it was tested over the next few months."

Further evidence that sweat plays a role in the incorporation of drugs into hair may be gained from the time of administration and appearance of drugs in the hair. Cone[61] and Püschel et al.[58] separately reported that codeine could be detected in beard hair 24 h after administration. A 5- to 7-d period would be necessary before the beard hair, which was forming during the period of maximum blood levels of the codeine, emerged above the skin. Cone[61] suggested that codeine secretion in sweat and its incorporation into hair could account for its initial appearance in the beard hair.

The data suggest that, given the similarity between the *in vitro* studies reported here and sweat, drugs could be easily incorporated into hair from sweat prior to appearance of the hair above the surface of the skin. The concentrations of cocaine, benzoylecgonine, and PCP analogs employed in our *in vitro* studies were well within the range of concentrations found in sweat.

---

* Püschel and co-workers[58] reported levels of morphine of 1.5 to 1.8 ng/mg hair for medicinal morphine users and 0 to 3.2 ng/mg hair for opiate users. Offidani et al.[57] reported morphine levels of 0.1 to 10.9 ng/mg hair for morphine abusers. Marigo and colleagues[56] reported 1 to 15 ng/mg hair for 22 heroin addicts. Klug[55] reported values ranging between 1 to 10 ng/mg hair for drug-induced overdose cases. More recently, Mangin and Kintz[59] reported levels of 0.62 to 27.1 ng/mg hair in 12 fatal heroin overdose cases.

F. Can Passive Exposure be a Vector for Incorporation of Drugs into Hair?

Yes. In situations where drugs are known to be present in the environment, it is easy to demonstrate that passive exposure can produce positive hair analysis results. In a study by Haley and Hoffmann[62] of the nicotine and cotinine concentrations in the hair of smokers and nonsmokers, there appeared to be a higher average of nicotine in the unwashed hair of smokers (average 8.75 ng/mg). However, nonsmokers also had an appreciable level of nicotine (average 2.42 ng/mg) which overlapped that of smokers. In contrast, cotinine (the nicotine metabolite) does appear to be a marker of tobacco use in this population. More recently, Kintz and co-workers[63] and Kintz[64] proposed a cutoff level of 2 ng of nicotine/mg of hair to eliminate nonsmoking individuals. Even at this level, some nonsmokers would be positive.

Like nicotine, drugs of abuse are also present in the environment. Smith et al.[65] examined the children and spouses of cocaine users. In this study, the children lived in a family where cocaine was used and thus present in the environment, although the quantity present was not known. This study assumed that children 3 to 10 years of age are unlikely to be self-administering cocaine, so that any cocaine in their hair must have come from passive exposure. Skin wipes were obtained from the children by wiping their foreheads with a cotton swab to assess external exposure. All of the skin wipes in the experimental population (N = 29) were positive for cocaine, indicating extensive surface contact. Saliva samples were also taken to assess ingestion of the cocaine. Most of the children's saliva samples were negative (2 of 23), indicating that ingestion of cocaine in the environment was not a likely source of the cocaine in the hair. In the active adult-using population, 80% were positive for cocaine in their hair. Of the children in this study, 85% were positive for cocaine. The distribution of the concentrations of cocaine in the hair of these two groups was also similar. Analysis of the data showed that several children of cocaine users had both cocaine and benzoylecgonine in their hair in varying quantities compared to the adult users. These amounts may be less than, greater than, or equal to the drug-using parent (Figure 5). The concentrations of cocaine within a family group varied greatly, which would be consistent with passive exposure being a random event (Figure 5d). This study may be the first demonstrating that passive exposure occurs in household settings and that certain metabolites (benzoylecgonine) may not be an indicator of drug use. The presence of benzoylecgonine in the environment and in hair has been noted before.[66,67]

Contrast the apparent ease in generating false positives for cocaine in hair analysis with several studies showing the difficulty in obtaining a positive urinalysis result due to passive exposure to cocaine.[68,69] A urine positive can result from the unusual practice of unprotected handling of kilogram quantities of cocaine.[70] Even in this case, the urine positive was for a short duration. These studies again reinforce the view that the body has metabolic and excretory mechanisms to eliminate cocaine and that the ingested cocaine is diluted into large blood and other fluid compartments.

Not *all* positive hair analysis results are due to passive exposure. Certainly, many positive results are due to ingestion of drugs. This section clearly illustrates that interpretation of the data must take passive exposure into consideration to determine the source of drugs in any given positive result.

G. Which Physicochemical Variables Affect Drug Incorporation into Hair?

The amount of drugs incorporated into hair depends on a variety of factors: (1) the concentration of drugs in hair increases with the concentration of the exposure solution and the time of exposure; (2) the concentration of drugs in hair decreases

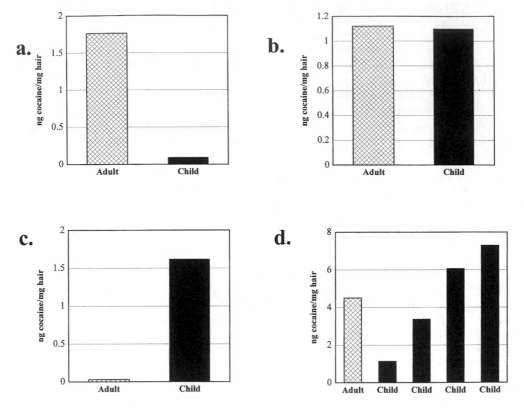

Figure 5
Concentrations of cocaine in the hair of family groups. Cocaine was used by all adults.

in the presence of competing cations, such as sodium; (3) the concentration of drugs in hair shows complex effects depending on the pH of the exposure solution and the chemical structure of the drug; and (4) the prior treatment history of the hair.

1. Concentration

The relationship between the concentration of cocaine in the exposure solution and the amount of cocaine in the final extract for two types of hair is shown in Figure 6. Concentrations employed for this study were from 1 µg/mL cocaine plus radiotracer to 500 µg/mL cocaine plus radiotracer in phosphate-buffered saline (PBS), and all samples were exposed for 1 h and dried as previously described.[33] Even though this figure represents data obtained from a series of several experiments, a linear plot is easily fitted to the data. However, the amount incorporated varies depending upon the hair type, reflecting a matrix bias that is explored in more detail below.

2. Duration of Exposure

The effect of varying the time of exposure of hair to externally applied cocaine and the amount of cocaine incorporated in the hair is shown in Figure 7a. Increasing the duration of exposure increases the uptake of drug, although the relationship is not linear, especially for the Asian hair sample. Even exposure times as short as 5 min produced specimens with detectable levels of drugs. Exposure times greater than 1 d were attempted (Figure 7b), but cocaine (and presumably the radiotracer) in the

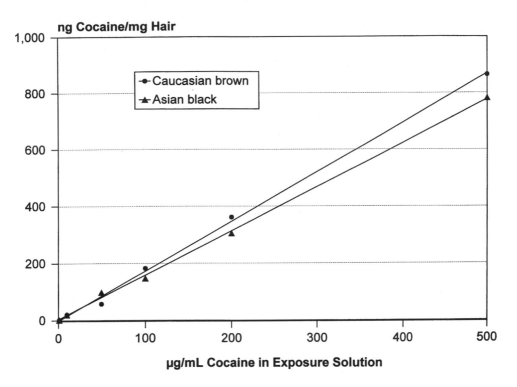

Figure 6
Effect of concentration on the incorporation of cocaine into hair. Two types of hair were exposed to cocaine for 1 h and were dried overnight before extraction. (From Blank, D. L. and Kidwell, D. A., *Forensic Sci. Int.*, 63, 145, 1993.)

exposure solution decomposed substantially after 24 h at 37°C, affecting the concentration of cocaine in the exposure solution and therefore decreasing the amount of drug available for uptake.* Furthermore, cocaine bound to the hair matrix during the first several hours of exposure could be hydrolyzed during continued exposure, resulting in conversion of the radioactive label to benzoic acid, a material that is not expected to bind well to the hair (see below).

3. Competing Cations

The effect of concentration of cations in the externally applied cocaine solutions on the uptake of cocaine into hair is shown in Figure 8. Cocaine plus radiotracer was added (final concentration, 10 µg/mL) to each of four solutions with different sodium chloride concentrations ranging from 0 to 500 mM. Two sets of samples were studied — each set containing either Caucasian brown hair or Asian black hair. The first set was dried after exposure and prior to decontamination/analysis for cocaine. The second set was carried directly through the decontamination procedure. As seen

---

* The rate of degradation of cocaine depends upon temperature and pH. The higher the pH, the more rapid the degradation. For example, solutions of cocaine in phosphate-buffered saline were incubated at 37°C and pH 8.5. Aloquotes were taken, concentrated to dryness, derivatized by silylation, and analyzed by GC/MS. The ratio of benzoylecgonine:cocaine = 40%, methyl ecgonine:cocaine = 4.5%, and ecgonine:cocaine = 1.9% after 1 h; benzoylecgonine:cocaine = 115%, methyl ecgonine:cocaine = 9.1%, and ecgonine:cocaine = 7.3% after 2 h. If the cocaine incorporated into the hair was similarly affected, most of the radioactive label (cocaine, 3,4-³H-benzoyl) would have been converted to radioactive benzoylecgonine or radioactive benzoic acid.

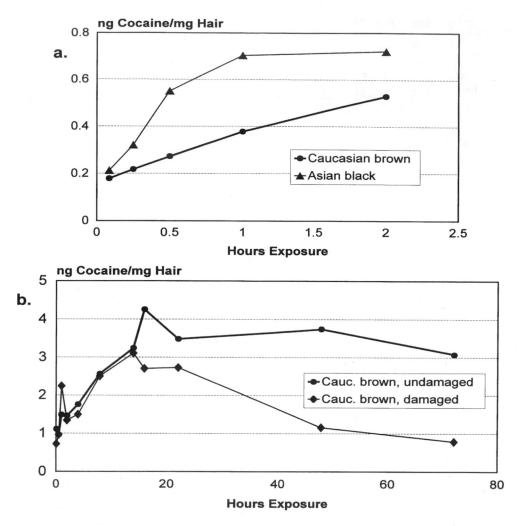

Figure 7
Effect of exposure time on the incorporation of cocaine into hair. The hair samples in (a) were exposed to a 1-µg/mL solution of cocaine containing a radioactive tracer. The damaged hair samples in (b) were produced by cosmetically treating hair with Clairol® Nice and Easy® Shampoo-in hair color #97 (natural extra light beige blond) for 20 min. The decrease in cocaine incorporation for longer exposure intervals was likely due to hydrolysis of cocaine during long exposures.

in Figure 8, as the sodium ion concentration increases, the amount of drug incorporated into the hair decreases. What underlying mechanism could account for these data? If incorporation of drugs into hair is through ionic bonding, then increasing the concentration of cations in the exposure solution would increase the competition for the available binding sites on the hair. Mass action would then dictate that the drug concentration would be lower.

4.   The pH of the Exposure Solution and Charge State of Drug

The effect of pH on the uptake of tritiated cocaine and tritiated benzoylecgonine from exposure solutions (5 µg/mL of drug) is shown in Figure 9. Raising the pH to more basic values causes an increase in the incorporation of all the drugs into hair, but at different rates. No correction was made for any hydrolysis of either cocaine

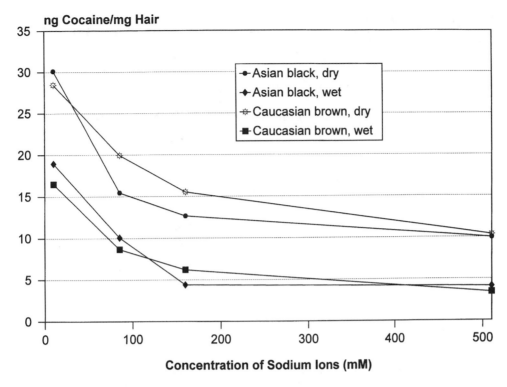

Figure 8

Effect of cation concentration on the incorporation of cocaine into hair. The hair was exposed to cocaine (10 µg/mL) in phosphate buffer (10 mM, pH 5.6) with increasing amounts of sodium chloride present in the solution. The final concentration of sodium ions included the contribution from the buffer. Two sets of samples were employed for this study, each set containing either Caucasian brown hair or Asian black hair. The first set (dry) was air dried prior to decontamination and analysis for cocaine and the second (wet) was carried directly through the decontamination procedure following cocaine exposure. (From Blank, D. L. and Kidwell, D. A., *Forensic Sci. Int.*, 63, 145, 1993.)

or benzoylecgonine in the exposure solution at the higher pH. A physicochemical model for this effect is discussed below.

5. *Prior pH History of the Hair Specimen*

The uptake of cocaine to hair after preliminary exposure to various acidic and basic solutions is shown in Figure 10. Three sets of experiments were performed. In the first experiment, hair was exposed to base (0.1 M sodium carbonate, pH 12) for 3 h and then washed extensively with water until the water was neutral. This was then followed by two phosphate buffer rinses at pH 5.3 for 10-min periods. The hair was then exposed to 5 µg/mL of cocaine, containing a radioactive tracer, in phosphate buffer at pH 5.3 for 1 h. After exposure the excess cocaine was removed with three water rinses of 10 min each and the hair was dried. Then the hair was decontaminated with ethanol and six phosphate washes, digested, and the cocaine concentration determined by scintillation counting as per previously described procedures.[26] Approximately twice as much cocaine was incorporated into the hair pretreated by exposure to base as compared to hair not subjected to any pretreatment. In the second experiment, hair was pretreated with 0.1 M HCl, then treated as in the first experiment. In this case approximately one sixth as much cocaine was incorporated as the control. In the third experiment, hair was pretreated with 5% acetic acid and treated as in the

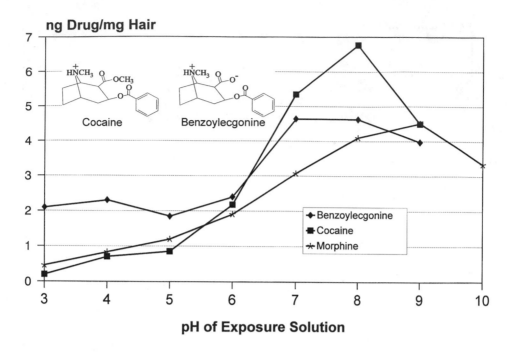

Figure 9

Effect of pH on the incorporation of cocaine, benzoylecgonine, and morphine into hair. Caucasian brown hair was exposed to the drug (5 µg/mL) containing a radioactive tracer in 10 m*M* phosphate buffer at the indicated pH for 1-h. The hair was rinsed three times with water and dried. A 30-min ethanol decontamination was followed by six phosphate decontamination washes. The hair was then digested with sodium hydroxide and the drug concentration measured by scintillation counting.

other two experiments. This specimen bound approximately 25% of the cocaine as the hair with no pretreatment. Thus, there is a 12-fold difference in incorporation of cocaine in the same hair type, depending on which hair preexposure solution was employed. Attempts were made to reverse the effect of the pretreatments. Surprisingly, a short (3- to 24-h) exposure to buffer *does not* restore the original binding capacity of the hair to cocaine. However, a much longer exposure (5 d) to buffer did return the hair to the original binding capacity, indicating that the change was not permanent (restoration in Figure 10).

Carboxylic acids in the amino acid side chains of the hair matrix can be esterified. Esterification removes the potential for a negative charge on the carboxylic acids and thereby reduces the ion exchange capacity of the hair. Esterification was performed by treating hair with methanolic HCl for 24 h at 60°C. Because an acid was used, the binding capacity of the hair for cocaine would be reduced regardless of whether or not the carboxyls were esterified. Therefore, the hair was also treated with buffer for 5 d and the binding capacity was measured. Only about half as much capacity was restored, indicating that the methanol did esterify some of the carboxyl functionalities. Because sulfate groups occur in hair due to oxidation of cystine, not all the binding capacity would be expected to be removed. Also, the esterification may have been incomplete or some of the esters may have been removed during the buffer exposure. As a control to demonstrate that the hair was not irreversibly changed by the methanolic HCl, esterified hair was treated with base to hydrolyze the esters back to carboxyl functionalities. The original binding capacity was restored. These experiments on esterification of hair are consistent with binding to the hair matrix being an ion exchange process with carboxyl functionalities participating.

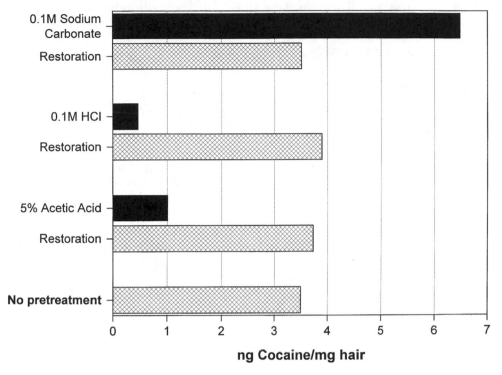

Figure 10

Effect of acid/base pretreatment on cocaine incorporation into Caucasian brown hair. The hair was exposed to the solutions listed for 3 h at 37°C and then washed with water until the water was pH neutral. The hair was rinsed two times with 10 mM phosphate buffer, pH 5.6. The hair was then exposed to cocaine (5 µg/mL), decontaminated, and analyzed. To "restore" the hair, each sample was soaked in 10 mM phosphate buffer, pH 5.6, for 5 d before exposure and analysis.

H. Model of Drugs Binding to Hair

A model for incorporation of drugs into hair that is consistent with the effects of competing ions, pH, and preexposure is shown schematically in Figure 11. Hair contains three major functionalities whose characteristics can vary as a function of pH, i.e., side chains of aspartic acid, side chains of glutamic acid, and sulfate groups produced by oxygen degradation of the amino acid cysteine. The sulfate groups would be expected to be more prevalent in damaged or chemically treated hair. This is supported by the observation of Pohl and co-workers[71] that the isoelectric point of hair decreases when the hair is damaged or when the hair ages. Hair is predominantly neutral at low pH near the isoelectric point of hair (approximately 4).[72] At these acidic pH, drugs could bind via weak hydrophobic interactions (Figure 11a). Thus, drugs should bind equally well at low pH provided they are similar in hydrophobicities and charge. Drugs which are positively charged at low pH would bind less well than those neutral or negatively charged. At pH greater than the isoelectric point of the hair, the side chains of aspartic and glutamic acids are deprotonated, and hair becomes negatively charged (Figure 11b). Therefore, the more positively charged the drug, the greater the likelihood of binding to hair. At still higher pH, more and more side chain carboxyl groups are deprotonated and hair becomes increasing negatively charged (Figure 11c). In short, there are three effects that a change in pH can produce: (1) a change in charge state of the hair, (2) a change in charge state of the drug, and (3) a change in the hydrophobicity of both the drug and the hair. The interactions of these three variables will determine the relative uptake of drugs into hair.

Cocaine has a pK_b of 8.4 to 8.69 and should remain predominately positively charged below pH 8.7.[73] Cocaine is represented by Drug A in Figure 11. The exact charge state of the drug is complicated by zwitterionic species, such as benzoylecgonine (see structure in Figure 9 and Drug C in Figure 11), because zwitterionic materials change from positive to negative charge, depending upon the pH. Negatively charged drugs (represented by Drug B in Figure 11) should bind poorly to the hair matrix. In this model, a negatively charged drug such as biotin (a weak acid with a pK_a about 3.5) would be repelled from the hair matrix at all pH studied, and would bind quite poorly to hair compared to cocaine. The binding of biotin was compared to the binding of cocaine, both followed by radiotracer analysis. When hair was exposed to 1 µg/mL of biotin at pH 7 in PBS for 1 h, only 12% as much biotin as cocaine bound to Asian black hair and 4% as much for Caucasian brown hair. Likewise, the observations that hair from marijuana users contains only trace quantities of negatively charged 11-nor-9-carboxy-Δ9-tetrahydrocannabinol (THC carboxylic acid)[74]* and that aspirin (pK_a = 3.49) also binds poorly to hair[75] would be quite consistent with the model presented here.**

The preferential binding of drugs to the hair matrix varies with pH, and therefore may be dependent upon the pH of the sweat. The pH of sweat varies considerably between individuals. Sweat has been reported to have an average pH of 5.82.[76] A range of 6.1 to 6.7 has also been reported,[77] but a pH of 4.5 is not uncommon.[78] The average ratio of cocaine:benzoylecgonine in sweat is approximately 4.4:1, but is highly variable.[78a] Likewise, the ratio of cocaine:benzoylecgonine in hair extracts varies considerably among individuals. By analyzing the data given in Cone et al.,[36] an average ratio of 6.7:1 can be calculated. Examination of the data of Henderson et al.[79] yielded an average ratio of 7:1 before decontamination washing and 5.4:1 after washing. Assuming the sweat of these individuals was near the average pH of 5.8, one could calculate from the *in vitro* data given in Figure 9 that the cocaine:benzoylecgonine ratio in hair should be approximately 1:1. This ratio would depend upon the hair type and assumes that the sweat contained equal concentrations of cocaine and benzoylecgonine. Because sweat does not contain equal amounts of cocaine and benzoylecgonine, the ratios must be adjusted by the drug ratios found in sweat.*** Accordingly, when these adjustments are made, the ratio of cocaine:benzoylecgonine in hair should likewise be approximately 4.4:1, if sweat is a vehicle by which drugs appear in hair. Given the uncertainties of drug ratios in sweat, sweat pH, and natural biovariability in all the measurements, this calculated value of 4.4:1 comes remarkably close to the ratios observed by others of 5.4 to 7:1.

The model of ionic binding was also evaluated by comparing the binding of fluorescein and rhodamine to hair, two molecules of similar shape, size, and fluorescent quantum yield. Rhodamine, being positively charged, is far more efficient at binding to hair compared to fluorescein.[29] The disparity in binding of materials to

---

* Hayes et al.[74] reported that in 200 samples of hair, the average value of THC-carboxylic acid was 4 pg/10 mg hair digest. This concentration appeared to be substantially less than the average value of 6 pg/10 mg of hair when the melanin fraction was examined.

** More recently, Kintz (personal communication) has attempted to analyze for aspirin in hair after taking 500 mg/d for two weeks. Approximately 1 ng of aspirin/mg of hair was observed. A similar user of cocaine would be considered a heavy cocaine user. If aspirin binding to hair was as strong as cocaine, 10 to 100 times more aspirin would have been expected.

*** This calculation is as follows: (cocaine/benzoylecgonine) in sweat from Schoendofer × (cocaine/benzoylecgonine) incorporation ratio from our study = predicted cocaine/benzoylecgonine ratio in hair extract.

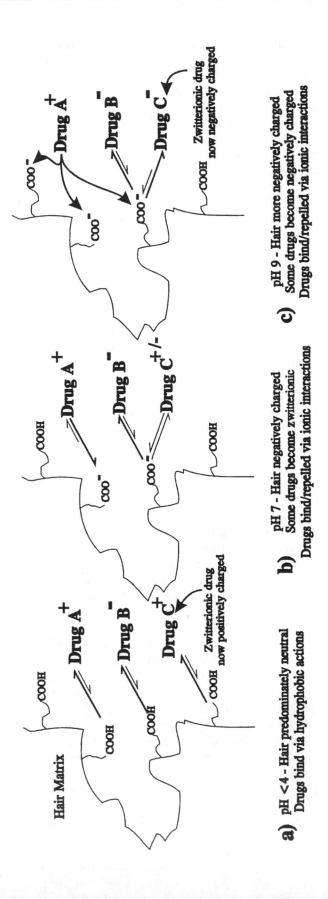

Figure 11

A model of drug incorporation into hair. Cocaine would be represented by Drug A. Biotin would be represented by Drug B. Benzoylecgonine would be represented by the zwitterionic Drug C, which would be positive, negative, or neutral, depending on pH. Likewise, any of the compounds, Drugs A through C, could represent a competing cation that could occupy a binding site on the hair and block a drug molecule from binding. (From Kidwell, D. A. and Blank, D. L., in *Hair Testing for Drugs of Abuse: International Workshop on Standards and Technology*, E. J. Cone, M. J. Welch, and M. B. Grigson Babecki, Eds., National Institutes of Health Publication #95-3727, Superintendent of Documents, U.S. Government Printing Office, Washington, D.C., 1995, pp. 19-90.)

hair is quite well known in the cosmetic industry.[72] Goddard[80] reported that negatively charged hair care products bind poorly to hair, while positively charged materials bind tightly. His results should hold equally well for drugs of abuse as it does for the compounds used in hair care products.

Hair may be thought of as an ion exchange resin with limited access to ion exchange sites further removed from the external environment. Commercial ion exchange resins frequently must be preconditioned before use to generate the ion exchange sites. Similarly, if hair is exposed to base, a larger number of ion exchange sites would be expected to be generated because of deprotonation of the carboxylic acid side chains. Exposure of hair to acid (as discussed in the section on prior pH history) would be expected to generate a lower number of sites because the carboxylic side chains would become protonated. The time required to restore the binding capacity depends upon the base and acid used in the preexposure and their diffusion rates into the hair matrix. For the acetic acid exposure, a short buffer soaking could restore the binding capacity, but for hydrochloric acid it could not. A likely explanation for the delay in reversibility is the concentration of buffer (pH 5.6) being 0.01 M compared to the preexposure solution being 0.1 M. This concentration difference affects the diffusional rate of the solutions into the hair matrix; the hair must be exposed to the buffer for a longer period of time to reach all the carboxyls on the interior of the hair. The delay in reaching interior binding sites is consistent with the data presented in Figure 10 and modeled in Figure 11.

I. Is there a Dose-Response Correlation for Drugs of Abuse?

Sometimes. In a data set going back as far as 1986, a poor correlation between self-reported drug use histories of military members in drug treatment and the amount of drugs found in hair was observed.[33] More recently, we reanalyzed data collected by Baumgartner and Hill.[81] The selected data set contained a sufficient number of subjects (N = 45) for statistical analyses. The correlation coefficient between cocaine found in hair and self-reported use was 0.59 and was significant ($p < 0.01$). The regression line for this data set yielded a Y intercept of 4.9 ng/mg hair. Statistically, the Y intercept would predict the quantity of the drug in hair given a zero dose. Therefore, a positive Y intercept could be interpreted in two ways: (1) individuals using drugs underreported their use; or (2) environmental exposure and incorporation of drugs occur in individuals exposed to cocaine, which increase the apparent amount of cocaine in their hair. Though statistically significant, the data reported by Baumgartner and Hill[81] had substantial scatter, with only 35% of the variance accounted for by an assumption of a linear dose response.

In a controlled dosage study with cocaine, Henderson and co-workers[37] did not find a predictable relationship between dose of cocaine administered and the amount of drug eventually incorporated into hair. Overall, the correlation between the mean amount found in hair and the administered dose was 0.03. Using a drug surrogate, methoxymethamphetamine, Nakahara et al.[82] found that both the quantity of the drug appearing in the hair and its location along the hair shaft were correlated with ingestion. However, they also found poor correlations between individuals with identical dosages. In contrast, for haloperidol, Matsuno et al.[83] found a good correlation (r = 0.83, n = 59) between dose administered and the amount of haloperidol and its metabolite in hair. These studies suggest that the concentrations of drugs found in hair are not simply related to the quantity of drugs ingested, but may vary depending on the structure of the drug and certain other factors.

J. Do Different Physical Classes of Hair and Hair Care Habits Affect Drug Uptake?

Yes. A poor correlation between self-reported drug use histories of military members in drug treatment and the amount of drugs found in hair was observed when hair color was taken into account.[84] The variability of drug incorporation along racial lines also has been observed by several other authors[36,39,48] for a limited population of individuals. Several variables such as hair texture and hair care habits could account for these differences.

In an attempt to examine this critical issue, the *in vitro* uptake of drugs for different hair types was extensively studied. The results shown in Figure 12 make this difference apparent. Ten hair samples were exposed to cocaine or morphine (5 µg/mL, 1 h) and then thoroughly decontaminated. The quantities of drugs incorporated were measured by radiotracer techniques.[47] In general, African black hair incorporated more cocaine or morphine than other hair types, perhaps indicative of prior hair history. However, variables other than color must be considered, as among the Caucasian hair types, blond hair incorporated similar amounts of cocaine and morphine compared to other more pigmented hair types. To emphasize that hair color is not the only determining factor, six hair samples of varying length from female subjects were all exposed simultaneously to a solution of cocaine (5 µg/mL) containing a radiotracer for 2 h. Each section from a single individual had different, but unknown prior treatment histories. To address some concerns about drug uptake at the cut ends, the ends were sealed in wax prior to exposure. The hair was washed five times with distilled water to remove the relatively high concentration of cocaine in the exposure solution and the hair was then allowed to dry. The samples varied in length from 3 to 8 cm. After drying, each hair type was sectioned into pieces 1 cm long. The ends sealed in wax were discarded. Then each section was weighed, individually decontaminated,[33] and the cocaine tracer measured by liquid scintillation counting. The hair samples from the four Caucasian females showed relatively constant uptake of cocaine in all segments (Figure 13). The African black hair sample exhibited a decrease in uptake in the more distal segments, whereas the Korean hair sample exhibited an increase in uptake in all but the last segment. The ratio of the average amount of cocaine incorporated into all segments of African black hair to all segments of Caucasian hair was 2.9:1, whereas the ratio of cocaine incorporated for all segments of Korean hair to Caucasian hair was 6.8:1. There is no simple hypothesis that can account for these data. For example, the hair diameters of all segments were measured and were not related to cocaine uptake. Likewise, the Korean individual did not report any hair treatment other than normal hygiene. As suggested in the model outlined in Figure 11, it might be expected that the charge state of the proteins of various hair types may be different. Studies are under way in this laboratory to examine this possibility.

In real-life scenarios, the uptake of drugs from the external environment and loss of the drugs incorporated into the hair would be influenced by the hair type and hair care habits of the individual. Preliminary data collected from this laboratory and literature reports[85] suggest that the frequency of hair washing is highly variable among individuals. For example, drugs deposited into the hair matrix from external contamination or from sweat would be removed at rates influenced by normal hygiene and hair care. A survey of a convenience sample of 74 middle-class high school students asked "How often do you wash your hair?" and showed marked differences along racial lines (Figure 14). Thus, if all variables of exposure to drugs were constant, an African female would likely show more drug incorporation from

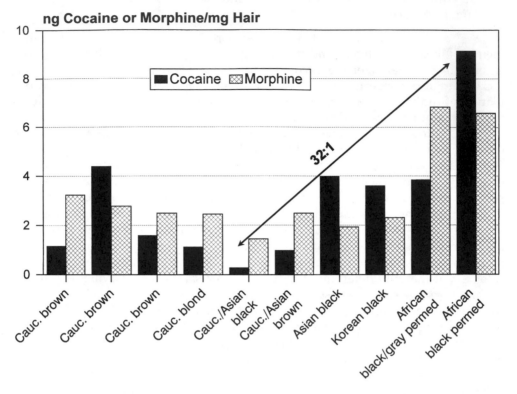

Figure 12
The binding of cocaine or morphine to ten different hair types. Hair type is not a well-defined term. However, it reflects hair color, hair history, race, and ethnicity of its donor.

external sources because her hair would be exposed to the drug longer, permitting greater uptake and fewer occasions on which drugs could be hygienically removed.

The *in vitro* studies of Henderson and co-workers[37] show a similar hair type bias to that depicted in Figures 12 and 13. The rank order of cocaine incorporation into hair exposed to cocaine vapor parallels our data, i.e., the amount of cocaine incorporated into Asian hair > African hair > Caucasian hair. Likewise, the *in vitro* studies of Reid and colleagues[86] also show a similar differential binding of benzoylecgonine depending on the hair color. Although their incorporation conditions for benzoylecgonine were severe (pH 4.0, 60°C, 42 h), these authors observed benzoylecgonine ratios of 7:2:1 for black, brown, and blond hair types, respectively.

Henderson and colleagues,[37] using a limited sample (N = 14), calculated the half-life of drugs in hair and found that for Caucasians, d5-cocaine (the administered tracer) had an apparent half-life of 2.1 months whereas for non-Caucasians the half-life was 1.6 months. The observed difference in half-life of cocaine may *not* be accurate because of the limited number of non-Caucasian subjects. The initial amounts of drugs must be considered with the half-life in determining the window of detection. Significantly, all the non-Caucasian hair had originally two to three times as much drug as the Caucasian hair samples, suggesting a longer window of detection for the small doses administered in this study.

While the *in vivo* studies cited above suggest bias among hair types, these studies contained too few subjects to reach statistical significance, or these studies were based upon self-reported use, which may be inaccurate. Mieczkowski and Newel[87] have

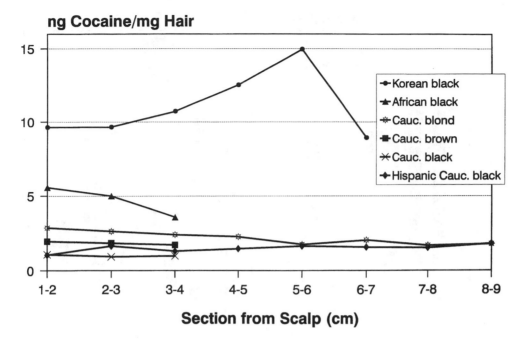

Figure 13
The binding of cocaine to six different types of female hair. (From Kidwell, D. A. and Blank, D. L., in *Hair Testing for Drugs of Abuse: International Workshop on Standards and Technology,* E. J. Cone, M. J. Welch, and M. B. Grigson Babecki, Eds., National Institutes of Health Publication #95-3727, Superintendent of Documents, U.S. Government Printing Office, Washington, D.C., 1995, pp. 19-90.)

raised such issues for these reasons. Mieczkowski and Newel[87] compared self-reported use, urinalysis, and hair analysis data and found that hair analysis detected significantly more presumptive "drug users" than either urinalysis or self-report. They also compared the patterns of self-reported use with the patterns of urinalysis and hair analysis. From this comparison they concluded that there was no apparent bias among hair types. However, to observe bias, i.e., the differential incorporation or removal of drugs for different hair types at different doses, it is necessary to know if an individual has used drugs and, if so, the dosage of those drugs. For example, if two individuals with different hair types consumed the same dosage of a drug and one individual was detected by hair analysis and the other was not, then a conclusion of bias could be drawn. Similarly, if two individuals were passively exposed to the same quantity of drug and one individual was detected by hair analysis and the other was not, then again a conclusion of bias could be reached. In other words, both users and those exposed to environmental drugs may be accused or exonerated depending upon their hair type. Bias could occur if one type of hair incorporated drugs more readily, incorporated them at higher levels, or released them more slowly making detection easier.

K. Is Uptake of Drugs Different for Cosmetically Treated Hair?

Yes. As Baumgartner and colleagues[28] have noted, drug levels in hair vary depending on the damage caused by cosmetic treatments. The magnitude of this effect was studied for two types of hair. Hair was treated by exposure to Clairol® Nice and Easy® Shampoo-in hair color #97 (natural extra light beige blond) for

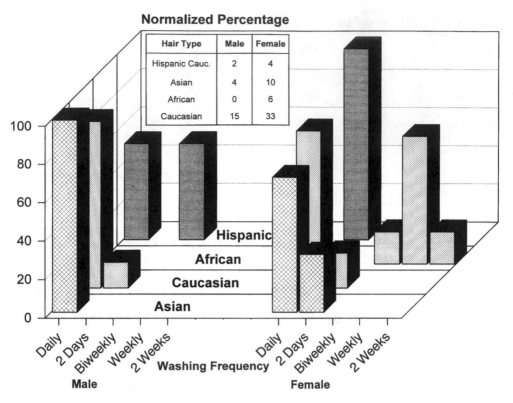

Figure 14
Hair washing survey. Results of a convenience sample of 74 middle-class high school students who were asked "How often do you wash your hair?". The distributions were normalized for each sex and ethnicity. The total number of individuals of each type is given in the inset table. (Data from Blank, D. L. and Kidwell, D. A., *Forensic Sci. Int.*, 70, 13, 1995.)

varying lengths of time ranging from 20 min to 80 min. The manufacturer's instructions on the application of this product noted an application time of 20 min for normal results. However, the customer may vary this application time or may perform several applications to achieve the color desired. The Caucasian brown hair was easier to color than was the Asian black hair and therefore was colored for the normal 20 min. Following cosmetic treatment, hair was exposed to a cocaine solution of 1 µg/mL plus radiotracer in PBS for 1 h and carried through the normal prewashing and decontamination procedures noted above. The data reveals that for Asian black hair, the longer the treatment was applied, the greater the incorporation of cocaine into the hair sample (Figure 15a). For Caucasian brown hair, the results were different. The treated hair showed less uptake of cocaine compared to the untreated hair as measured by radiotracer techniques (Figure 15a). Figure 15b shows another set of Caucasian brown hair treated by several methods. Depending upon the cosmetic treatment, this hair type may bind less or more externally applied drug (morphine in this example, 5 µg/mL, 1 h) as compared to the untreated control.

Henderson and co-workers[37] demonstrated an even more pronounced effect on the retention of cocaine after a permanent wave solution was applied. The amounts of cocaine incorporated from crack cocaine smoke prior to treatment were as follows: Asian, 52.6 ng/mg hair; Caucasian brown, 22.7 ng/mg hair; and Caucasian blond, 14.9 ng/mg hair. After treatment, these concentrations drop to: Asian, 8.2 ng/mg hair; Caucasian brown, 2.2 ng/mg hair; and Caucasian blond, 0.8 ng/mg hair. The percentage decreases from the untreated hair were similar: Asian, 84%; Caucasian

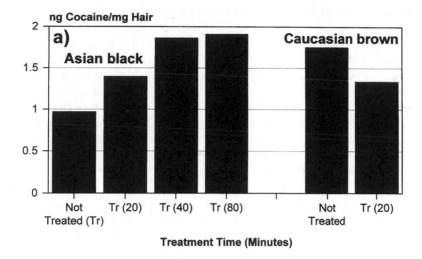

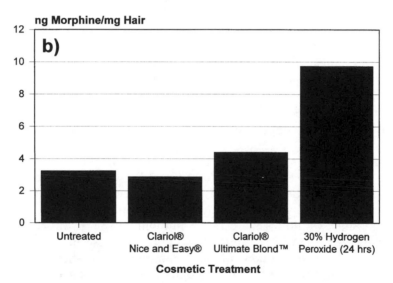

Figure 15
Effect of cosmetic treatments on the incorporation of cocaine into hair. Two types of hair: (a) Asian black hair and Caucasian brown hair were treated with Clairol® Nice and Easy® Shampoo-in hair color #97 (natural extra light beige blond) for varying times as shown in parentheses. Following treatment, hair was exposed to a cocaine solution (1 μg/mL) plus radiotracer in PBS for 1 h and carried through our normal analysis procedure. (b) Caucasian brown hair was either treated with a commercial hair coloring solution according to the manufacturer's recommendations or to 30% hydrogen peroxide for 24 h. The treated hair was exposed to cocaine (5 μg/mL) plus radiotracer in 10 mM phosphate, pH 5.6, for 1 h and carried through our normal analysis procedure.

brown, 90%; and Caucasian blond, 95%. Permanent wave solutions are among the most severe treatments applied to hair because they are both basic and contain materials that break the disulfide bonds in hair that give hair its texture.[72] Consequently, perming solutions would be expected to cause more hair damage and loss of drugs than would hygienic washes or hair-coloring solutions.

Marques and colleagues[88] also suggest the loss of drugs from the hair matrix due to cosmetic treatments. In studying the concentration of cocaine in the hair of mother/infant pairs, they found that it was possible to increase the correlation between the amount of cocaine in the mother's hair and the infant's hair by eliminating "damaged hair" from the analysis. Unfortunately, by doing so, data for approximately one half of the sample pairs had to be dropped.

As demonstrated in Figure 10, prior hair history affects the uptake of drugs from the external environment. Also, prior shampoo use can affect drug uptake,[48] perhaps by leaving shampoo residues on the hair.[89] African hair is more likely to be subjected to straightening preparations. Such treatments often employ solutions of a high pH.[72] Because relatively long (days) equilibration is necessary to restore the original binding capacity after strong base treatments, straightened hair might incorporate more drug just because of this history rather than an intrinsic binding capacity.

In our opinion, the binding capacity of hair is not static. The hair varies in its binding capacity from time to time depending upon environmental conditions. Chemical treatment, mechanical abrasion, and ultraviolet exposure will all contribute to the uptake of drugs. Therefore, a single measurement of the binding capacity may be a poor predictor of drug uptake. Not all treatments make the hair more porous. Thus, methylene blue staining[50] may not reveal the amount of "alteration" that may have been produced by treatments prior to environmental exposure and testing.

L. Are there Inaccessible Regions in Hair, and Why Are They Important?

Inaccessible regions are not evident in our data, but they are critical for the theories of decontamination. The entrapment model assumes that drugs are inaccessible after the hair matures. If all the hair matrix can be accessed by the external environment, then it can be changed by the external environment. Several years ago, we began a series of experiments to determine if there were inaccessible regions in hair that sequester drugs of abuse.[33] Two molecules were selected for study. A fluorescent compound was attached to PCP to produce DANSYL-PCP and to aniline to produce a control (DANSYL-aniline). DANSYL-PCP was synthesized to have only one potential cationic functionality, which was absent in DANSYL-aniline. The DANSYL group does contain a dimethylamino substituent, which, if protonated when bound to hair, would form a cationic ammonium salt. If this compound were retained in the hair, the DANSYL group would not be fluorescent.

Hair was exposed to solutions of these compounds, washed, embedded, and sectioned for microscopic observation. Although hair soaked in a solution of DANSYL-aniline showed some fluorescence, its intensity was much less than that of the labeled PCP. This could indicate that either DANSYL-aniline did not bind as well as the labeled PCP or that when it was bound, it was protonated and therefore not fluorescent. Because the source and history of the hair were known, it was clear that the hair was not damaged by bleaching or perming. Such damage could allow access of the fluorescent compounds to the interior. In many cross sections of the hair shaft the DANSYL-PCP was only on the surface. However, the fluorescence was observed throughout the hair in a few cases. In other sections, the fluorescence was both in the center and on the outside. These differential results were obtained on the same strand of hair indicating that the hair was not resistant to penetration of the drug derivative. Because a single strand of hair produced such varying results, even greater variability among hair samples from different individuals might be expected.

Further exploration of "inaccessible regions" by DeLauder and Kidwell[89a] used fluorescent materials (rhodamine 6G and fluorescein) as drug surrogates. Eleven types of hair were exposed to these materials for varying lengths of time at 37°C with varying concentrations of fluorescent compounds (1 to 10 µg/mL). The exposed hair was washed, mounted, and sectioned. Under a fluorescent microscope, cross sections in almost every hair type showed incorporation of rhodamine throughout the hair.

Some specimens required either longer exposure times or greater concentrations of fluorescent dye to incorporate it throughout the hair. In only a Korean hair sample was uniform rhodamine staining throughout the hair shaft not achieved.

At low rhodamine concentrations, the rate of diffusion into the hair shaft was sufficiently slow to monitor. A cross section of Caucasian hair exposed to rhodamine at 1 µg/mL is shown in Figure 16. As exposure time was increased, more rhodamine reached the interior of the hair. At 120 min, even though the surface was much brighter than the interior, substantial fluorescence was observed in the interior when compared with unexposed hair or to the 30-min exposure. At a rhodamine exposure of 10 µg/mL for 2 h the rhodamine had penetrated throughout the hair sample (Figure 16).

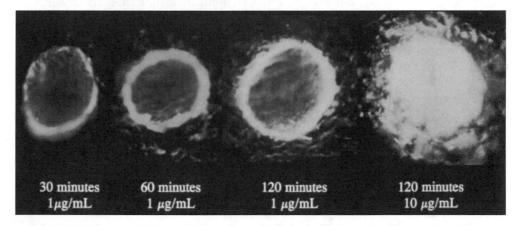

Figure 16
Caucasian hair exposed to rhodamine 6G for varying lengths of time. Whole Caucasian brown hair was exposed to rhodamine in 10 mM phosphate, pH 5.6, at 37°C for the times indicated. The excess was removed, the hair dried, mounted, and sectioned. The sections were photographed using a fluorescent microscope with appropriate excitation and barrier filters. The exposures were held constant to allow comparison of fluorescent intensities. The photographs were scanned into a computer, fluorescence intensity converted to a gray scale, and the gamma curve adjusted to reflect the relative intensities of the original photographic image.

These data suggest that, at least at the micron level (using 20× or 40× objectives), there are no inaccessible regions. Incorporation at the molecular level is not ruled out by these data, although stoichiometric considerations may suggest otherwise.*

M. Does the Hair Matrix Protect Previously Deposited Materials from Modification by the External Environment? Are Metabolites Adequate to Determine Usage?

No. In certain situations, such as low moisture, drugs have been shown to remain in hair for long periods.[90] However, this condition is unrealistic for humans using

---

* For example, assume that one protein molecule can sequester one drug molecule. The molecular weight of most proteins is approximately 100 kDa and that of most drugs approximately 300 Da. Therefore, the molecular weight ratio of protein to drug is approximately 300:1. Further assume that drugs are preferentially sequestered over other species in the blood at a 1000:1 ratio. Cocaine has a maximum blood concentration of approximately 1 µg/mL of blood or 1 ppm. Thus, the maximum amount of cocaine that could be sequestered would be 30 ng/mg hair (1 mg hair[amount of protein] × 1000[binding ratio]/(300[protein/drug ratio] × 1,000,000[1 ppm])). This amount is less than some drug users have in their hair. Furthermore, we believe that the preferential sequestration of drugs over other blood components is quite unrealistic, and dedication of that much protein for this purpose seems biologically unreasonable.

normal hygienic practices. Nakahara and colleagues[82] have shown a loss of 50% of methoxymethamphetamine in five individuals after five months. Similarly, Henderson and co-workers[37] have shown a loss of deuterated cocaine in hair for individuals administered this drug and have suggested a half-life of approximately two months for cocaine in hair.

For some drugs such as amphetamines, PCP, and marijuana, the metabolites are either in low concentrations in the hair relative to the parent drug or are nonexistent. For cocaine, the metabolites can be identical to hydrolysis products that occur *in vitro*. These degradation products are also present in street samples of cocaine.[66] We call these compounds "pseudometabolites" as they are chemically identical to metabolites, but are created by chemical rather than metabolic processes. Thus, an individual could be exposed to these "pseudometabolites" without having ingested them, or they could be created in their hair by environmental degradation of passively incorporated cocaine. Five samples of street cocaine, furnished by the Drug Enforcement Administration and the U.S. Customs Service, were analyzed by GC/MS. The amounts of cocaine-related materials relative to cocaine for five samples are shown in Table 2. Some samples, such as crack cocaine, have up to 13% benzoylecgonine. Depending on a variety of circumstances, hair exposed to this impure cocaine could produce ratios of cocaine-to-benzoylecgonine identical to that observed in specimens of cocaine users' hair (see discussion on sweat above).

TABLE 2.

Amounts of Cocaine Metabolites in Street Samples of Cocaine

Cocaine source	Benzoylecgonine/cocaine	Methylecgonine/cocaine	Ecgonine/cocaine
Cocaine-HCl #1	0.86	0.57	1.76
Cocaine-HCl #2	2.39	3.41	17.3
Free base #1	4.96	0.07	5.41
Free base #2	3.63	0.10	5.45
Free base #3	12.7	0.09	29.7

Note: Values are relative to cocaine and were determined by GC/MS analysis of the samples.

Baumgartner and co-workers[50] argue that the presence of drugs in archaeological artifacts demonstrate that drugs may remain in hair for long periods of time. Springfield and colleagues[90] have analyzed cocaine in the hair of 12 ancient coca chewers (Peruvian mummies dating from A.D. 1000). They found more cocaine metabolites than cocaine (ratios of cocaine/benzoylecgonine of 0.04 to 0.19:1 compared to ratios of about 5.2:1[79] or 2.9:1[38] in current-day coca chewers). Springfield and co-workers[90] suggested that this significant difference in ratios may be due to degradation *in situ*.

Metabolites of cocaine could be produced in hair *in vitro* by exposing cocaine-contaminated hair to various situations. Cocaine was incorporated into two types of hair by exposure to cocaine solutions at 37°C for 1 h. After cocaine incorporation, the hair was either exposed to UV light to mimic sunlight or exposed to bicarbonate to mimic basic hair treatments, such as shampoos. Table 3 shows the percent of pseudometabolites relative to cocaine for each type of hair as determined by GC/MS analysis. Only percentages which were at least two times the control values are presented. Clearly, mild bases, such as bicarbonate, can produce significant amounts of benzoylecgonine and ecgonine in hair from the cocaine present. Likewise, Nakahara and Kikura[92] concluded from a study of rats injected with cocaine that some of the cocaine is hydrolyzed to benzoylecgonine and methyl ecgonine. From these studies, it appears that the hair matrix does not present an insurmountable barrier to environmental degradation. The data, presented in Smith and co-workers,[65] shows that several children of cocaine users have both cocaine and benzoylecgonine in their hair in quantities greater than the drug-using parent. This study again reinforces the presumption that benzoylecgonine is not an absolute indicator of cocaine use.

TABLE 3.

Degradation of Cocaine in Hair

	UV, 45°C, 21 h		Bicarbonate, room temp., overnight		Bicarbonate, 37°C, overnight	
	Brown	Black	Brown	Black	Brown	Black
Benzoylecgonine/cocaine	NS[a]	NS	13%	8.4%	9.7%	24%
Methyl ecgonine/cocaine	3.7%	NS	NS	NS	NS	NS
Ecgonine/cocaine	NS	NS	NS	NS	11%	27%

Note: After exposure to the indicated environmental conditions, the "metabolites" were extracted from the hair with acid, the acid evaporated, the extract silylated, and the derivatives analyzed by GC/MS. Unexposed controls were run to account for any degradation of cocaine during the incorporation and extraction stages.

[a] NS = not significant if less than 2 times the control.

Furthermore, sweat contains nonspecific esterases[93,94] and other enzymes[95,96] which may allow for degradation of the drug on the surface of the hair before incorporation into the hair matrix. These mechanisms for the production of metabolites in hair have not been considered in hair analysis. The plethora of hair treatments and environmental chemicals available that might be capable of influencing the production of "pseudometabolites" in hair would argue against automatically accepting the presence of metabolites in hair as the *sine qua non* of human ingestion.

N. Do Drugs Migrate Along the Hair Shaft?

Not quickly. Püschel et al.[58] found that cut human hair placed in a solution containing codeine showed diffusion of the drug to the distal portions of the hair within 1 h.* It is possible that our *in vitro* contamination procedure would not mimic passive exposure *in vivo* if diffusion of drugs along the hair shaft was possible. To test the possibility of diffusion along the hair shaft, two groups of hair (Asian black and Caucasian brown) were examined. The first group was sealed at both ends by immersion of the hair in hot paraffin. The second group was cut into 1-cm sections and not sealed. The hair was exposed to cocaine plus radiotracer, washed, and analyzed as previously described,[33] except the drying ethanol prewash was eliminated. As can be seen in Figure 17, no differences in the incorporation of cocaine were observed. In another study, the possibility of diffusion of materials along the hair shaft was tested by placing the ends of eleven different hair types in 10 µg/mL of rhodamine for 1 h at 37°C. The remainder of the hair was suspended above the liquid level. Following removal of the hair from the solution, the excess rhodamine was washed away in a flowing stream of water. The samples were cut near the liquid

---

* This is the only report in the literature mentioning diffusion along the hair shaft. It lacks mention of the amount of drug found in the distal ends, the concentration of the soaking solution, or the decontamination procedure. Furthermore, the assay for the codeine consisted only of a RIA.

level and both parts were mounted and sectioned so that the sections covered the part near the liquid level. A comparison of the two sections showed large amounts of rhodamine in the portion that was in contact with the solution, and no rhodamine in the hair portion outside the exposure solution. We concluded that, if migration of drugs occurs along the hair shaft, it would be only over a short distance. To test long-term diffusion, a series of experiments were conducted using the set-up shown in Figure 18. A single hair was placed in approximately 1 mL of phosphate buffer. A wax seal was formed by quickly pippeting hot wax onto the surface of the buffer. Then, 1 mL of the same buffer was placed on the top containing 50 µg/mL of either fluorescein or rhodamine 6G. The upper layer was sealed with wax. The cell was read at the initial time and at varying intervals afterward by placing it in a spectrometer and measuring the absorption at the wavelength maximum of each dye. A number of experiments were conducted with a series of phosphate buffers from pH 3 to 9 and two hair types, Asian black and Caucasian brown. No dye appeared in the lower buffer layer for at least a month. Some cuvettes had no dye migration even after two months. There did not appear to be any correlation between the buffer pH, dye structure, or hair type and the rate of migration. Because the hair sections were approximately 3 cm in length, a maximum longitudinal diffusion rate of approximately 1 mm/d can be estimated. If the diffusion was not entirely along the length of the hair strand, then the diffusion rate would be much slower. If the cross-sectional diffusion rate of these dye molecules were similar to the longitudinal diffusion rate, then it would take 0.7 h to diffuse into a 30-µm cross section of hair. This is in agreement with the microscopic measurements, cited above, given that the concentrations of dye in this study are a factor of 5 to 50 greater than in the cross-sectional studies.

In several controlled *in vivo* studies, Henderson and co-workes,[37] Cone[61] and Nakahara et al.[82] have shown that the thickness of the drug band in the hair shaft is greater than would be expected for the amount of hair growth and time that the drugs are in the bloodstream. Whether this represents movement of the drug along the hair during/after the formation process or incorporation of the drugs from sweat has not been determined.

O. Does Melanin Play a Role in Drug Binding?

Perhaps. Harrison and co-workers[97] suggest a specific structural requirement for drug incorporation to take place into melanin.[75] Compounds structurally similar to the L-DOPA precursors of melanin (such as the amphetamines) may be incorporated into the synthesis of the melanin contained in the hair. On the other hand, Ishiyama and colleagues[98] demonstrated that methamphetamine binds to hair of albino animals in amounts that are similar to those found in pigmented animals. They suggest that drugs need not be incorporated into the melanin pigment. In contrast to these results, Forrest et al.[99] found differing rates of incorporation of chlorpromazine into the black and white areas of Dutch Belt rabbits. They do not offer an explanation for this difference, such as varying fur growth rates. Forrest and colleagues[99] only monitored the incorporation of chlorpromazine through radioactivity measurements. It is possible that the moiety with the radioactive label could have been metabolized to another compound which subsequently was used in the synthesis of fur pigments. Sato and colleagues[100] also found substantial differences between the concentrations of chlorpromazine in different colors of hair in humans with both white and black hair. Likewise, Uematsu and co-workers[101] found substantial differences between the

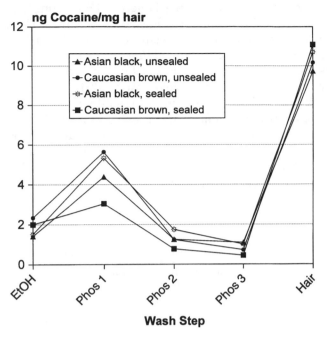

Figure 17
Effect of sealing the ends of the hair on the uptake of cocaine. The hair was exposed to cocaine, decontaminated, and analyzed. (From Kidwell, D. A. and Blank, D. L., in *Hair Testing for Drugs of Abuse: International Workshop on Standards and Technology*, E. J. Cone, M. J. Welch, and M. B. Grigson Babecki, Eds., National Institutes of Health Publication #95-3727, Superintendent of Documents, U.S. Government Printing Office, Washington, D.C., 1995, pp. 19-90.)

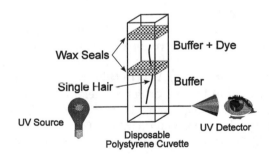

Figure 18
Schematic of device employed to determine long-term diffusion of dyes through hair. A similar set-up could be used for measurement of drugs employing radioactive tracers by placing the drug in the lower chamber. Removing aloquots from the top at regular intervals and measuring the radioactivity would then determine the breakthrough time.

amount of haloperidol in white vs. black hair in both rats and humans. In humans with some gray hair, the gray hair had only 10% as much haloperidol as the black hair. In all of their experiments, the haloperidol was extracted from the hair matrix after digestion of the hair with sodium hydroxide. Because no separation of the melanin fraction was performed, a clear demonstration of the association of haloperidol with melanin was never made.

To examine the role of melanin for the incorporation of cocaine, two types of hair were exposed to cocaine plus radiotracer. The amount of cocaine present in the melanin fraction was not substantial (see Table 4). The sodium hydroxide digestion procedure was used to replicate previous digestion procedures. These data are shown in the first two columns of Table 4 for Asian black and Caucasian brown hair. For this procedure, very little cocaine was found in the melanin fraction alone. A potential criticism of this digestion methodology is that sodium hydroxide would degrade the cocaine by converting the tritiated benzoyl group to tritiated benzoic acid. Because benzoic acid may not bind to melanin in the same manner as cocaine, the sodium hydroxide digestion may not reflect the true binding of cocaine to melanin. To address this concern, the enzymatic procedure of Baumgartner and Hill[91] was adopted. These results are shown in the third and fourth columns of Table 4. Again, little cocaine was observed in the melanin fraction. We concluded that regardless of hair type and digestion procedure, only small amounts of cocaine were associated with melanin.

TABLE 4.

Incorporation of Cocaine into Melanin as Measured by Digestion Techniques

	Sodium hydroxide digestion		Enzymatic digestion	
	Asian Black	Caucasian Brown	Asian Black	Caucasian Brown
Hair digest without melanin	9.98	5.97	7.49	6.09
Melanin fraction alone	0.69	0.27	0.15	0.20

Note: Numbers are in units of ng cocaine/10 mg hair. The hair was exposed to 2.5 µg/mL of cocaine containing tracer for 1 h, rinsed two times with distilled water, and dried. After centrifugation, all the digested hair in the supernatant could not be separated from the melanin pellet. Therefore, some residual amount of the drug remained associated with the melanin pellet. This amount is estimated to be between 2 to 10% of the total. Thus, the amount of cocaine actually associated with the melanin fraction is even less than that cited above.

Nakahara[102] has measured the uptake of a number of drugs into rat fur. He has suggested three factors that determine the incorporation of drugs into the fur from the bloodstream alone, as rats do not sweat: (1) the bioavailability of the drug, as measured by the area under the curve for the drug in blood (the time and concentration plots for a drug in the bloodstream); (2) the oil-water partition coefficient; and (3) the melanin binding affinity. We consider melanin to be a poor binding site for drugs. However, the binding of drugs to melanin is likely to be a similar ion exchange mechanism as to hair protein. The binding of drugs to hair protein was compared to melanin and, for a given quantity of hair, the protein fraction binds at least 58 times more drug than does the melanin fraction (Table 5). The amount of drug bound to each fraction per mg of original hair sample was corrected for the separation step and the amount that fraction corresponds to in the original hair specimen. The digestion process was varied in an attempt to isolate a representative protein fraction. However, all these procedures have their deficiencies in that the soluble proteins that result do not necessarily reflect those originally present in the intact matrix. Regardless of preparation procedure, the protein fraction bound substantial quantities of cocaine and morphine.

We have also employed melanin from synthetic sources and *Sepia officinalis*. However, synthetic melanin has a structure distinctly different[103] from naturally occurring melanins, and therefore it makes a poor model for hair melanin. Based on these results, melanin may play a role in some drug binding, especially for highly lipophilic drugs, but this role may be overshadowed by that played by the hair protein matrix.

TABLE 5.

Binding of Cocaine and Morphine to Hair Protein and Melanin

Drug used	μg of Drug/equivalent mg of hair						
	Melanin #1	Protein #1	Melanin #2	Protein #2	Melanin #3	Protein #3	Protein #4
Cocaine	0.0155	1.09	0.017	0.99	0.010	1.03	1.03
Morphine	0.0017	0.73	0.0017	0.69	0.0018	0.45	0.65
Ratio of binding (protein:melanin)							
Cocaine	70:1		58:1		103:1		
Morphine	429:1		406:1		250:1		

Note: Hair protein was isolated by each of four procedures. In procedure #1, hair was digested with sodium hydroxide and separated into melanin and protein fractions by centrifugation. Each fraction was extensively dialyzed against phosphate buffer to remove the sodium hydroxide and excess salts. Procedure #2 consisted of sodium hydroxide digestion, neutralization with hydrochloric acid, dialysis, and then separation of the melanin. Procedure #3 consisted of sodium hydroxide digestion, separation of the melanin, precipitation of the protein fraction with trichloroacetic acid, and then dialysis. Procedure #4, a modification of several literature methods, consisted of digestion of the protein with 8 *M* urea containing 10 mM dithiothreitol and 100 mM 2-mercaptoethanol at 50°C for 2 to 3 d with rapid stirring.[117-119] Procedure #4 did not digest all the hair. The undigested hair was removed by centrifugation and the protein solution extensively dialyzed.

The protein and melanin fractions were exposed to either cocaine or morphine at 5 μg/mL, containing a radioactive tracer. The excess drug not bound was removed by passage of the solution through Amicon® Centricon® 30 concentrators. The concentrated solution was washed once more with phosphate buffer (10 mM, pH 7) and reconcentrated. The drug concentration in all solutions was determined by scintillation counting. Because the solutions were colored, quenching corrections were performed as were done for hair samples.[47] The amount of proteins removed by the Centricon® 30 concentrators was quantitated by ultraviolet spectroscopy and the data in Table 5 corrected for this removal, which varied, depending upon the procedure employed to prepare the protein. The amount of drug bound is calculated relative to 1 mg of hair protein or the equivalent amount of melanin in 1 mg of Caucasian brown hair.

III. ANALYTICAL CONSIDERATIONS IN THE ANALYSIS
OF DRUGS IN HAIR

A. How Effective are Decontamination Procedures in Removing External Contamination?

Not very. Decontamination procedures vary substantially. For example, Martz[104] substitutes a methanol wash for the more widely employed anhydrous ethanol wash. He then examines the concentration of the drug in the methanol wash and compares it to the amount found in the hair to determine if the hair has been contaminated. The effectiveness of a methanol/ethanol wash in decontaminating hair has been studied in our laboratory.[29,47] In more recent experiments, a sample of Asian black hair was exposed to cocaine solution (5 µg/mL) plus radiotracer in PBS, pH 7, washed three times with water, and air dried prior to decontamination with either ethanol or methanol. For this hair type (Figure 19a) methanol removes more of the externally applied drug than ethanol but still did not remove more than 19%, even after two 30 min alcohol washes. The second methanol wash removed far less of the remaining cocaine (4.6%) than did the first wash. A false sense of security could be generated if the second methanol extract were analyzed to determine the extent of external contamination. The quantity of cocaine extracted into the methanol, being only a fraction of that present, may be below the instrumental detection limit. Consequently, the methanol may be considered negative, but the hair, with substantially more cocaine, could still be positive.

Other investigators have reached similar conclusions. If an individual's hair was exposed to an external source of drugs, a small percentage would be incorporated into the hair and the rest would remain on the surface. If a hair specimen was taken at that time, the decontamination kinetics would likely indicate that the hair was contaminated. However, after several normal hygienic washings, all of the surface-bound drug would be removed and only the drug in the interior would remain.* If a hair specimen was taken at that time, the decontamination kinetics would be substantially flatter and the specimen would not likely be considered contaminated. Baumgartner et al.[31] showed that even nine normal hygienic washings were unable to completely decontaminate hair exposed to crack cocaine smoke. Others have shown that hair soaked in solutions of drugs or exposed to the vapors are not completely decontaminated.[32,36]

Detergents appear to be among the worst solutions for decontaminating the hair. Figure 19c compares 1-h decontaminations using a number of commercial surfactants and hair care products. None of these remove more than 35% of the externally applied cocaine. A false sense of security could be generated if detergents are employed as decontaminating solutions.

The early work from this laboratory[23] has been extended to include a large number of solvents which could be used to decontaminate hair externally contaminated with cocaine (Figures 19b and c). After exposure to cocaine (5 µg/mL in 10 mM phosphate, pH 5.6), the Caucasian brown hair was washed five times with water and then dried. Quantitation was done using radiotracers, as previously described.[47] Each sample was decontaminated with the experimental solution for a period of 1 h. The remaining cocaine in the hair was determined after digestion with sodium hydroxide. The solvents most efficient in removing cocaine were 0.1 M HCl and dimethylformamide. However, even these solutions left approximately 30% of the externally applied cocaine in the hair.

---

* For example, see Figure 2 of Baumgartner and Hill (1992).[31] After two brief washings with Prell® shampoo the kinetic wash data are essentially flat. However, 1.68 ng (0.014% times 12,000, from Baumgartner's data) of cocaine/mg of hair still remain in the subsequent analysis.

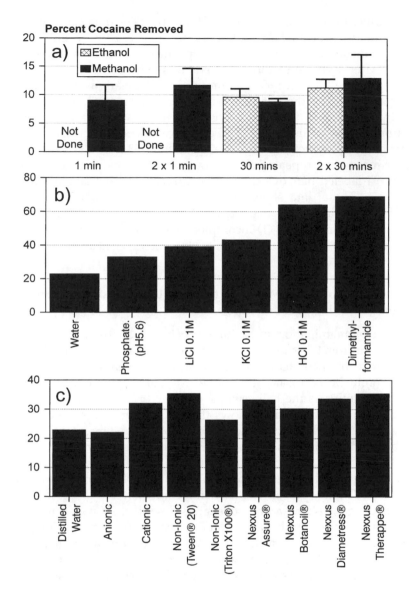

Figure 19
Total percent cocaine removed by various solutions.

B. Extraction Considerations

The difference between decontamination (removal of exogenous drugs and other interfering substances) and extraction (removal of endogenous drugs for analysis) is often blurred. Some investigators[39] employ the same extraction solution as was used for decontamination except applying it for a longer period or at a higher temperature. Although obvious, with respect to extraction of drugs from undigested hair, it is important to remember that extraction efficiency should parallel the decontamination efficiency for the various solvents. Therefore, the effectiveness of extraction solutions will be in proportion to their decontamination effectiveness.

The two most prevalent types of extraction procedures are (1) digestion of the hair matrix and extraction of the drug from the digest with liquid-liquid or solid-phase extraction procedures, or (2) extraction of the drug from the intact hair matrix.

These techniques have been reviewed by Chiarotti.[105] The data exploring the removal of cocaine and radiotracer by acid extraction from *intact* hair are shown in Figure 20. Cocaine was incorporated into hair by exposure to cocaine solutions containing radiotracer. After decontamination, the hair was extracted with 0.1 M HCl and the radioactivity in the HCl extract was measured. The hair was then digested with sodium hydroxide and the residual radioactivity measured. The ratio of cocaine extracted by HCl to the total amount of cocaine in the sample for different durations of extraction are shown in Figure 20a. For the two types of hair studied, extraction efficiency varied by extraction time and ranged from 53 to 86% in the 42-h extraction. This experiment was repeated for extraction of benzoylecgonine (Figure 20b) employing ^{125}I-labeled benzoylecgonine as a tracer. The extraction efficiency never exceeded 41%. These data show that extreme caution must be exercised in the analysis of intact hair samples extracted by HCl. Data resulting from acid extraction may be biased in three ways: (1) incomplete extraction, (2) different extraction efficiencies across hair types, and (3) different extraction efficiencies across drug moieties.

C. Are There Quantitative Measures that can Distinguish Between Endogenous and Exogenous Exposure?

No. All decontamination procedures are based on the *assumption* that drugs deposited in hair from the external environment are loosely bound to either the surface of the hair or to the hair matrix. If this assumption were correct, then only gentle treatments might be necessary to remove externally applied drugs. For example, Martz[104] decontaminates the hair using successive brief methanol extracts and tests each extract until the last extract shows that no further drugs have been removed. Henderson and co-workers[79] rinse hair samples with a detergent, followed by water, followed by methanol prior to digestion and extraction. Cone et al.[36] rinse hair samples briefly with methanol, then extract the drugs by incubating with methanol at 40°C overnight. Mangin and Kintz[106] decontaminate hair using a dichloromethane wash for 15 min at 37°C. They presumed this method sufficient because a second wash was negative for drugs. Nakahara and colleagues[107] wash their hair samples with 0.1% sodium dodecyl sulfate and water. Moeller et al.[108] attempt to eliminate external contamination with a 5-min warm-water wash followed by a 5-min acetone wash. A number of different decontamination procedures were tested to remove external contamination produced by solutions of drugs. These results are discussed above and shown in Figures 19 and 20. None were sufficient to remove all drugs.

Baumgartner and Hill[91] have adopted a more rigorous and consistent approach to decontamination. First, the hair is evaluated for damage by measuring the uptake of methylene blue dye. The severity of the decontamination procedure is varied, depending on this evaluation, and is made less severe when greater damage is observed in the hair sample. An undamaged hair sample would be washed with anhydrous ethanol to remove surface contamination, oils, and dirt. More recently, an isopropanol wash is employed for grease removal.[108a] Then a well-defined decontamination using three phosphate buffer washes is employed. The hair is then digested and analyzed. According to Baumgartner and Hill,[91] drugs loosely adhering to the surface are removed with the initial alcohol wash. If the methylene blue staining indicates porous or damaged hair, the phosphate buffer washes are replaced by varying mixtures of ethanol and water, 5 to 15%. Unfortunately, the objective criteria for evaluating porosity are not reported. Each of the phosphate washes is analyzed for drug concentration and the data analyzed for external contamination by the criteria cited in Table 6. The empirically determined cutoff values cited in

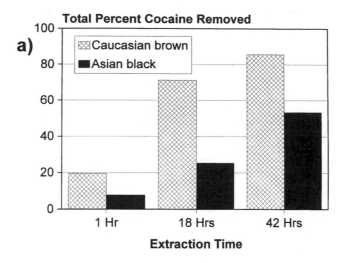

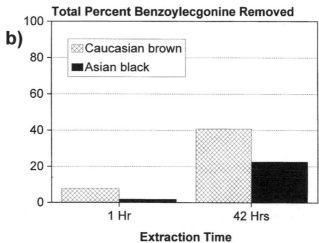

Figure 20

Removal of drugs by HCl extractions. (a) Cocaine or (b) [125]I benzoylecgonine. (From Kidwell, D. A. and Blank, D. L., in *Hair Testing for Drugs of Abuse: International Workshop on Standards and Technology*, E. J. Cone, M. J. Welch, and M. B. Grigson Babecki, Eds., National Institutes of Health Publication #95-3727, Superintendent of Documents, U.S. Government Printing Office, Washington, D.C., 1995, pp. 19-90.)

Table 6 for cocaine may vary for other drugs (see Baumgartner and Hill).[91] If a specimen fails any of the three cutoff criteria by having a value less than that listed in Table 6, it would be considered externally contaminated.

Baumgartner and Hill's decontamination procedure[91] was evaluated on more than 500 different hair samples that were externally contaminated in this laboratory. In the initial phases of this study we used four phosphate washes. In later phases, based upon the comments in Baumgartner and Hill,[109] our procedure was modified to use between three and ten phosphate washes. However, in the radioactive tracer experiments we typically perform, it is possible to account for all radioactivity; the sample is digested and no radioactivity is lost from any extraction step. Therefore, it is possible to calculate the effect of either three or four phosphate washes on the decontamination of the hair and on the various calculated criteria. No remarkable differences were found between our four-phosphate extraction process and the recalculated values for the equivalent three-phosphate extraction process.

TABLE 6.

Definitions of Kinetic Wash Criteria and Values
for Cocaine

$$\text{Rew} = \frac{\text{amount of drug in digest}}{\text{amount of drug in last phosphate wash}}$$

$$\text{Rsz} = \frac{\text{amount of drug in digest}}{\text{amount of drug in all the phosphate washes}}$$

$$\text{Rc} = \frac{\text{amount of drug in the three phosphate washes}}{3 \times \text{amount of drug in last phosphate wash}}$$

Cutoff values for cocaine. Specimens with values below
these numbers are considered contaminated. Otherwise,
the individuals are considered users of cocaine.

Rew	10
Rez	0.33
Rc	1.3

Definitions from Baumgartner and Hill[91] and reproduced
from Kidwell and Blank.[26]

The purpose of any objective kinetic wash criterion is to assess how effective the decontamination procedure has been. Both the extended wash ratio (Rew) and the safety zone ratio (Rsz) measure effectiveness of decontamination. However, the curvature ratio (Rc) does not appear to measure this variable. For example, the Rc for samples prepared by three different methods was examined. A sample of Asian hair was exposed to 50 μg/mL of cocaine plus radiotracer for 1 h and not prewashed with water prior to alcohol and phosphate decontamination. The second sample of Asian hair was exposed to 1 μg/mL of cocaine plus radiotracer for 1 h and was also not prewashed prior to decontamination. The third sample of Asian hair was also exposed to 1 μg/mL of cocaine plus radiotracer for 1 h, but it included three water prewashes before decontamination. Examples of the decontamination kinetics for these samples are shown in Figures 21 a, b, and c, respectively. As expected, samples that did not include a prewash exhibited a greater "curvature", i.e., a greater decrease in the amount extracted by successive phosphate washes than did the samples that included water prewashing. In other words, the greater the contamination, the greater the value of Rc and the more easily it would pass the cutoff criteria. Because this term does not seem to have any special predictive value and the amount of drug in all phosphate washes and the amount of drug in the last phosphate wash appears in calculations of Rew and Rsz, Rc is not reported here. The calculated values for Rew and Rsz for 110 specimens are shown in Figure 22. The data are plotted in order of increasing values for Rew.

Of the 110 samples examined, only a few fell below the cutoff values for these externally contaminated samples and then by only marginal amounts. A criticism of these data may be that many samples only pass the cutoff values by trivial percentages and that these empirically determined cutoffs could be adjusted to take the data into account.[109] Cutoff values of at least 214 for Rew would exclude all the samples in Figure 22 as contaminated, and at least 137 would exclude 95% of the samples as contaminated. Cutoff values this high would also preclude analysis of many, if not all, user samples. The frequency distributions of Rew and Rsz values for the specimens shown in Figure 22, that pass the respective cutoffs, are plotted in Figure 23 and compared to Baumgartner and Hill's distributions.[91] The distributions are remarkably similar.

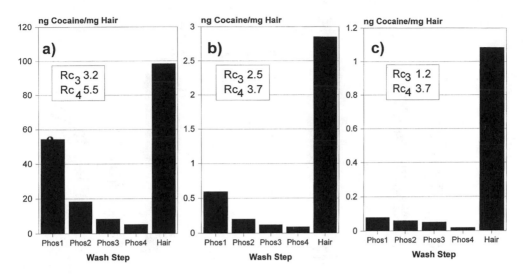

Figure 21
Decontamination curves for three sets of Asian black hair. The Rc values were taken as either $Rc_3 = [(Phos1 + Phos2 + Phos3)/(3 \times Phos3)]$ or $Rc_4 = [(Phos1 + Phos2 + Phos3 + Phos4)/(3 \times Phos4)]$. (From Kidwell, D. A. and Blank, D. L., in *Hair Testing for Drugs of Abuse: International Workshop on Standards and Technology*, E. J. Cone, M. J. Welch, and M. B. Grigson Babecki, Eds., National Institutes of Health Publication #95-3727, Superintendent of Documents, U.S. Government Printing Office, Washington, D.C., 1995, pp. 19-90.)

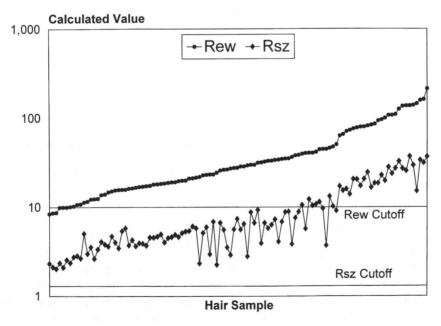

Figure 22
Rank ordered plot of Rew and Rsz values for 110 externally contaminated samples. Ten different types of hair were exposed to cocaine in PBS at 5 µg/mL for varying lengths of time. (From Kidwell, D. A. and Blank, D. L., in *Hair Testing for Drugs of Abuse: International Workshop on Standards and Technology*, E. J. Cone, M. J. Welch, and M. B. Grigson Babecki, Eds., National Institutes of Health Publication #95-3727, Superintendent of Documents, U.S. Government Printing Office, Washington, D.C., 1995, pp. 19-90.)

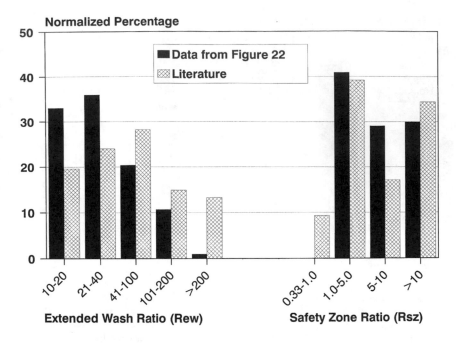

Figure 23

Distributions of Rew and Rsz from the data in Figure 22 and the literature. The three samples not meeting the Rew cutoff were excluded. The literature values were adapted from Baumgartner and Hill.

More recently, we have employed an extensive, ten-phosphate decontamination paradigm over a 5-h period before analysis in an attempt to remove the externally applied drugs. Even with this extensive and laborious washing, substantial quantities of drugs continue to be sequestered in the hair (Figure 24). In fact, some of the quantitative criteria, such as Rew (61:1 and 18:1), becomes *more "user-like" the more extensive the decontamination*. These results strongly suggest that regardless of the decontamination procedure employed, *external contamination cannot be fully removed*.

IV. IMPLICATIONS OF DATA AND THEORETICAL FRAMEWORK

The data suggest that the mechanism for incorporation of drugs into the hair matrix is a complex process. A schematic diagram of possible routes of incorporation of drugs into hair is shown in Figure 3. If an individual is a drug user, drugs may enter the hair by one or more of three different routes. The first is via sweat. Ingested drugs and drug metabolites will be excreted in sweat and sebum, bathe the hair shaft, and be incorporated into the hair matrix, possibly by an ion-exchange mechanism. The second is via external exposure. It is likely that the environment of a drug user will contain drugs of abuse. If the drugs came in contact with hair, they would likewise be incorporated, perhaps mediated by solutions such as sweat or normal wetting of hair by hygienic practices. This route includes transfer of drug vapors or powders directly to the hair or physical transfer from drugs deposited on surfaces, e.g., money[110] or clothing, to hair via the hands. The third route of transfer is via the blood, with drugs being transferred into the growing portion of the hair shaft from the capillary bed engulfing that structure.

As Figure 3 further illustrates, there is usually some passage of time between ingestion and hair analysis. During that time, drugs loosely bound to the surface of the hair could be washed away by normal hygienic hair care. The removal of drugs

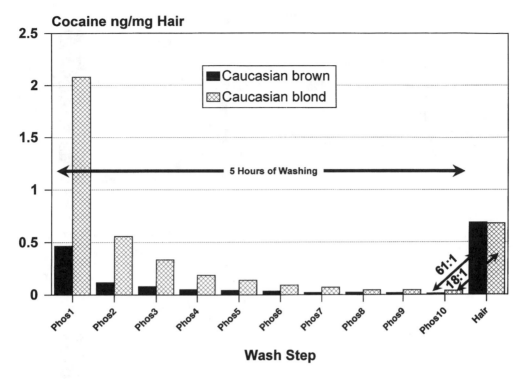

Figure 24
Removal of externally applied cocaine by extensive washing.

will depend on several variables, not the least of which are the characteristics of the solutions used to wash or treat the hair. In fact, one might conceptualize these practices as an *in vivo* extraction of drugs of abuse and contrast them with laboratory extraction procedures.

Baumgartner and Hill[91] proposed the concept of inaccessible regions primarily based on the experimental extraction kinetics. The means by which drugs enter the hair shaft may not be revealed by hair analysis procedures. For example, if the model shown in Figure 1 is assumed correct, drugs present in the bloodstream could be entrapped into inaccessible regions of the hair during the hair growth process. A cross section of such a hair shaft is modeled in Figure 25. The model in Figure 25 assumes that the concentration of the drug in the hair shaft would be greater in the interior layers of the hair. Because the hair shaft is considered relatively impermeable to contamination by drugs in the external environment (or alternatively, external contamination could be washed away), drugs in the external layers are less concentrated than in the core. Analyzing a hair sample whose cross-sectional concentration profile is depicted in Figure 25 should yield the decontamination and analysis curve shown. We should reemphasize that this concept is only a hypothesis that appears to explain the experimental decontamination kinetics.

In our dye experiments discussed earlier, we demonstrated that either a uniform cross-sectional drug profile or one with more in the exterior layers could be generated by passive exposure. These two profiles are modeled in Figure 26. The decontamination kinetics for the model described in Figure 25 should be distinctly different from that shown in Figure 26 — *but for one problem:* people wash their hair. Normal hygiene will remove the easily removable drugs and convert either profile with more drugs concentrated on the outside, or one with a uniform concentration, to a profile mimicking that presumed for the inaccessible region model shown in Figure 25. Thus,

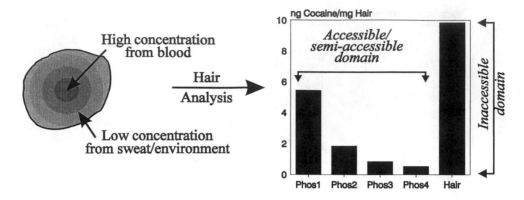

Figure 25
Model of drug distribution consistent with inaccessible regions. For clarity, the drug concentration in the hair cross section is modeled as discrete steps of concentration. A more appropriate model may be a continuous gray scale representing a continuous concentration gradient of the drug.

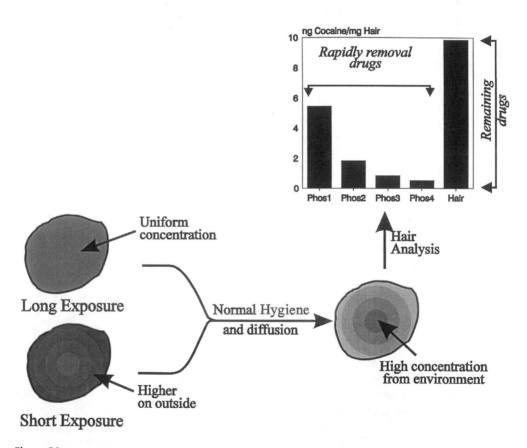

Figure 26
Alternative model for producing a drug distribution that would mimic inaccessible regions. Diffusion of drugs from the hair matrix due to normal hygiene would change the initial drug distribution to one with less drug on the exterior portions. Laboratory decontamination curves of such a hygienically washed hair would generate identical curves to that of suspected drug users modeled in Figure 25.

we believe that most decontamination curves only reflect rapidly removed drug vs. the remaining drug concentration — not the origin of the drug in the hair.

As noted earlier, drugs can be incorporated into hair over rather short periods of time, and the extraction kinetics of these externally deposited drugs mimic those from the hair of a drug user. The extent to which this would be observed in a nonuser would depend upon the exposure to drugs from the environment in which that individual lives. For example, Smith et al.[65] have shown that cocaine can be found in hair samples collected from spouses and children of cocaine users. In some of these cases (such as young children 3 to 8 years of age), intentional ingestion of cocaine would be highly unlikely. It is well known that cocaine is found on circulating currency,[110] but the extent to which cocaine could be transferred from currency to hands to hair is not known. However, it must be emphasized that conditions on the head of an individual could favor the incorporation of drugs into the hair matrix. For example, if drugs are transferred to head sweat, their concentration would increase as the sweat evaporates, thus favoring incorporation into the hair matrix. In addition, the infrequent washing of hair or the application of hair care products by some individuals could also favor incorporation.

The profound effects of inadvertent exposure to drugs are illustrated in Figure 27. This hypothetical example assumes that two nonusing individuals are exposed to drugs. The first (A) inadvertently ingests 80 to 100 mg of cocaine while the second (B) was inadvertently exposed to a few micrograms of cocaine transferred to his hair. Figure 27 further assumes that both A and B are subjected to a random screening program (A, urinalysis; B, hair analysis) where either could be tested once on any day in a 120-d period. Individual B has a standard military hairstyle with hair 3 cm in length and obtains a haircut every two weeks. At the above-mentioned dosage, assume that cocaine can be detected in A's urine for 3 d following ingestion, and the amount of cocaine that can be detected in B's hair decreases gradually over a 120-d period; the decrease in concentration due to hygiene is accelerated by the frequent military haircuts. In this hypothetical example, it would be 37 times more likely for individual B to be positive on a hair test than individual A on a urinalysis.* Clearly, given

---

* The following events are defined below:

TP = Testing positive
DP = Drugs present in system — for individual A this means ingestion of 80 to 100 mg cocaine into the body. For individual B this means hair is dusted with microgram quantities of cocaine.
EX = Exposure to cocaine — this means physically in the presence of cocaine.

In general, the probability of a positive test can be expressed as the product: probability of testing positive = (probability of testing positive given drugs are present) × (probability of drugs being present given a certain exposure) × (probability of the exposure). Mathematically this is represented by:

$$P(TP) = P(TP \mid DP) \, P(DP \mid EX) \, P(EX)$$

For individual A, we have:
Probably of A testing positive = (days which drugs are detectable in urine) × (probability of A being given a certain exposure) × (probability of exposure) or mathematically:

$$P_A(TP) = (3/120) \, P_A(DP \mid EX) \, P_A(EX)$$

and for individual B:
Probability of B testing positive = (days which drugs are detectable in hair) × (probability of B being given a certain exposure) × (probability of exposure) or mathematically:

$$P_B(TP) = (110/120) \, P_B(DP \mid EX) \, P_B(EX)$$

If the probabilities of exposure are the same [$P_A(EX) = P_B(EX)$], it would be 37 times more likely for individual B to be positive on a hair test than for individual A on urinalysis (110/120 ÷ 3/120).

inadvertent exposure, the window for false accusation by urinalysis is far shorter than for that of hair analysis. This problem would be exacerbated by the fact that individual A could have noticed a physiological or psychological effect of inadvertent ingestion and could document that event in his defense, whereas individual B would have no knowledge of being passively exposed until that event was revealed by hair analysis. Given the longer window of detection for hair analysis, this would put the burden of proof on the accused to document their exact whereabouts over a long period of time. The inadvertent ingestion of 80 to 100 mg of cocaine does not intuitively seem to be as likely as passive exposure of hair to 8 to 10 µg of cocaine. The former would most likely require an intentional act of another individual, whereas the latter would only require that an individual be in an environment where cocaine is present.

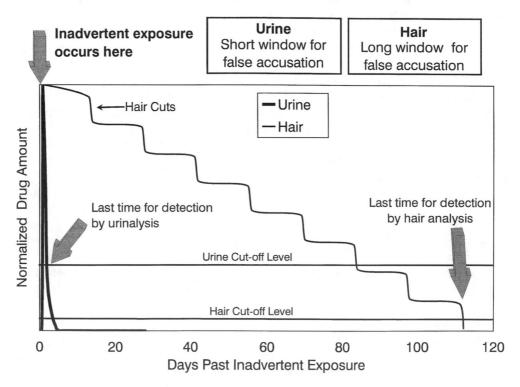

Figure 27
Hypothetical window of detection for urinalysis and hair analysis after an inadvertent exposure to cocaine. Losses of cocaine from the hair come mainly from haircuts rather than the gradual loss by normal hygiene. (From Kidwell, D. A. and Blank, D. L., in *Hair Testing for Drugs of Abuse: International Workshop on Standards and Technology*, E. J. Cone, M. J. Welch, and M. B. Grigson Babecki, Eds., National Institutes of Health Publication #95-3727, Superintendent of Documents, U.S. Government Printing Office, Washington, D.C., 1995, pp. 19-90.)

Another example might help to elucidate this issue. Suppose someone attended a party on Friday night and was given cocaine in a mixed drink, without his/her knowledge. On Monday they were tested by urinalysis and, by the end of the week, urinalysis results would indicate that they tested positive for cocaine. Prior to a judicial proceeding, that individual would be able to recall the inadvertent exposure event and, by so doing, present data on that incident in his/her defense. Furthermore, witnesses could be produced that would substantiate the claim that cocaine was being used at the Friday night party and that the accused did not appear to participate in the use of cocaine. Contrast this situation to reconstructing these events 120 d later and being unable to know when or where the inadvertent exposure had occurred. The credibility of any defense witnesses would be impaired by time and uncertainty when measured against the hard fact (as determined by hair analysis) that cocaine exposure had occurred. Perhaps, until better decontamination procedures are established, only short hair lengths near the scalp should be tested to limit environmental exposure, as was suggested for metal analysis in hair.[14] Of course, such a limitation would reduce the potential advantage of hair analysis to provide long-term drug use history for an individual.[111]

V. CONCLUSIONS

Current findings suggest that cocaine and benzoylecgonine (and likely other drugs) may be readily incorporated into hair from environmental exposure and not removed by any one of several decontamination techniques. The radiotracer procedures outlined in this chapter allow a large number of specimens to be analyzed with high precision under a variety of conditions. Any new decontamination procedure could be tested through the use of these techniques. Several authors have proposed wash ratios as one criteria to distinguish active from passive exposure. This chapter reviewed data from a number of externally contaminated samples and found that the suggested criteria were inadequate to identify the samples as contaminated. Several decontamination scenarios employing detergents and solvents were also examined, and none completely decontaminated the hair. For some drugs, metabolites in hair may distinguish active use from passive exposure. However, for cocaine most of the metabolites may be produced by means other than metabolic processes secondary to drug ingestion.

Several variables must be considered in the analysis of hair besides the decontamination procedure. The most important is the method by which drugs are to be extracted from the hair matrix. Because no extraction solution can remove all of the drug present, dissolution of the hair matrix before extraction of the drug appears essential to achieve complete removal for analytical purposes.

The evidence from this laboratory and from others reviewed in this chapter reinforces and extends the serious concern that external contamination of hair by drugs of abuse can easily occur. Any interpretation of hair analysis data should consider the prospect that the sample could have been externally contaminated. The pharmacokinetics of the incorporation of drugs into many tissues has been well elaborated.[112] However, substantial additional information on the mechanisms for incorporation of drugs into hair, the decontamination of hair, the differentiation between exposure to exogenous and endogenous drugs, and the meaning of the presence of metabolites in the hair are needed before hair analysis can be employed in many forensic applications.

Given the current level of understanding, hair analysis could be useful as an adjunct to population surveys, as suggested by a recent U.S. General Accounting Office report.[113] But even in this case, the intended application of the data must be carefully considered. If trends in drug use in a large population are generated from a small cross section of the population by hair analysis, the inclusion of a few false positives may seriously skew the data to an overly high use rate. If these data were being used to make policy decisions, then a large or increasing use rate could prompt the government to divert resources, personnel, or enact legislation in a wasteful manner.

The lack of a firm scientific basis was partially responsible for the 1990 consensus option of the Society of Forensic Toxicology (SOFT) that stated "Hair may be a useful specimen in forensic investigations when supported by other evidence of drug use..."[114,115] This review extends the SOFT consensus option that hair analysis should not be used (even in preemployment testing) without corroborative evidence of drug use because the source of drugs in hair cannot be firmly established at the present state of knowledge in hair testing.

As a familiar American proverb says, "The Devil is in the details." There are numerous details left unanswered concerning hair analysis. It is time to pay the Devil his due and address these details before hair testing is widely employed.

ACKNOWLEDGMENTS

The authors would like to thank Mehret Mandefro for obtaining the hair care survey results, and Saundra DeLauder and Mehret Mandefro for their work in some of the fluorescein, rhodamine, and radioactivity incorporation experiments. We would also like to thank Wendy Lavern and Janel Holland for their work on many of the radioactivity incorporation experiments. The numerous suggestions for improvement of this chapter by Dianne Murphy, David Venezky, and Gayle Kupcho were very helpful. Support for part of this work from the Office of Naval Research, the Bureau of Naval Personnel, and the Office of National Drug Control Policy is also gratefully acknowledged.

REFERENCES

1. O'Conner, C. and Miller, M., The military says 'No', *Newsweek*, Nov. 10, 1986, p. 26.
2. Bray, R. M., Kroutil, L. A., Wheeless, S. C., Iannacchione, V. G., Anderson, D. W., Marsden, M. E., Bailey, S. L., Fairbank, J. A., and Hartford, T. C., *1995 Department of Defense Survey of Health Related Behaviors Among Military Personnel*, Research Triangle Institute, RTI/6019/06-FR, December, 1995.
3. Spolar, C., How lab results can change lives, *Washington Post*, Dec. 6, 1988, front page.
4. Holstein, W. L., The other side of drug testing, *Chemtech*, 523, September 1992.
5. Beers, N., Navy's drug test goes overboard, Letter to the Editor, *Philadelphia Inquirer*, Dec. 13, 1986.
6. Ambre, J., The urinary excretion of cocaine and metabolites in humans: a kinetic analysis of published data, *J. Anal. Toxicol.*, 9, 241, 1985.
7. ElSohly, M. A., Stanford, D. F., and ElSohly, H. N., Coca tea and urinalysis for cocaine metabolites, *J. Anal. Toxicol.*, 10, 256, 1986.
8. Siegel, R. K., ElSohly, M. A., Plowman, T., Rury, P. M., and Jones, R. T., Cocaine in herbal tea, *JAMA*, 255, 40, 1986.
9. Fenton, J. W., Navy's drug testing won't snag 'innocent bystanders', Letters to the editor, Los Angeles *Herald Examiner*, Dec. 4, 1986, p. A16.
10. Knudson, M., Non-users test positive, fuel drug screening furor, *Baltimore Sun*, Nov. 11, 1986, p. 2F.
11. Cone, E. J. and Johnson, R. E., Contact highs and urinary cannabinoid excretion after passive exposure to marijuana smoke, *Clin. Pharmacol. Ther.*, 40, 247, 1986.
12. Cone, E. J., Johnson, R. E., Darwin, W. D., Yosefnejad, D., Mell, L. D., Paul, B. D., and Mitchell, J., Passive inhalation of marijuana smoke: urinalysis and room air levels of delta-8-tetrahydrocannabinol, *J. Anal. Toxicol.*, 11, 89, 1987.
13. Lenihan, J., *Measuring and Monitoring the Environment*, J. Lenihan and W.W. Fletcher, eds., Academic Press, New York, 1978, pp. 66–86.
14. Wilhelm, M., Ohnesore, F. K., Lombeck, I., and Hafner, D., Uptake of aluminum, cadmium, copper, lead, and zinc by human scalp hair and elution of the adsorbed metals, *J. Anal. Toxicol.*, 13, 17, 1989.
15. Valković, V., *Human Hair Fundamentals and Methods for Measurement of Elemental Composition*, Volume I, CRC Press, Boca Raton, FL, 1988.
16. Valković, V., *Human Hair Trace Element Levels*, Volume II, CRC Press, Boca Raton, FL, 1988.
17. Hopps, H. C., The biologic basis for using hair and nail for analysis of trace elements, *Sci. Total Environ.*, 7, 71, 1977.

18. Chatt, A. and Katz, S. A., *Hair Analysis: Applications in the Biomedical and Environmental Sciences*, VCH Publishers, New York, 1988, pp. 14–16 and pp. 77–81.

19. Manson, P. and Zlotkin, S., Hair analysis — a critical review, *Can. Med. Assoc. J.*, 133, 186, 1985.

20. Harkey, M. R. and Henderson, G. L., Hair analysis for drugs of abuse, *Adv. Anal. Toxicol.*, 2, 298, 1989.

21. Barrett, S., Commercial hair analysis: science or scam?, *JAMA*, 254, 1041, 1985.

22. Hambridge, K. M., Hair analysis: worthless for vitamins, limited for minerals, *Am. J. Clin. Nutr.*, 36, 943, 1982.

23. Kidwell, D. A., Analysis of drugs of abuse in hair by tandem mass spectrometry. *Proceedings of the 36th American Society of Mass Spectrometry Conference on Mass Spectrometry and Allied Topics*, ASMS, San Francisco, CA, June 6–10, 1988, pp. 1364–1365.

24. Blank, D. L. and Kidwell, D. A., Screening of hair for drugs of abuse — is passive exposure a complication, 41st Annual Meeting of the American Academy of Forensic Sciences, Las Vegas, NV, February 1989.

25. Kidwell, D. A. and Blank, D. L., Deposition of drugs of abuse into human hair. Society of Forensic Toxicology Conference on Hair Analysis, May 26, 1990.

26. Blank, D. L. and Kidwell, D. A., External contamination of hair by drugs of abuse; an issue in forensic interpretation, *Forensic Sci. Int.*, 63, 145, 1993.

27. Henderson, G. L., Harkey, M. R., and Jones, R., Hair analysis for drugs of abuse. Final Report on Grant Number NIJ 90-NIJ-CX-0012 to National Institutes of Justice, September 1993.

28. Baumgartner, W. A., Hill, V. A., and Blahd, W. H., Hair analysis for drugs of abuse, *J. Forensic Sci.*, 34, 1433, 1989.

29. Kidwell, D. A. and Blank, D. L., Mechanisms of incorporation of drugs into hair and the interpretation of hair analysis data. In: *Hair Testing for Drugs of Abuse: International Workshop on Standards and Technology*, E. J. Cone, M. J. Welch, and M. B. Grigson Babecki, eds., National Institutes of Health Publication #95-3727, Superintendent of Documents, U.S. Government Printing Office, Washington, D.C., pp. 19-90.

30. Koren, G., Klein, J., Forman, R., and Graham, K., Hair analysis of cocaine: differentiation between systemic exposure and external contamination, *J. Clin. Pharmacol.*, 32, 671, 1992.

31. Baumgartner, W. A. and Hill, V. A., Hair analysis for drugs of abuse: decontamination issues, in I. Sunshine, ed., *Recent Developments in Therapeutic Drug Monitoring and Clinical Toxicology*, Marcel Dekker, New York, 1992, pp. 577–597.

32. Welch, M. J., Sniegoski, L. T., Allgood, C. C., and Habrum, M., Hair analysis for drugs of abuse: evaluation of analytical methods, environmental issues, and development of reference materials, *J. Anal. Toxicol.*, 17, 389, 1993.

33. Kidwell, D. A. and Blank, D. L., Hair Analysis: Techniques and Potential Problems. Presented at Recent Developments in Therapeutic Drug Monitoring and Clinical Toxicology, Barcelona, Spain, 1991. In: I. Sunshine, ed., *Recent Developments in Therapeutic Drug Monitoring and Clinical Toxicology*, Marcel Dekker, New York, 1992, pp. 555–563.

34. Baumgartner, A. M., Jones, P. F., and Black, C. T., Detection of phencyclidine in hair, *J. Forensic Sci.*, 26, 576, 1981.

35. Sramek, J. J., Baumgartner, W. A., Tallos, J. A., Ahrens, T. N., Heiser, J. F., and Blahd, W. H., Hair analysis for detection of phencyclidine in newly admitted psychiatric patients, *Am. J. Psychiatry*, 142, 950, 1985.

36. Cone, E. J., Yousefnejad, D., Darwin, W. D., and Maguire, T., Testing human hair for drugs of abuse. II. Identification of unique cocaine metabolites in hair of drug abusers and evaluation of decontamination procedures, *J. Anal. Toxicol.*, 15, 250, 1991.

37. Henderson, G. L., Harkey, M. R., and Jones, R., Hair analysis for drugs of abuse. Final Report on Grant Number NIJ 90-NIJ-CX-0012 to National Institutes of Justice, September 1993.

38. Möller, M. R., Fey, P., and Rimbach, S., Identification and quantitation of cocaine and its metabolites, benzoylecgonine and ecgonine methyl ester, in hair of Bolivian coca chewers by gas chromatography/mass spectrometry, *J. Anal. Toxicol.*, 16, 291, 1992.

39. Cone, E. J., Darwin, W. D., and Wang, W.-L., The occurrence of cocaine, heroin and metabolites in hair of drug abusers, *Forensic Sci. Int.*, 63, 55, 1993.

40. Mieczkowski, T., Barzelay, D., Gropper, B., and Wish, E., Concordance of three measures of cocaine use in an arrestee population: hair, urine, and self-report, *J. Psychoactive Drugs*, 23, 241, 1991.

41. Harkey, M. R., Henderson, G. L., and Zhou, C., Simultaneous quantitation of cocaine and its major metabolites in human hair by gas chromatography/chemical ionization mass spectrometry, *J. Anal. Toxicol.*, 15, 260, 1991.

42. Fritch, D., Groce, Y., and Rieders, F., Cocaine and some of its products in hair by RIA and GC/MS, *J. Anal. Toxicol.*, 16, 112, 1992.

43. Kidwell, D. A., Analysis of drugs of abuse in hair by tandem mass spectrometry. *Proceedings of the 36th American Society of Mass Spectrometry Conference on Mass Spectrometry and Allied Topics*, ASMS, San Francisco, CA, June 6–10, 1988, pp. 1364–1365.

44. Kidwell, D. A., Analysis of phencyclidine and cocaine in human hair by tandem mass spectrometry, *J. Forensic Sci.*, 38, 272, 1993.

45. Kidwell, D. A. and Blank, D. L., Deposition of drugs of abuse into human hair. Society of Forensic Toxicology Conference on Hair Analysis, May 26, 1990, Washington, D.C.

46. Blank, D. L. and Kidwell, D. A., External contamination of hair by drugs of abuse; an issue in forensic interpretation, *Forensic Sci. Int.*, 63, 145, 1993.

47. Blank, D. L. and Kidwell, D. A., Decontamination procedures for drugs of abuse in hair. Are they sufficient?, *Forensic Sci. Int.*, 70, 13, 1995.

48. Sellers, J. K., The Effects of Hair Treatment on Cocaine Contamination from External Exposure, M.S. thesis, University of Alabama, Birmingham, 1994.

49. Baumgartner, A. M., Jones, P. F., Black, C. T., and Blahd, W. H., Radioimmunoassay of cocaine in hair, *J. Nuclear Med.*, 23, 790, 1982.

50. Baumgartner, W. A., Hill, V. A., and Blahd, W. H., Hair analysis for drugs of abuse, *J. Forensic Sci.*, 34, 1433, 1989.

51. Smith, F. P. and Liu, R. H., Detection of cocaine metabolite in perspiration stain, menstrual bloodstain and hair, *J. Forensic Sci.*, 31, 1269, 1986.

52. Balabanova, S., Schneider, E., Wepler, R., Bühler, G., Hermann, B., Boschek, H. J., Schneitler, H., and Jenzmik, H., Capacity of the eccrine sweat glands to store cocaine, *Dermatol. Monatsschr.*, 178, 89, 1992.

53. Balabanova, Von, S. and Schneider, E., Nachis von drogen im schiβ. *Beitr. Gerichtl. Med.*, 48, 45, 1990.

54. Ishiyama, I., Nagai, To., Nagai, Ta., Komuro, E., Momose, T., and Akimori, N., The significance of drug analysis of sweat in respect to rapid screening for drug abuse, *Z. Rechtsmed.*, 82, 251, 1979.

55. Klug, E., Zur morphinbestimmung in kopfhaaren, *Z. Rechtsmed.*, 84, 189, 1980.

56. Marigo, M., Tagliaro, F., Polesi, C., Lafisca, S., and Neri, C., Determination of morphine in the hair of heroin addicts by high performance liquid chromatography with fluorimetric detection, *J. Anal. Toxicol.*, 10, 158, 1986.

57. Offidani, C., Carnevale, A., and Chiarotti, M., Drugs in hair: a new extraction procedure, *Forensic Sci. Int.*, 41, 35, 1989.

58. Püschel, K., Thomash, P., and Arnold, W., Opiate levels in hair, *Forensic Sci. Int.*, 21, 181, 1983.

59. Mangin, P. and Kintz, P., Variability of opiate concentrations in human hair according to their anatomical origin: head, axillary and pubic regions, *Forensic Sci. Int.*, 63, 77, 1993.

60. Kintz, P., Tracqui, A., and Mangin, P., Opiate concentrations in human head, axillary, and pubic hair, *J. Forensic Sci.*, 38, 657, 1993.

61. Cone, E. J., Testing human hair for drugs of abuse. I. Individual dose and time profiles of morphine and codeine in plasma, saliva, urine, and beard compared to drug-induced effects on pupils and behavior, *J. Anal. Toxicol.*, 14, 1, 1990.

62. Haley, N. J. and Hoffmann, D., Analysis for nicotine and cotinine in hair to determine cigarette smoker status, *Clin. Chem.*, 31, 1598, 1985.

63. Kintz, P., Ludes, B., and Mangin, P., Evaluation of nicotine and cotinine in human hair, *J. Forensic Sci.*, 37, 72, 1992.

64. Kintz, P., Gas chromatographic analysis of nicotine and cotinine in hair, *J. Chromatogr.*, 580, 347, 1992.

65. Smith, F. P., Kidwell, D. A., and Cook, L. F., Children of cocaine users show passive drug incorporation in their hair. Paper presented at the 2nd International Meeting on Clinical and Forensic Aspects of Hair Analysis, Genova, Italy, June 1994. In: E.J., Cone, ed. *Proceedings of the 2nd International Meeting on Clinical and Forensic Aspects of Hair Analysis*, National Institute on Drug Abuse Research, DHHS Publ. No. Superintendent Documents, U.S. Goverment Printing Office, Washington, D.C., in press.

66. Casale, J. F. and Waggoner, R. W., Jr., A chromatographic impurity signature profile analysis for cocaine using capillary gas chromatography, *J. Forensic Sci.*, 36, 1312, 1991.

67. Janzen, K., Concerning norcocaine, ethylbenzoylecgonine, and the identification of cocaine use in human hair, *J. Anal. Toxicol.*, 16, 402, 1992.

68. ElSohly, M. A., Urinalysis and casual handling of marijuana and cocaine, *J. Anal. Toxicol.*, 15, 46, 1991.

69. Zieske, L. A., Passive exposure of cocaine in medical personnel and its relationship to drug screening tests, *Arch. Otolaryngol. Head Neck Surg.*, 118, 364, 1992.

70. Le, S. D., Taylor, R. W., Vidal, D., Lovis, J. J., and Ting, E., Occupational exposure to cocaine involving crime lab personnel, *J. Forensic Sci.*, 37, 959, 1992.

71. Pohl, S., Hnatchenko, M., and Feinland, R., Hair conditioning compounds and method for use, U.S. patent 4, 507, 280.

72. Corbett, J. F., The chemistry of hair-care products, *J. S. D. C.*, 285, Aug. 1976.
73. Polášek, M., Gaš, B., Hirokawa, T., and Vacík, J., Determination of limiting ionic mobilities and dissociation constants of some local anesthetics, *J. Chrom.*, 596, 265, 1992.
74. Hayes, G., Scholtz, H., Donahue, T., and Baumgartner, W., in an abstract of the 39th meeting of the American Society for Mass Spectrometry and Allied Topics held in Nashville, TN, on May 19–24, 1991.
75. Harrison, W. H., Gray, R. M., and Solomon, L. M., Incorporation of d-amphetamine into pigmented guinea-pig hair, *Br. J. Dermatol.*, 91, 415, 1974.
76. Lentner, C., ed. *Geigy Scientific Tables*, Volume 1, Medical Education Division, Ciba-Geigy Corporation, West Caldwell, NJ, 1981, p. 108.
77. Doran, D., Tierney, J., Varano, M., and Ware, S., A study of the pH of perspiration from male and female subjects exercising in the gymnasium, *J. Chem. Ed.*, 70, 412, 1993.
78. Vree, T. B., Muskens, A. Th. J .M., and Van Rossum, J. M., Excretion of amphetamines in human sweat, *Arch. Int. Pharmacodyn.*, 199, 311, 1972.
78a. Schoendofer, D., unpublished data.
79. Henderson, G. L., Harkey, M. R., and Zhou, C., Cocaine and metabolite concentrations in the hair of South American coca chewers, *J. Anal. Toxicol.*, 16, 199, 1992.
80. Goddard, E. D., Substantivity through cationic substitution, *Cosmetics Toiletries*, 102, 71, 1987.
81. Baumgartner, W. A. and Hill, V. A., Hair analysis for drugs of abuse: no evidence for racial bias. Technical memorandum to the National Institutes of Justice, June 1991, data also appearing in: Hair analysis for drugs of abuse: forensic issues. *Proceedings of the International Symposium on Forensic Toxicology*, Federal Bureau of Investigation, June 15–19, 1992, Quantico, Virginia, in press.
82. Nakahara, Y., Shimamine, M., and Takahashi, K., Hair analysis for drugs of abuse. III. Movement and stability of methoxyphenamine (as a model compound of methamphetamine) along hair shaft with hair growth, *J. Anal. Toxicol.*, 16, 253, 1992.
83. Matsuno, H., Uematsu, T., and Nakashima, M., The measurement of haloperidol and reduced haloperidol in hair as an index of dosage history, *Br. J. Clin. Pharmacol.*, 29, 187, 1990.
84. Kidwell, D. A. Caveats in testing for drugs of abuse. National Institute on Drug Abuse conference on Methodological Issues in Epidemiological, Prevention and Treatment Research on the Effects of Prenatal Drug Exposure in Women and Children, July 25–26, 1990, in M. M. Kilbey and K. Asghar, eds., *Methodological Issues in Epidemiological, Prevention and Treatment Research on Drug Exposed Women and Their Children*. National Institute on Drug Abuse Research Monograph #117, Superintendent of Documents, Government Printing Office, Washington, D.C., 1992.
85. Purpura-Tavano, L., Elements of style: beauty-conscious women use electric-hair care products to stay fashionable, *HFD — The Weekly Home Furnishings Newspaper*, 66, Dec. 7, 1992.
86. Reid, R. W., O'Connor, F. L., and Crayton, J. W., The *in vitro* binding of benzoylecgonine to pigmented human hair samples, *Clin. Toxicol.*, 32, 405, 1994.
87. Mieczkowski, T. and Newel, R., An evaluation of patterns of racial bias in hair assays for cocaine: black and white arrestees compared, *Forensic Sci. Int.*, 63, 85, 1993.
88. Marques, P. R., Tippetts, A. S., and Branch, D. G., Cocaine in the hair of mother-infant pairs: quantitative analysis and correlations with urine measures and self-report, *Am. J. Drug Alcohol Abuse*, 19, 159, 1993.
89. Andrasko, J. and Stocklassa, B., Shampoo residue profiles in human head hair, *J. Forensic Sci.*, 35, 569, 1990.
89a. DeLander, S. F. and Kidwell, D. A., in preparation.
90. Springfield, A. C., Cartmell, L. W., Aufderheide, A. C., Buikstra, J., and Ho, J., Cocaine and metabolites in the hair of ancient Peruvian coca leaf chewers, *Forensic Sci. Int.*, 63, 269, 1993.
91. Baumgartner, W. A. and Hill, V. A., Sample preparation techniques, *Forensic Sci. Int.*, 63, 121, 1993.
92. Nakahara, Y. and Kikura, R., Hair analysis for drugs of abuse. VII. The incorporation rates of cocaine, benzoylecgonine and ecgonine methyl ester into rat hair and hydrolysis of cocaine in rat hair, *Arch. Toxicol.*, 68, 54, 1994.
93. Herrmann, W. P. and Habbig, J., Immunological demonstration of multiple esterases in human eccrine sweat, *Br. J. Dermatol.*, 95, 67, 1976.
94. Ryhanen, R. J. J., Pseudocholinesterase activity in some human body fluids, *Gen. Pharmacol.*, 14, 459, 1983.
95. Herrmann, W. P. and Habbig, J., Immunologishe darstellung einer β-glucuronidase im ekkrinen scheiβ, *Arch. Derm. Res.*, 255, 63, 1976.
96. Loewenthal, L. J. A. and Politzer, W. M., Alkaline phosphatase in human eccrine sweat, *Nature*, 195, 902, 1962.
97. Harrison, W. H., Gray, R. M., and Solomon, L. M., Incorporation of L-DOPA, L-α-methylDOPA and DL-isoproterenol into guinea pig hair, *Acta Dermatovener (Stockholm)*, 54, 249, 1974.

98. Ishiyama, I., Nagai, T., and Toshida, S., Detection of basic drugs (methamphetamine, antidepressants, and nicotine) from human hair, *J. Forensic Sci.*, 28, 380, 1983.

99. Forrest, I. S., Otis, L. S., Serra, M. T., and Skinner, G. C., Passage of ^{3}H-chlorpromazine and ^{3}H-Δ-tetrahydrocannabinol into the hair (fur) of various mammals, *Proc. West. Pharmacol. Soc.*, 15, 83, 1972.

100. Sato, H., Uematsu, T., Yamada, K., and Nakashima, M., Chlorpromazine in human scalp hair as an index of dosage history: comparison with simultaneously measured haloperidol, *Eur. J. Clin. Pharmacol.*, 44, 439, 1993.

101. Uematsu, T., Sato, R., Fujimori, O., and Nakashima, M., Human scalp hair as evidence of individual dosage history of haloperidol: a possible linkage of haloperidol excretion into hair with hair pigment, *Arch. Dermatol. Res.*, 282, 120, 1990.

102. Nakahara, Y., Effect of the physicochemical properties on the incorporation rates of drugs into hair, paper presented at the 2nd International Meeting on Clinical and Forensic Aspects of Hair Analysis, Genova, Italy, June 6–8, 1994.

103. Prota, G., Progress in the chemistry of melanins and related metabolites, *Med. Res. Rev.*, 8, 525, 1988.

104. Martz, R. M., The identification of cocaine in hair by GC/MS and MS/MS, *Crime Lab. Dig.*, 15, 67, 1988.

105. Chiarotti, M., Overview on extraction procedures, *Forensic Sci. Int.*, 63, 161, 1993.

106. Mangin, P. and Kintz, P., Variability of opiate concentrations in human hair according to their anatomical origin: head, axillary and pubic regions, *Forensic Sci. Int.* 63, 77, 1993.

107. Nakahara, Y., Takahashi, K., and Konuma, K., Hair analysis for drugs of abuse. VI. The excretion of methoxphenamine and methamphetamine into beards of human subjects, *Forensic Sci. Int.*, 63, 109, 1993.

108. Moeller, M. R., Fey, P., and Wennig, R., Simultaneous determination of drugs of abuse (opiates, cocaine and amphetamine) in human hair by GC/MS and its application to a methadone treatment program, *Forensic Sci. Int.*, 63, 185, 1993.

108a. Baumgartner, W. A., private communication.

109. Baumgartner, W. A. and Hill, V. A., Comments on the paper by D. L. Blank and D. A. Kidwell: "External contamination of hair by cocaine: an issue in forensic procedures", *Forensic Sci. Int.*, 63, 157, 1993.

110. Hudson, J. C., Analysis of currency for cocaine contamination, *Can. Soc. Forensic Sci.*, 22, 203, 1989.

111. Holden, C., Hairy problem for new drug testing method, *Science*, 249, 1099–1100, 1990.

112. Hawks, R. L. and Chaing, C. N., eds., *Urine Testing for Drugs of Abuse*, National Institute on Drug Abuse Research Monograph 73, DHHS Publ. No. (ADM)87-1481 Superintendent of Documents, U.S. Government Printing Office, Washington, D.C., 1986.

113. U.S. General Accounting Office, Drug use measurement: strengths, limitations and recommendations for improvement, Report to the Chairman, Committee on Government Operations, House of Representatives, report GAO/PEMD-93-18, U.S. Government Printing Office, Washington D.C., 1993.

114. Consensus opinion summarizing the current applicability of hair analysis to testing for drugs of abuse, Society of Forensic Toxicologists, Inc. report to the National Institute of Drug Abuse summarizing a conference on hair testing that took place on May 26–27, 1990, Washington, D.C. This report is discussed in more depth in Mieczkowski, T., New approaches in drug testing: a review of hair analysis, *Ann. Am. Acad. Pol. Soc. Sci.*, 521, 132, 1992.

115. Mieczkowski, T., New approaches in drug testing: a review of hair analysis, *Ann. Am. Acad. Pol. Soc. Sci.*, 521, 132, 1992.

116. Baer, J. D., Baumgartner, W. A., Hill, V. A., and Blahd, W. H., Hair analysis for the detection of drug use in pretrial, probation, and parole populations, *Fed. Probation*, LV, 3, 1991.

117. Said, H. M., Newsom, A. E., Tippins, B. L., and Mathews, R. A., Reversed-phase high-performance liquid chromatography of some ultra-heterogeneous and covalently-modified proteins from human hair, *J. Chromatogr.*, 324, 65, 1985.

118. Menefee, E., Component distributions in keratins and their estimation from amino acid analysis, *J. Soc. Cosmet. Chem.*, 36, 17, 1985.

119. Wiewióra, A. and Buszman, E., Amino acid composition of human hair melanoproteins, *Acta Biochem. Pol.*, 39, 81, 1992.

Chapter 3

THE POTENTIAL FOR BIAS IN HAIR TESTING FOR DRUGS OF ABUSE

Edward J. Cone and Robert E. Joseph, Jr.

CONTENTS

I. INTRODUCTION

Recent advances in drug testing have led to the development of new drug testing methodologies for the detection and measurement of drugs of abuse in hair. An important consideration in implementing a new drug testing methodology is that the method should provide an equitable assessment of drug use in all populations. The unexplained presence of drug in a biological sample is often considered *prima facia* evidence of drug use. Therefore, the reliability of the testing procedure is of considerable importance due to the seriousness of drug charges that may arise from positive test results. Drug testing methods must be free of bias to ensure that different populations do not suffer unduly from the application of such tests. Bias in drug testing could occur with a specific test procedure or be due to particular characteristics of the test matrix. For example, ethnic bias in urine testing for cannabinoids was alleged to occur because of higher urinary excretion rates of melanin and its metabolites among Africoids in comparison to Caucasoids.[1,2] The higher concentration of melanin and its metabolites in Africoid urine specimens was suggested to produce false positives by interfering with enzyme-based cannabinoid screening assays. However, no bias was found in further studies of the interference of melanin with cannabinoid screening reagents.

Recently, the potential for ethnic bias in hair testing for drugs of abuse has arisen.[3] Hair analysis involves clipping strands of hair from the head or other regions of the body and extracting drug from the hair matrix. If an illicit drug is present, a presumption of drug use is often made. The most significant advantage of hair testing for drugs appears to be its unusually long detection window in comparison to other biological samples. Once drug is deposited in hair, it remains detectable for a period of months to years. However, if drug deposition is influenced by those properties of hair attributed to a particular ethnic group, e.g., hair color or structure, then certain ethnic groups may test positive more frequently than other ethnic groups who may accumulate less drug in hair.

Hair testing is currently applied in a variety of situations in the workplace, in courtroom cases, in cases assessing gestational drug exposure, and as a means of monitoring drug use by parolees. However, the mechanisms of drug accumulation in hair are unknown, and there is only limited information on the physicochemical characteristics of how drugs bind to hair. Some evidence suggests that drug incorporation into hair may depend on ethnic hair type, hair color, and hair treatments. Further research into these issues is clearly needed since the use of hair testing to detect drug use in the workplace is rapidly increasing.

This review surveys morphological, ultrastructural, and chemical differences between ethnic hair types which may affect the binding of drugs of abuse to hair. Selected studies are discussed regarding the binding of metals, chemicals, and drugs of abuse to different ethnic hair types and the effects of hair color and chemical treatments on drug disposition in hair. A mechanistic discussion of chemical and metal binding to different hair types is included since these processes appear to be similar in manner to those by which drugs bind to hair.

II. HAIR ANATOMY

Hair is a complex tissue whose structure and biology are only partially understood. There are basic structural similarities among all hair types regardless of hair color and ethnicity. Hair is an epidermal outgrowth from a hair follicle. Figure 1 illustrates the structural components of a hair follicle and hair shaft. A bulb at the base of the follicle contains matrix cells which give rise to layers of the hair shaft including the cuticle, cortex, and medulla.[4,5] The cuticle is the outermost layer of hair, and the cortex is located between the cuticle and the innermost region, the medulla.[5,6] There are multiple layers of the cuticle which serve to protect the underlying cortex from damage and fraying. The cuticular cells and the inner root sheath of the hair follicle are interdigitated, and the inner root sheath is degraded as hair emerges from the follicle, which gives the cuticle an appearance of shingles.[4] A cell membrane is the outermost layer of the cuticle and a cell membrane complex is present between overlapping cuticular cells. The cell membrane complex also forms a network throughout the hair shaft which interwinds between the cuticle and cortex. In the cuticle, the alpha (A) layer is located beneath the cell membrane, and the exocuticle is situated between the A-layer and the endocuticle. The cortex forms the bulk of the hair shaft and is located immediately beneath the cuticle.[6] The medulla is the innermost region of hair, and consists of scattered cells and hollow space.[6,7] It is small in relation to the cortex and may be continuous, discontinuous, or absent in adult hair.

III. PHYSIOLOGY OF HAIR

Matrix cells at the base of the hair follicle undergo morphological and structural changes as they move upward during growth to form different layers of the hair shaft. Proteins are synthesized within cells which ultimately determine the durability and strength of the hair shaft which emerges from the follicle. Matrix cells also may acquire pigment as they differentiate to form individual layers of hair, and pigment present in hair cells determines the color of the hair shaft. The individual layers of hair often can be distinguished by differences in the type and quantity of proteins and pigment. These differences are important, since they may determine which structural components and layers of hair are responsible for binding drugs.

A. Keratin

Hair is composed of approximately 65 to 95% protein, 1 to 9% lipid, and small quantities of trace elements, polysaccharides, and water.[6,8,9] The majority of hair protein is often referred to as keratin, which is a general term used to describe aggregates of protein with low or high sulfur content.[10] These proteins are synthesized in the keratogenous zone of the hair follicle as matrix cells move upward from the hair bulb to form layers of the hair shaft.[7,11] The cuticle, cortex, and medulla are comprised largely of keratin, although it is structurally different in each layer.[11,12] Keratin in the exocuticle contains a high concentration of cysteine, which forms disulfide bonds which link the A-layer to the exocuticle, and this makes the cuticle

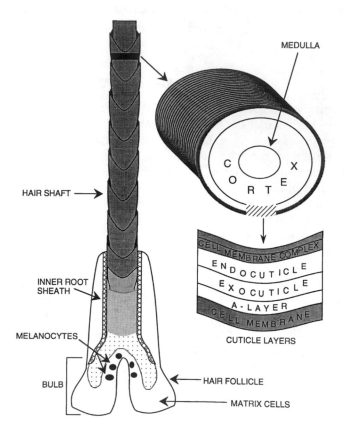

Figure 1
Anatomy of a hair follicle and hair shaft.

a tough and resilient structure capable of protecting the underlying cortex from being damaged.[8,9] The endocuticle contains less cysteine than the A-layer and is therefore not as hard or durable.[8,9] Basic and acidic amino acids such as arginine and glutamate are present in the endocuticle.

Keratin in the cortex comprises 85% or more of the mass of the hair shaft. Cortical keratin is composed of two types of structural proteins, matrix proteins and fibrous proteins.[12-15] Matrix proteins have a high sulfur content and contain polypeptides with a molecular weight of approximately 10 to 28 kDa. Fibrous proteins are embedded in matrix proteins and are characterized by a low sulfur content. They have a molecular weight of approximately 40 to 58 kDa. Also, matrix proteins have a nonhelical structure and are readily soluble at pH 4.5 in 0.5 M KCl, whereas fibrous proteins exhibit a helical structure and are insoluble in this same solution.[12]

The structure, organization, and ratio of matrix and fibrous proteins contribute to the physiochemical properties of keratinous tissues.[16,17] For example, a primary difference between hair and nails is the arrangement of fibrous proteins and the concentration of matrix proteins present in each tissue.[17] In cells destined to form the cortex of hair, fibrous proteins are oriented to form filaments which cluster to form fibrils. In the keratogenous zone, fibrils undergo lateral fusion to ultimately produce the cortex.[11] The medulla also contains keratin which has been characterized as a collection of irregular fibrous proteins.[12] Fibrous proteins form a trabecular framework comprising 95% of the medulla, and medullary proteins are less resistant to chemical degradation than proteins in the cortex.[12] The cell membrane complex, which courses throughout the hair shaft, is reported to consist of lipids, polysaccharides, and protein.[8,9,18] The cysteine content of the cell membrane complex is small in relation to the other components of the cuticle, although basic amino acids like lysine and glycine are present in higher concentrations relative to the A-layer.

B. Pigmentation

Melanins are among the most frequently occurring pigments in nature.[19,20] They are responsible for tegumentary pigmentation in chordates, coloration of insect cuticles, and the color of hair.[19] Hair cells which form the hair shaft do not produce melanin.[21] Melanin is synthesized in melanosomes which are discrete organelles located within melanocytes.[19] Melanocytes, which are derived from the neural crest, are present in the upper portion of the hair bulb.[19] Melanocytes are the only cells capable of melanin synthesis in mammals, and melanocytes transfer melanosomes to differentiating cortical and medullary cells during growth as these cells move upward from the hair bulb to form the hair shaft.[19] Cells which form the cortex and medulla contain melanosomes and melanin, whereas cuticular cells contain little or no melanin.[21]

Enzymes present in melanosomes synthesize two types of melanin, eumelanin and pheomelanin.[19,21,22] Figure 2 illustrates the proposed biosynthetic pathways of eumelanin and pheomelanin. The synthesis of eumelanin requires tyrosinase, an enzyme located in melanosomes. Tyrosinase catalyzes the conversion of tyrosine to dopa, which is further oxidized to dopaquinone. Through a series of enzymatic and nonenzymatic reactions, dopaquinone is converted to 5,6-indole quinone and then to eumelanin, a polymer. This polymer is always found attached to proteins in mammalian tissues, although the specific linkage site between proteins and polymers is unknown.[19] Polymers affixed to protein constitute eumelanin, but the exact molecular structure of this complex has not been elucidated. Pheomelanin is also synthesized in melanosomes.[19,21,22] The initial steps in pheomelanin synthesis parallel eumelanin synthesis, since tyrosinase and tyrosine are required to produce dopaquinone. Dopaquinone then combines with cysteine to form cysteinyldopa, which is oxidized and polymerized to pheomelanin.[22] The exact molecular structure of pheomelanin also has not been determined.

There are distinct chemical properties of eumelanin and pheomelanin which affect the physiochemical properties of hair.[22,23] Pheomelanin granules are smaller and less resistant to chemical degradation than eumelanin granules.[22] Pheomelanin is soluble in dilute alkali in comparison to eumelanin, which is insoluble in almost all solvents.[22] Pheomelanin and eumelanin also differ in sulfur content. Pheomelanin is high in sulfur content (9 to 12%) in contrast to eumelanin, which contains only 0 to 1% sulfur.[23]

Color is determined by the quantity of pheomelanin and eumelanin present in hair.[21,22] Black and brown hair contain more eumelanin than red and blond hair.[24,25] The major difference between black and brown hair is that black hair contains a higher concentration of eumelanin.[23,25] The melanosomes in black hair are also frequently larger and more numerous than in brown hair.[24,26] In red hair, the concentration of pheomelanin is often greater than eumelanin, depending on the intensity of color.[27] Blond hair contains eumelanin and possibly pheomelanin, although the concentration of eumelanin in blond hair is considerably less than in brown hair.[21,22,25] Gray hair is considered to be an intermediate transition color in the aging process.[21] Gray hair bulbs contain melanocytes, but there is little melanogenic activity and melanosomes are few in number. Hair bulbs which produce white hair contain few

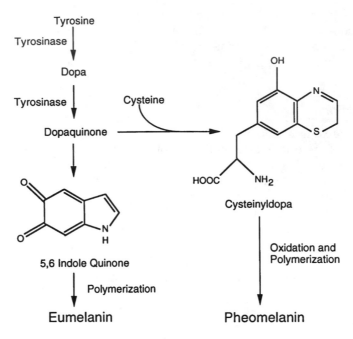

Figure 2
Proposed biosynthetic pathways for eumelanin and pheomelanin.

melanocytes, and tyrosinase activity is minimal, which suggests that melanin is not being actively synthesized.[21] Melanin is not present in the shaft of white hair.

IV. COMPARISON OF ETHNIC HEAD HAIR TYPES

There are differences in the chemical and physical attributes of ethnic hair types together with considerable intraethnic variation. Table 1 provides a compilation of differences between hair types of three major ethnic groups. Unfortunately, many of these findings are based on small numbers of subjects and are in need of further validation. Despite these limitations, differences between ethnic hair types warrant special attention because of their importance in the binding of drugs to hair.

A. Morphology and Ultrastructure

Studies of ethnic hair types in the 19th century lead investigators to propose hair as the primary racial characteristic for anthropological classification.[28] In 1827, Bory de Saint Vincent proposed the races of mankind be classified according to straight hair and woolly hair.[28] In the 1940s, anthropologists became dissatisfied with racial characterization based on these hair forms because they could not be accurately quantified. However, these early observations of hair form prompted investigators to examine ultrastructural features of hair from different ethnic groups and report differences between ethnic hair types.[29-31]

Hrdy[31] conducted a quantitative study of head hair form among Africoid, Caucasoid, and Mongoloid populations. Hair diameter, cuticular scale count, curvature, kinking, and the degree of medullation were determined for head hair collected from 30 Africoid, Caucasoid, and Mongoloid subjects (Table 1). The diameter and degree of medullation were significantly greater for Mongoloid hair in comparison to Africoid

TABLE 1.

Comparison of Ethnic Head Hair Types

Hair	Ethnicity[a]			Ref.
	Caucasoid	Africoid	Mongoloid	
Structure[b]				31
Cuticular scales	15.07 ± 1.80	17.95 ± 2.04	15.47 ± 1.46	
Diameter (μm)	79.17 ± 11.27	76.10 ± 15.83	95.53 ± 10.35	
Medullation	0.33 ± 0.55	0.70 ± 0.657	1.23 ± 0.63	
Form[c]				31, 33
Kink	0	0.10 ± 0.31	0	
Curvature	0.19 ± 0.13	7.75 ± 1.78	0.07 ± 0.4	
Crimp	0.03 ± 0.06	0.95 ± 0.57	0	
Cross section	Oval or round	Flat	Round	
Pigmentation				21, 22, 25
Color	Various intensities of blond, brown, red	Black	Brown-black	
Melanin[d]	Varies with hair color: (blond, eumelanin (+); brown, eumelanin (++); red, pheomelanin (++) and eumelanin (+)	Eumelanin (+++)	Eumelanin (+++)	
Melanosome morphology	Varies with hair color: blond, elliptical; brown, elliptical; red, spherical and elliptical	Elliptical	Elliptical	
Melanosome distribution	Complex	Single	Single/complex	
Protein (mg/0.1 g hair)				34
Fibrous	9.2 ± 0.7	8.6 ± 0.8	14.3 ± 0.8	
Matrix	31.9 ± 1.3	46.0 ± 2.7	32.9 ± 2.7	
Fibrous/matrix	0.29 ± 0.02	0.18 ± 0.02	0.45 ± 0.03	
Follicular form	Helical and curved or straight	Helical, curved	Straight	33

[a] Ethnic groups were defined as follows: Mongoloid, primarily Japanese descent; Caucasoid, European descent; Africoid, African descent.

[b] Structure: cuticular scales were counted over a distance of 0.52 mm. Medullation was determined by scoring hair samples as follows: 0 = no medullation; 1 = discontinuous medullation; 2 = continuous medullation.

[c] Form: kink refers to a constriction in the hair shaft and was determined by scoring hair samples as follows: 0 = no kink; 1 = kink. Curvature was determined by placing hair samples between 2 glass slides and matching circles of known radius to the shape of hair. Crimp refers to the number of changes in direction of the hair shaft per unit length.

[d] The relative concentration of melanin was scored as follows: (+++) = high concentration; (++) = moderate concentration; (+) = low concentration.

and Caucasoid hair. Curvature, scale count, and kinking were significantly greater for Africoid hair in comparison to Caucasoid and Mongoloid hair.

Hair morphology is another attribute of hair which may differ between ethnic groups.[32] Hair morphology refers to the cross-sectional shape of hair and also the gross appearance of hair. Lindelöf et al.[33] examined cross sections of head hair collected from Africoid, Mongoloid, and Caucasoid subjects. Hair samples were described as follows: Africoid hair was curly, Mongoloid hair was straight, and Caucasoid hair was a combination of curly/straight. Africoid hair follicles demonstrated a helical and curved form which was typical of the curly hair produced by these follicles. Hair from Africoid subjects appeared flat in cross section. Mongoloid hair follicles extended straight down into skin, and hair produced by these follicles was straight and appeared round in cross section. Caucasoid hair follicles had a structure which was intermediate between Africoid and Mongoloid hair follicles, and hair produced by these follicles was described as moderately curly to straight. Hair from Caucasoid subjects appeared round and ovoid in cross section. Lindelöf et al.[33] indicated that the shape of the hair follicle determined the morphology of ethnic hair types.

B. Biochemistry

The amino acid content of head hair from the three major ethnic groups is similar, as indicated in Table 2.[32] However, the concentration and type of hair proteins may differ. Dekio and Jidoi[34] determined the concentration of fibrous proteins and matrix proteins in Africoid, Caucasoid, and Mongoloid hair. They collected head hair from five male and female volunteers from each ethnic group. Hair samples were treated with urea and mercaptoethanol to reduce disulfide bonds present in keratin. Hair was then homogenized and iodoacetic acid was added to homogenates which yielded S-carboxymethylated (SCM) protein derivatives. SCM matrix proteins and SCM fibrous proteins were separated by centrifugation. These results are included in Table 1. The mean concentration of fibrous proteins was greater in Mongoloid hair in comparison to Caucasoid and Africoid hair, but the mean concentration of matrix proteins was highest for Africoid hair. The ratio of fibrous proteins to matrix proteins differed considerably between Africoid, Caucasoid, and Mongoloid hair.

TABLE 2.

Amino Acid Content of Ethnic Head Hair Types

	Ethnicity		
Amino acid[a]	Africoid N = 3	Mongoloid N = 3	Caucasoid N = 3
Cysteic acid	3.3	3.0	2.3
Aspartic acid	58.5	58.0	53.6
Threonine	68.4	67.2	67.6
Serine	114.5	113.8	124.7
Proline	74.6	72.0	76.2
Glutamic acid	120.4	117.2	115.3
Glycine	56.2	56.5	59.1
Alanine	44.3	45.9	45.3
Cysteine	150.0	143.0	139.0
Valine	41.1	50.2	44.2
Methionine	1.7	2.6	4.6
Isoleucine	23.3	24.4	20.8
Leucine	54.9	58.2	55.8
Tyrosine	19.1	18.9	19.1
Phenylalanine	13.9	14.4	14.8
Lysine	16.9	18.7	23.7
Histidine	7.2	7.3	8.3
Arginine	52.1	57.2	55.9

[a] Data are expressed as μM of amino acid/g of hair. Reproduced from Gold, R. J. M. and Scriver, C. R., *Clin. Chim. Acta,* 33, 465, 1971. With permission.

C. Pigmentation

In 1925, Hausman[35] conducted one of the earliest comparative ethnic studies of pigment in head hair. He reported that races could not be distinguished based on pigmentary structures of head hair and that hair color, but not race, reflected differences in hair pigmentation. More recent reports described differences in the organization and structure of melanin in different ethnic hair types.[22,24,26,36-38] Many of these studies employed electron microscopy to distinguish differences in location, distribution, and transfer of pigment cells in ethnic hair types. However, there are important limitations associated with classification of ethnic hair types based on pigmentation. Definitive studies comparing ethnic hair types of similar color, e.g., Africoid black hair vs. Caucasoid black hair, need to be conducted so that an accurate comparison between ethnic hair types can be made. For the studies cited in this review, hair color cannot be excluded as the basis for many of the observed differences. There is strong selection for dark hair in Africoid and Mongoloid populations, whereas hair color selection is not strong among the Caucasoid population.[39,40]

1. Africoid Hair

Melanocytes are present in the outer root sheath of follicles and in the bulb of Africoid hair.[41] Elliptical and rod-shaped melanosomes synthesize eumelanin in Africoid hair which is predominantly black in color.[21,22,25] In the hair bulb, melanosomes are transferred from melanocytes to developing cortical and medullary cells. Cesarini[22] reported that each melanosome is transferred singly and does not form aggregates with other melanosomes. Melanosomes in Africoid hair bulbs also may be twice the size of those found in Caucasoid hair bulbs. However, Barnicot et al.[26] noted that some Caucasoid hair bulbs which produced dark brown hair contained melanosomes which were comparable in size and occasionally larger than those in Africoid hair bulbs.

Cesarini[22] examined the appearance of melanin in melanosomes of Africoid hair bulbs and reported that melanin granules were electron opaque and completely obscured the internal structures upon which melanin polymers are formed. This was due to extensive melanin synthesis within melanosomes. Barnicot et al.[26] compared melanin granules in Caucasoid and Africoid hair which were dark brown and black, respectively. They noted that melanin granules present in Caucasoid hair were similar in appearance to those in Africoid hair, but the number of melanin granules present in Africoid hair was greater in comparison to Caucasoid hair. In Africoid hair, Barnicot et al.[26] reported the size of eumelanin granules to be approximately 0.83 µm in length by 0.34 µm in width. Barnicot et al.[26] also noted that a dark surface shell was more often visible surrounding melanin granules in Africoid hair in comparison to Caucasoid hair, but Africoid hair samples examined in this study were reportedly darker in color than Caucasoid hair samples. Therefore, the presence of a surface shell surrounding melanin granules may be related to hair color and not necessarily to ethnicity.

2. Mongoloid Hair

Melanosomes in Mongoloid hair contain primarily eumelanin that is typical of all black and brown hair regardless of ethnicity. There may be subtle differences between melanosomes in Mongoloid and Africoid hair. Cesarini[22] described the appearance of melanosomes in Mongoloid hair as electron dense with round electron-lucent structures present at the periphery. Electron-lucent regions likely indicate an absence of melanin at these sites in contrast to the more heavily melanized melanosomes present in some Africoid hair samples. However, Cesarini[22] only examined one Mongoloid hair sample, and the differences noted between Africoid and

Mongoloid melanosomes again could be attributed to differences in hair color. Cesarini[22] also reported that melanosomes in Mongoloid hair are transferred individually or in aggregates to cells moving upward from the hair bulb to form the hair shaft.

3. Caucasoid Hair

a. Red and Blond

Bulbs of Caucasoid red hair are filled with premelanosomes, which are precursors to melanosomes.[22] Premelanosomes have not yet begun the process of melanization. As premelanosomes mature, pheomelanin primarily is synthesized in oval-shaped melanosomes and small quantities of eumelanin may also be synthesized in elliptical-shaped melanosomes. Cortical and medullary cells receive aggregates of two or three melanosomes which form complexes. The concentration of complexes in cortical and medullary cells determines the intensity of color of the hair shaft.[21,22] Bright red hair contains a higher concentration of complexes in comparison to hair which exhibits a darker shade of red.[21,22] In red hair, melanin granules appear to be more rod shaped and fewer in number than those present in brown and black hair.[26] Barnicot et al.[26] reported that granules in red hair measured approximately 0.54 µm in length by 0.24 µm in width. In the shaft of red hair, complexes of melanosomes appear to be rapidly digested, leaving polymorphous melanosomes.

Blond hair contains elliptical and oval melanosomes. The granules in melanosomes are small, few in number, and likely represent eumelanin, although pheomelanin also may be present in blond hair.[21,22,24] Ortonne and Prota[21] suggested that blond hair results from a quantitative decrease in the synthesis of melanin in comparison to black and brown hair. The main features which distinguish brown hair from blond hair are the presence of more melanosomes and a higher concentration of melanin in brown hair.[21,22,24] Melanosomes present in the shaft of blond hair also appear to be more susceptible to degradation than melanosomes in black hair. Cesarini[22] reported that melanosomes may not be present in the shaft of blond hair, possibly due to digestion of melanosomes by lysosomes.

b. Brown and Black

Caucasoid hair bulbs which produce brown hair contain ellipsoidal melanosomes which synthesize primarily eumelanin granules.[21,22] The size of melanosomes generally is greater in hair which is dark brown in comparison to light brown hair.[36] Melanosomes are transferred in aggregates from melanocytes to cortical and medullary cells in comparison to melanosomes present in Africoid hair, which are transferred singly.[22] Complexes formed by aggregates of melanosomes contain electron-dense matrices filled with melanin granules, but the size and number of these granules are usually less than in Africoid hair.[26] Cesarini[22] reported that melanosomes present in the shaft of brown Caucasoid hair may be degraded, whereas melanosomes present in Africoid and Mongoloid hair may be more stable. Differences in the size, stability, and concentration of melanosomes may distinguish ethnic hair types, but data suggest that there is a wide range of intraethnic variability in pigmentary structures in hair. As indicated earlier, the majority of hair studies reporting ethnic differences lacked appropriate controls for hair color. There are no reports which have compared black Caucasoid hair to Mongoloid or Africoid hair. Consequently, Barnicot et al.[26] suggested there may be little difference in the structure and distribution of melanin and melanosomes in ethnic hair types which are similar in color.

V. BINDING SITES IN HAIR

There are several potential binding sites in hair for drugs. Keratin and melanin contain many polar groups, and evidence strongly suggests that these substituents can serve as attachment points for binding drugs, chemicals, and metals. Accordingly, differences between hair types in the structure, organization, and concentration of melanin and keratin may affect the degree of binding. Lipids in hair may also serve as possible binding sites. However, the exact binding sites in hair have not been determined.

A. Keratin

Keratin in hair is comprised of numerous basic and acidic amino acid residues.[7] Acidic amino acid residues in keratin include aspartic acid and glutamic acid, and basic amino acid residues include arginine, lysine, and histidine. The number of acidic groups is slightly less than the number of basic groups. Potentially reactive constituent groups in keratin include free carboxyl, amide, phenolic hydroxyl, aliphatic hydroxyl, and sulfhydryl groups. Many hydrogen bonds are also present in keratin and may provide additional binding sites.

The polarity of chemical ligands appears to affect their affinity for binding sites in keratin. Chemicals such as benzene bind more extensively to hair than cyclohexane, a molecule which is similar in size, but considerably less polar than benzene. Breuer[42] proposed that peptide linkages in polypeptide chains of keratin are sites at which polarizable molecules bind to hair. A peptide bond in keratin consists of a carboxyl group bound covalently to an amino group, and this bond exhibits a large permanent dipole moment to which molecules may be attracted. This could result in dipole-dipole interactions between peptide bonds and polarizable molecules, which could hypothetically explain why polar molecules such as benzene bind more extensively than nonpolarizable molecules. Further, Breuer[42] indicated that water molecules bind to hair by a similar mechanism. When hair is exposed to water, water molecules appear to move along the polypeptide chains of hair. Water molecules which approach peptide bonds may form water-polypeptide bonds based on rapid hydrogen bond exchanges between water and carboxyl and amino groups present in keratin.

Cystine, which contains a disulfide bond, is reported to be the most numerous and reactive amino acid present in hair keratin.[43] Disulfide bonds in cystine are reduced by mercaptans and phosphines, and oxidized by perborates, bromates, and bleach. These reactions result in structural rearrangements within keratin which may affect the physiochemical properties of hair, since disulfide bonds in cystine contribute to the stability of hair.[43] For example, hydrogen peroxide bleaching of hair is an oxidative process which occurs readily in an alkaline medium. This results in the formation of perhydroxy anions which have been proposed to react with cystine to form cysteic acid residues. The process of bleaching results in the loss of approximately 15% of the cystine bonds originally present in keratin and may explain the increased permeability of bleached hair to chemicals.[43,44]

Disulfide bonds present in the keratin of hair are potential binding sites for many nucleophilic molecules. Tolgyesi and Fang[45] investigated the structural changes which occur in keratin as a result of alkaline treatments of hair. They reported that hydroxide ions initiated a β-elimination reaction, resulting in cleavage of the disulfide bonds of cystine to produce dehydroalanine intermediates. Cysteine and lysine may react with dehydroalanine to form new cross links in keratin. Nucleophilic amines such as ethylamine and n-pentylamine may also react with dehydroalanine to form cross-linking structures in keratin. Tolgyesi and Fang[45] indicated that these reactions modify the structural features of keratin.

Asquith and Carthew[46] investigated the alkaline degradation of cystine present in the keratin of wool fibers. They reported that alkyl amines form adducts with cystine, resulting in the production of β-alkyl amino acids. They compared the concentration of β-alkyl amino acids formed from the addition of ethyl, propyl, butyl, and amyl amines to wool. The highest concentration of β-alkyl amino acids resulted from the addition of amyl amine to wool. In comparison, the lowest concentration of β-alkyl amino acids resulted following the addition of ethyl amine to wool. The authors indicated that cystine present in wool fibers may have a greater affinity for long-chain amines. They suggested that amines with long aliphatic chains are capable of breaking hydrophobic bonds within keratin and are more available to react with cystine residues in comparison to amines with short aliphatic chains.

For drugs of abuse, the affinity of basic drugs for keratin appears to be substantially greater than that of acidic drugs.[47] Basic drugs such as cocaine, nicotine, and morphine are incorporated into hair to a much greater extent than acidic molecules such as aspirin and Δ9-tetrahydrocannabinol (THC). These observations are consistent with the findings of Ward and Lundgren,[7] who indicated that basic dyes have a greater affinity for hair, and presumably keratin, in comparison to acidic dyes.

B. Melanin

Although the structures of pheomelanin and eumelanin have not been resolved, melanin is one of several suspected binding sites in hair for metals, chemicals, and drugs of abuse. The quantity and type of melanin in hair should determine the extent to which drugs bind to hair. Melanin is present in several mammalian tissues including the brain, skin, hair, iris of the eye, vas deferens, and cochlea of the inner ear.[19,48-50] Since melanin is present in many human tissues, drug may bind to other bodily tissues as well as hair.

One of the earliest studies which may have inadvertently demonstrated the capacity for melanin to bind drugs was conducted by Chen and Poth[51] in 1929. They examined the mydriatic effects of topical solutions of ephedrine (10%), pseudoephedrine (10%), cocaine (4%), and eupthalamine (5%) in the eyes of Caucasoids, Mongoloids, and Africoids. Dilation of Caucasoid subjects' pupils was considerably more pronounced than that of Mongoloid and Africoid subjects. A definitive explanation for these findings was not offered by the investigators, although they suggested that differences in pigmentation may have contributed to the observed differences in dilation of the subjects' pupils. A possible explanation was that melanin in heavily pigmented irides of Africoid and Mongoloid subjects bound more drug, thus less drug was available to bind to adrenergic receptors with resultant dilation of the pupil.

Later, Patil[52] studied the binding of radiolabeled cocaine to pigmented and nonpigmented irides. Nonpigmented irides were obtained from albino guinea pigs and pigmented irides were obtained from guinea pigs with dark irides. Patil[52] observed that pigmented irides had a binding capacity for cocaine which was 18-fold greater than nonpigmented irides. Also, the majority of cocaine bound by pigmented irides remained attached in contrast to nonpigmented irides, which lost cocaine at an exponential rate. It was concluded that the major determinant for the accumulation of cocaine by irides was the degree of pigmentation.

Similar results were obtained by Patil et al.,[53] who investigated the binding of radiolabeled ephedrine by pigmented and nonpigmented irides. The binding of ephedrine by pigmented irides was approximately eightfold greater than nonpigmented irides.[53] It was determined that irides with moderate pigmentation bound more drug than nonpigmented irides but much less than heavily pigmented irides. In a separate study, Patil and Jacobowitz[54] investigated the binding of radiolabeled (*l*)-norepinephrine, (*d,l*)-epinephrine, and (*d,l*)-phenylephrine to pigmented and nonpigmented guinea pig irides. For these phenolic amines, there was only a twofold difference in accumulation of drug by pigmented irides in comparison to nonpigmented irides. However, the accumulation of radiolabeled β-phenethylamine, a nonphenolic lipophilic amine, was eight- to tenfold greater for pigmented irides in comparison to nonpigmented irides. These results were similar to those observed for the binding of cocaine by pigmented and nonpigmented irides.[53]

Shimada et al.[55] further investigated the nature of binding of drugs to melanin by measuring the binding of radiolabeled amphetamine, epinephrine, ephedrine, octopamine, norepinephrine, atropine, and cocaine to synthetic levodopa melanin. The investigators determined the binding capacity of melanin and the strength of drug binding (affinity). They also calculated net binding, the product of binding capacity and affinity. Their results indicated that binding capacity and affinity differed, but that net binding was similar for all drugs evaluated. These results differed from those obtained for the binding of similar drugs to pigmented irides which may be explained by the presence of lipophilic components in iride preparations. The authors indicated that lipophlic drugs such as ephedrine may cross lipid membranes in irides more readily, resulting in greater availability of drug for binding to melanin in comparison to less lipophilic drugs.

Melanins may be considered to behave like weak cationic exchange polymers and consequently form simple ionic bonds with metals and chemicals.[56-58] This may explain why drugs with cationic properties such as cocaine, phencyclidine, amphetamines, and opiates appear to be bound and retained by hair to a greater extent in comparison to anionic species such as aspirin and THC. Larsson and Tjalve[56] reported that cations may compete for binding sites in melanin. They determined that metal ions such as sodium, potassium, and calcium inhibited the binding of chloropromazine to melanin. However, anions in general had no impact on the binding of chloropromazine to melanin. Similarly, Kidwell and Blank[59] reported that the binding of cocaine to Caucasoid and Mongoloid hair decreased when the concentration of sodium ions was increased. Larsson and Tjalve[56,57] indicated that melanins also may be associated with charge transfer complexes which participate in the binding of metals and drugs by accepting electrons. They suggested that nonelectrostatic interactions such as Van der Waals forces may occur between aromatic nuclei of melanin and aromatic rings of molecules, and this may be an additional mechanism by which drug binds to melanin.

C. Lipids

Recently, Su et al.[60] demonstrated nonspecific binding of cocaine and morphine to hair homogenates. Nonspecific binding refers to binding of a molecule to components in tissue, e.g., lipids, which act as a virtual reservoir for accumulation of drugs. Although lipids constitute only a small portion of hair (1 to 9%), they are possible binding sites for drugs, and nonspecific binding could result from drug binding to lipid components in cell membranes of hair. For example, phosphatidylcholine, a zwitterionic fatty acid present in lipid membranes, contains negatively charged phosphate groups and positively charged choline groups which are capable of binding cationic and anionic drugs, respectively. Mohr and Struve[61] studied the binding of amphiphilic drugs to liposomes containing phosphatidylcholine. They observed that the binding of chlofibric acid anions to liposomes was considerably less in comparison to chlorphentermine cations. Both drugs are amphiphilic and have an identical aromatic ring system. The investigators attributed these differences in binding to the ability of anionic and cationic molecules to access internal hydrophobic regions of lipid membranes. They suggested that cationic molecules such as chlorphentermine bind to phosphate groups, which are closer to the hydrophobic region of lipid membranes than choline groups, which bind anionic molecules. When the pH of the incubation media was adjusted such that both drugs were uncharged, similar binding was observed for both neutral species. If drugs bind to lipids in cell membranes of hair, these findings may explain why the binding of cationic drugs such as cocaine to hair appears to be greater than anionic drugs.

VI. BINDING STUDIES WITH DIFFERENT HAIR TYPES

Presently, there is strong evidence to suggest that ethnicity, hair color, and hair treatments affect the binding of drugs to hair. Similar observations have been made in studies which investigated the binding of metals and chemicals other than drugs to different hair types. Although there is no clear physiochemical or molecular basis for these differences in binding between hair types, it is conceivable that differences in the structure and chemical composition of hair may affect the binding of drug to different hair types. The potential for bias in hair testing was evaluated in the following studies of the binding of trace elements, chemicals, and drugs of abuse to different hair types.

A. Trace Elements

Hair testing for trace elements was initially employed in forensic investigations as a measure of heavy metal poisoning and was also thought to be useful in clinical testing for the diagnoses of cystic fibrosis, celiac disease, geophagia, and phenylketonuria.[62-64] Between 1970 and 1980, hair analysis for trace elements received considerable public attention. Trace elements present in foodstuffs were considered to be present in hair solely as a result of diffusion from blood into hair follicles and hair cells. Companies promoted hair testing to the general public by indicating that analysis of hair would be useful for determining their overall health and nutritional status. However, the reliability of hair testing for trace elements was questioned by the scientific community, since it was not known whether or not variables other than diet could affect trace element concentrations in hair.[64]

A series of studies were conducted which sought to determine whether or not hair testing for trace elements was a valid approach for assessing metal exposure.[65-69] Sky-Peck[67] studied the effects of sex, hair treatments, age, ethnicity, and hair color on the incorporation of trace elements into hair by collecting head hair from Caucasoid, Africoid, and Mongoloid male and female volunteers. Table 3 illustrates the trace element content of hair samples collected from females from these ethnic groups. Africoid hair differed significantly ($p < 0.01$) from Caucasoid hair in sulfur, calcium, iron, nickel, chromium, manganese, arsenic, mercury, lead, and rubidium content. Sky-Peck[67] also determined that sex and age affected trace element concentrations in hair. Female hair contained significantly higher concentrations of calcium, nickel, copper, and zinc in comparison to male hair; and hair from young subjects (3 to 20 years) had higher levels of sulfur, calcium, zinc, selenium, and strontium in comparison to older subjects (58 to 70 years). Hair color and hair treatments significantly affected the trace element content in hair. Dark-colored female hair had significantly higher concentrations of iron than female blond hair, and natural, bleached, and permed hair differed significantly in trace element content. When it was realized that sex, age, hair color, ethnicity, and hair treatments considerably affected trace element concentrations in hair, the practice of analyzing hair to determine nutritional status was discontinued.

TABLE 3.

Trace Element Concentrations in Ethnic Head Hair Types

	Ethnicity[a]		
Trace element[b]	Africoid (N = 60)	Caucasoid (N = 86)	Mongoloid (N = 45)
Arsenic	1.10 ± 1.58	0.33 ± 0.44	0.15 ± 0.18
Bromine	2.49 ± 1.37	2.15 ± 1.80	1.87 ± 1.00
Calcium	1,550.00 ± 1,132.20	986.3 ± 642.8	468.2 ± 267.0
Chromium	1.48 ± 0.71	0.99 ± 0.42	0.88 ± 0.29
Copper	21.80 ± 11.20	21.4 ± 7.10	15.4 ± 5.40
Iron	20.9 ± 9.90	11.6 ± 5.40	7.5 ± 2.0
Lead	28.40 ± 32.40	5.51 ± 5.97	1.56 ± 1.47
Mercury	0.40 ± 0.44	0.87 ± 0.48	0.81 ± 0.45
Manganese	1.68 ± 1.05	1.13 ± 0.56	0.68 ± 0.36
Nickel	2.16 ± 1.50	0.70 ± 0.48	0.45 ± 0.19
Rubidium	0.51 ± 0.24	0.74 ± 0.41	0.66 ± 0.38
Selenium	0.62 ± 0.29	0.55 ± 0.18	0.65 ± 0.21
Strontium	6.73 ± 5.10	5.69 ± 4.00	3.3 ± 2.3
Sulfur	32.5 ± 8.70	36.8 ± 4.90	36.5 ± 4.30
Zinc	181.8 ± 68.90	194.6 ± 39.60	197.6 ± 33.00

[a] Ethnic groups were defined as follows: Mongoloid, Japanese descent; Caucasoid, European descent; Africoid, African descent.
[b] Data are expressed as μg of trace element/g of hair.
Reproduced from Sky-Peck, H. H., *Clin. Physiol. Biochem.*, 8, 70, 1990. With permission.

Whether or not variables such as sex, ethnicity, and hair color significantly affect drug incorporation into hair remains highly controversial. It has been suggested that similar mechanisms are involved in the binding of trace elements and drugs to pigmented tissues.[56,57] By analogy, if trace elements and drugs share similar binding sites in hair, then drug incorporation into hair may be affected by the same variables which affect trace element concentrations in hair.

B. Chemicals

Compounds such as phenols, resorcinol, catechol, sodium picrate, and trimethylamine-naphthimide are known to bind to hair.[42,70] The affinity of a chemical for hair is believed to be determined by the following molecular properties: size; electric charges; the ratio of hydrophobic to hydrophilic groups; and the presence of highly polarizable groups.[42] Small molecules with electron-rich aromatic rings and molecules with hydrophobic groups appear to have the highest affinity for hair.

The binding of chemicals to hair is also affected by hair type. Lotzsch et al.[71] studied the adsorption and desorption of ^{14}C distearyl-dimethylammonium chloride (DSDMAC) with three Caucasoid and three Mongoloid hair samples. DSDMAC is a cationic surfactant which is used in commercial cosmetic products such as setting lotions and rinses for hair treatment. For hair samples treated with DSDMAC, adsorption was twofold greater for Mongoloid hair (0.94 mg/g of hair ± 0.33) in comparison to Caucasoid hair (0.47 mg/g of hair ± 0.14). Lotzsch et al.[71] also investigated the binding of DSDMAC to Caucasoid hair treated with peroxide. Bleaching of hair with peroxide causes increased porosity, changes in the molecular structure of hair pigment, and damage to the cuticle of hair.[43,72-74] Lotzsch et al.[71] reported that peroxide-treated hair adsorbed greater amounts of DSDMAC in comparison to untreated hair, and this was likely due to chemical and structural changes from bleaching.

C. Drugs

One of the first reports of drug accumulation in hair was made by Goldblum et al.[75] in 1954. They identified phenobarbital in hair of guinea pigs that had received approximately 65 mg of phenobarbital. Since this report, numerous drugs have been identified in human hair including cocaine, codeine, heroin, phencyclidine, morphine, amphetamines, antipsychotic drugs, and antifungal drugs.[47,76-78] The mechanisms by which drugs enter hair and which component(s) of hair serve as binding sites have not been resolved. Henderson[79] proposed several pathways for drug entry into hair: (1) diffusion of drug from blood into the hair follicle and into hair cells; (2) deposition of drug in skin followed by leaching out by sweat or sebum, which bathe hair shafts and subsequently deposit drug onto hair; and (3) diffusion of drug from blood into sweat and sebaceous glands, followed by excretion onto hair shafts. Determining the mechanisms for drug deposition is crucial in interpreting hair test data. The accumulation of drugs by different hair types may be selectively affected by each mechanism for drug entry.

1. Hair Color

a. Animal Studies

Hair color varies widely among species and even among hairs found in a single individual or animal. Initial evidence for the effect of hair color on drug incorporation was provided in animal studies. Forrest et al.[80] conducted one of the first studies to determine the effect of color on drug incorporation. The investigators administered chronic doses of radiolabeled chlorpromazine to rabbits and determined the percentage of drug incorporated into black and white hair by measuring radioactivity of hair samples at one, two, four, and six weeks after the initial dose. All black hair samples contained a higher concentration of drug in comparison to white hair samples. The average amount of drug incorporated into black hair was approximately fourfold greater than white hair.

Uematsu et al.[76] studied the excretion of ofloxacin, an antimicrobial drug, in pigmented and nonpigmented hair obtained from dark Agouti rats (predominantly brown hair) and albino rats (white hair). They administered ofloxacin (3, 10, 30 mg/kg) to rats by the intraperitoneal route twice daily for five weeks. Hair samples were taken from rats after five weeks and the concentration of ofloxacin in dark Agouti hair and white hair was measured. Table 4 illustrates the concentration of ofloxacin in Agouti and white hair. Agouti hair contained significantly more drug ($p < 0.01$) in comparison to white hair. Uematsu et al.[77] conducted a similar study to measure the incorporation of haloperidol, an antipsychotic drug, in Agouti and white hair of rats. Haloperidiol (0.5 and 1.0 mg/kg) was administered twice daily to two groups of six albino rats and two groups of three dark Agouti rats for two weeks, and then hair was collected for analyses. The concentration of haloperidol in Agouti hair was considerably greater than in white hair obtained from rats receiving identical doses of haloperidol (Table 4).

TABLE 4.

Drug Incorporation into Brown Agouti and White Rat Hair

Drug	Dose	N	Conc (ng/mg)	
			White hair	Brown Agouti hair
Ofloxacin				
	6 mg/kg	3–5	2.27 ± 0.4	8.84 ± 4.25
	20 mg/kg	3–5	6.25 ± 0.90	36.16 ± 5.88
	60 mg/kg	3–5	28.86 ± 3.36	389.7 ± 175.8
Haloperidol				
	0.5 mg/kg	3–6	1.21[a]	22.40[a]
	1.0 mg/kg	3–6	2.75[a]	39.29[a]

[a] Standard deviation from the mean was not reported.
Data are cited from References 76 and 77.

Ishiyama et al.[81] studied the incorporation of methamphetamine into black and white hair obtained from mice which received 100 µg of methamphetamine hydrochloride daily for a period of three weeks during which hair was periodically sampled and analyzed for drug. At five, eight, and fourteen days, the drug concentration in black hair was not substantially greater than white hair. However, three weeks following the initial dose the concentration of drug in black hair was twofold greater than in white hair.

b. Human Studies

The effect of hair color on drug incorporation into human hair is similar to that observed with animals. Uematsu et al.[76] investigated the incorporation of ofloxacin into white and black hair obtained from each of five subjects who had used ofloxacin within five months prior to hair collection. The investigators determined that the ofloxacin concentration was greater in black hair in comparison to white hair obtained from the same subject. Ofloxacin concentrations ranged from 17.1 to 71.1 ng/mg of black hair in comparison to 0 to 1.9 ng/mg of white hair. Uematsu et al.[77] also performed a similar study in which they measured the concentration of haloperidol in white and black hairs from each of ten male and ten female subjects. These subjects had received daily maintenance doses of haloperidol (3 to 5 mg per day) for at least one month prior to hair collection. The concentration of haloperidol in white hair was determined to be less than 10% of the haloperidol concentration in black hair collected from the same subject. These studies strongly suggested that hair color was responsible for the greater accumulation of haloperidol and ofloxacin in black hair in comparison to white hair. Unfortunately, there are few studies which have investigated the effect of hair color on the incorporation of drugs of abuse.

Reuschel and Smith[82] conducted a study in which they measured the concentration of benzoylecgonine in black, brown, blond, and red hair collected from jail detainees. They tested 48 hair samples for drug and 22 tested positive. The concentration of benzoylecgonine in black and brown hair was 5.6 ± 5.2 ng/mg of hair in comparison to that in blond and red hair, which was 1.39 ± 1.34 ng/mg of hair. However, Reuschel and Smith[82] did not elicit information about prior drug use from individuals, and these results could be attributable to differences in the frequency and extent of drug use.

Reid et al.[83] performed *in vitro* studies to determine the effect of color on the incorporation of benzoylecgonine into hair. Black, brown, and blond hair were obtained from drug-free subjects. Hair samples were placed in a solution containing 2.0 µg of benzoylecgonine. After 42 h, samples were removed and washed with 1.0% sodium dodecyl sulfate followed by 100% ethanol. Hair samples were then digested with 0.5 M sulfuric acid and analyzed for benzoylecgonine. Significant differences ($p < 0.001$) in benzoylecgonine concentrations were found in black, brown, and blond hair. The mean concentration of benzoylecgonine in black hair was sevenfold greater in comparison to blond hair and twofold greater in brown hair in comparison to blond hair. Further, Reid et al.[83] reported that benzoylecgonine present in blond hair was totally removed from hair washed once with sodium dodecyl sulfate and ethanol, but two washes were required to remove drug from black and brown hair. These findings suggested that benzoylecgonine may have a lower affinity for blond hair; consequently, benzoylecgonine may be lost from blond hair at a faster rate than black and brown hair.

Hair color also appears to affect the accumulation and retention of drugs such as cocaine and phencyclidine in hair. Kidwell and Blank[47] reported the concentration of cocaine in black and light brown hair obtained from subjects with self-reported cocaine use. The concentration of cocaine in black hair was greater than brown hair for subjects reporting similar cocaine use. Kidwell and Blank[47] also investigated the extraction efficiency of phencyclidine from black and light brown hair. The hair samples were soaked in solutions of phencyclidine and then drug was extracted from hair. The PCP concentration in black hair was three- to sevenfold greater than in brown hair samples, but the extraction efficiency was lower for hair samples which were dark brown or black in comparison to light brown hair.

2. Ethnic Hair Types

Some investigators have suggested that differences may exist in drug incorporation into ethnic hair types. Cone et al.[84] reported the occurrence of cocaine, heroin, and metabolites in hair obtained from subjects previously enrolled in an outpatient maintenance and detoxification study. They analyzed arm hair and head hair from nine Caucasoid subjects and eleven Africoid subjects for cocaine, benzoylecgonine, 6-acetylmorphine, and morphine. The results are presented in Table 5. Africoid head and arm hair samples had significantly higher levels of cocaine ($p < 0.05$) in comparison to Caucasoid head and arm hair samples, although no significant differences were found in the concentration of benzoylecgonine, 6-acetylmorphine, and morphine. The authors indicated that these findings could be explained by difference in the extent and frequency of drug use between the two groups or by ethnic differences in drug deposition and retention.

TABLE 5.

Drug Concentrations in Hair of Africoid and Caucasoid Subjects with Self-Reported Drug Use

	Ethnicity	N	COC ± SD	BE ± SD	6-AM ± SD	MOR ± SD
Arm hair	Caucasoid	9	137.9 ± 138.5	8.1 ± 9.1	1.7 ± 4.3	0.0
	Africoid	11	399.4 ± 413.6	23.9 ± 36.9	4.1 ± 9.4	0.0
Head hair	Caucasoid	9	60.8 ± 60.0	18.3 ± 28.4	2.1 ± 3.0	0.3 ± 0.8
	Africoid	11	235.5 ± 233.8	27.7 ± 45.3	1.9 ± 2.6	1.3 ± 3.1

Note: Abbreviations are as follows: COC, cocaine; BE, benzoylecgonine; 6-AM, monoacetyl-morphine; MOR, morphine.
Data are cited from Reference 84.

Mieczkowski and Newel[85] evaluated the potential for ethnic bias in hair testing for benzoylecgonine, a metabolite of cocaine. They analyzed urine and head hair collected from 1224 Africoid and Caucasoid arrestees with self-reported cocaine use. Data collected from analyses of hair and urine samples for drug indicated Africoid arrestees had a higher incidence of positives both in urine and hair assays for benzoylecgonine in comparison to Caucasoid arrestees. Africoid arrestees also had a higher percentage of self-reported use during the last 48 h, 30 d, and 60 d prior to arrest in comparison to Caucasoid arrestees. The percentage of Africoid and Caucasoid arrestees reporting cocaine use within 48 h was consistent with the number of subjects in each category who tested positive by urine and hair. Mieczkowski and Newel[85] concluded that there was no evidence of bias in hair testing and the higher rate of positive cocaine hair tests was likely due to greater drug use by Africoids.

The potential for ethnic bias in hair testing for cocaine was recently evaluated by Henderson et al.[86] They recruited 32 volunteers and recorded subject information including hair color, sex, and prior hair treatment. There were 28 Caucasoid subjects and 4 subjects described as non-Caucasoid (two Africoid subjects and two of East Indian, Eurasian, and Hispanic descent) who participated in the study. The investigators administered various doses of deuterated cocaine (0.6 to 1.2 mg/kg) by the intranasal and intravenous routes and periodically collected hair from subjects. Subjects were administered deuterated cocaine in order to distinguish it from cocaine present in subjects' hair from previous use. The results are illustrated in Figure 3. Non-Caucasoid subjects had between 2- and 12-fold more cocaine in their hair in comparison to Caucasoid subjects who received identical doses. Unfortunately, the sample size for non-Caucasoid subjects was limited to four. It was noteworthy, however, that cocaine concentrations in hair samples from all non-Caucasoid subjects were greater than those from Caucasoids.

Kidwell and Blank[59] performed an *in vitro* study of the binding of cocaine to Korean, Africoid, Caucasoid (brown and black), Hispanic, and Italian hair. The ends of cut hair samples were sealed in wax to avoid cocaine uptake. The hair samples were then soaked in a 5-µg/mL solution of radiolabeled cocaine for 2 h and washed repeatedly in deionized water to remove unbound cocaine. These samples were cut into segments (1.0 cm) and analyzed for cocaine. The investigators observed that the ratio of drug bound by Africoid to Caucasoid hair (including Hispanic) was 2.9 and the ratio for Korean to Caucasoid hair was 6.8. The authors suggested that their findings were due to ethnic differences in the protein structure of hair.

3.   *Ethnicity, Hair Color, and Chemical Treatments*

Drug incorporation into hair may be influenced by a combination of variables including hair color, ethnic hair type, and chemical treatments of hair. Henderson et al.[86] investigated the incorporation of drug into ethnic hair types which differed in hair color and hair treatments. They tested samples which included black Africoid hair, brown Caucasoid hair, blond Caucasoid hair, and bleached Caucasoid hair. These hair samples were exposed to cocaine vapor by placing samples in a Plexiglass™ box and vaporizing 10 mg of cocaine-free base. After 5 min, hair samples were removed and either subjected to direct analysis for drug or washed in sodium dodecyl sulfate, deionized water, and methanol prior to analysis. Caucasoid blond hair which had been bleached in a solution of 30% hydrogen peroxide accumulated considerably more drug than other hair types. This may be explained by the increase in porosity of hair which results from bleaching.[42] However, bleached hair samples which were washed prior to analysis for drug had the lowest concentration of cocaine

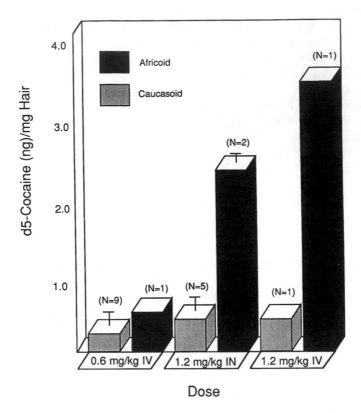

Figure 3
Concentrations of deuterated cocaine in Africoid and Caucasoid head hair samples collected at monthly intervals following drug administration. Hair samples were collected until drug was no longer detectable in subjects' hair. Bars represent the standard deviation from the mean. (Data are illustrated from Reference 86.)

in comparison to other samples. The removal of cocaine from bleached hair may have been enhanced due to the high porosity of bleached hair. For other hair types, Africoid hair accumulated more drug than blond and brown Caucasoid hair regardless of whether hair samples were washed or not washed prior to analysis for drug. The concentration of cocaine in Mongoloid hair was approximately twofold greater in comparison to other hair types. Henderson et al.[86] also investigated the effect of different cosmetic treatments on drug present in hair by perming, lightening, and straightening the same hair samples which had been contaminated with cocaine. Virtually all cosmetic treatments resulted in substantial loss of cocaine from hair.

Some preliminary evidence provides insight into the effect of chemical treatments on the accumulation of drug by different hair types. Skopp et al.[87] studied the diffusion of opiates in chemically treated hair (bleached or coated with nail polish) and untreated hair. One end of a bundle of hair strands was placed in water and the other end was placed in an aqueous solution containing 1 mg/mL of morphine, codeine, or dihydrocodeine. The investigators reported that radial and axial diffusion of these drugs occurred in hair samples in the presence of water, and that the diffusion of drug in bleached hair was dramatically greater in comparison to untreated hair. For hair samples that were coated with nail polish, diffusion of drug did not occur over a period of one week, which suggested that drug diffusion in hair was determined by the permeability of hair to water.

Additional evidence indicates that hair treatments affect drug incorporation into hair. Baumgartner and Hill[88] investigated the accumulation of cocaine in cosmetically treated hair and untreated hair. Porous treated hair (permed), Caucasoid hair, and Mongoloid hair were soaked in solutions of cocaine hydrochloride (5 µg/mL) for 3.5 h at 37°C. After soaking, hair samples were washed with either phosphate buffer or ethanol to remove drug from hair. Analysis of samples indicated that cocaine uptake by hair was greatest for porous hair samples. The concentration of cocaine was highest in permed hair in comparison to untreated Caucasoid and Mongoloid hair samples which were less porous. The concentration of cocaine in Caucasoid hair also was greater in comparison to Mongoloid hair, which was reportedly due to greater porosity of Caucasoid hair.

There is also evidence to suggest that shampoo treatments have a considerable effect on the absorption and binding of cocaine by hair. Sellers et al.[89] recently investigated the effects of different shampoos on cocaine absorption by ethnic hair types. They collected drug-free hair from Africoid, Caucasoid, and Mongoloid subjects. Hair samples were washed in 12 commercially available shampoos and then soaked in a solution of 1 µg/mL of cocaine in 0.1 M phosphate buffer (pH 6.0). Hair samples were washed in methanol, followed by a phosphate buffer wash, and then analyzed for drug. The investigators observed that the concentration of drug in Africoid, Mongoloid, and Caucasoid hair was dependent upon the brand of shampoo used to wash samples. For example, hair samples washed in Head and Shoulders® (fine/oily) had lower concentrations of cocaine in comparison to samples washed with Pert® (oily/fine) and Prell® (normal and normal oily). Sellers et al.[89] also reported that the concentrations of cocaine were highest in Africoid hair, and that Africoid and Mongoloid hair bound more cocaine in comparison to Caucasoid hair regardless of the brand of shampoo which was used to wash samples.

VI. CONCLUSIONS

A primary concern in drug testing is that the methodology should provide an objective means for assessing drug use by an individual. Drug testing methods are biased if particular ethnic groups are predisposed to test positive more often in comparison to other groups. Presently, there is mounting evidence which suggests that bias exists in hair testing for drugs of abuse due to selective accumulation of drug by particular hair types. Evidence indicates that the binding of drug to Africoid and Mongoloid hair is substantially greater in comparison to Caucasoid hair. Hair color and cosmetic treatments also affect the binding of drugs to different hair types. Present data indicate that black hair binds more drug in comparison to brown and blond hair, and that bleached hair binds more drug in comparison to untreated hair.

There is currently no clear explanation for selective binding of drug by particular hair types, but differences in ultrastructure, morphology, and protein structure between hair types may be the basis for the observed differences. For instance, the binding of drugs may be greater to heavily pigmented black hair such as Mongoloid and Africoid hair in comparison to blond and brown Caucasoid hair, because dark hair contains a higher concentration of melanin, a likely binding site for drugs. Structural differences could also account for the greater accumulation of drugs by brown and black hair. A recent explanation for the observed differences was provided by Kalasinsky et al.,[90] who hypothesized that the binding of drugs to hair occurs largely in medullated sections. The degree of medullation in Mongoloid hair is reported to be greater than Caucasoid hair, and this could explain greater accumulation of drug by Mongoloid hair. For bleached hair, the binding of drugs may be greater than for untreated hair since bleached hair is highly permeable, but this also could result in greater loss of drugs with solvent washes. The findings of Skopp et al.[87] offer the possibility that drug in a drug user's sweat or sebum could be deposited at a greater rate in bleached hair due to rapid diffusion through bleached hair in comparison to untreated hair.

Although there are strong indications for selective accumulation of drugs by particular hair types, hair testing is currently applied in a variety of situations to assess drug use by an individual. Drug charges based on the identification of a drug of abuse in hair have been made in child custody cases and have also resulted in denial and loss of employment. The use of hair testing to make these types of decisions may be unjust if selective accumulation of drugs by different hair types occurs which predisposes certain populations to test positive. There is definite need of further evaluation of the potential for bias in hair testing before this testing procedure can be considered as a viable alternative to more widely accepted and clinically validated drug testing methodologies.

REFERENCES

1. Elsohly, M. A., Jones, A. B., Elsohly, H. N., and Stanford, D. F., Analysis of the major metabolite of delta-9-tetrahydrocannabinol in urine. VI. Specificity of the assay with respect to indole carboxylic acids, *J. Anal. Toxicol.*, 9, 190, 1985.
2. Ricci, M. J., Benoit, C. A., Lertora, J. J. L., and George, W. J., Melanin and other indole compounds do not cause false-positive cannabinoid assay results, *Clin. Chem.*, 35, 1809, 1989.
3. Joseph, R., Su, T. P., and Cone, E. J., Possible ethnic bias in hair testing for cocaine, presented at *TIAFT/SOFT Joint Congress*, Tampa, Oct. 31–Nov. 4, 1994.
4. Parakkal, P. F., The fine structure of anagen hair follicle of the mouse, in *Advances in Biology of Skin: Hair Growth*, IX, Montagna, W. and Dobson, R. L., Eds., Pergamon Press, Braunschweig, Germany, 1969, 441.
5. Montagna, W. and Van Scott, E. J., The anatomy of the hair follicle, in *The Biology of Hair Growth*, Montagna, W. and Ellis, R. A., Eds., Academic Press, New York, 1958, 39.
6. Harkey, M. R., Anatomy and physiology of hair, *Forensic Sci. Int.*, 63, 9, 1993.
7. Ward, W. H. and Lundgren, H. P., The formation, composition, and properties of the keratins, in *Advances in Protein Chemistry*, Anson, M. L., Bailey, K., and Edsall, J. T., Eds., Academic Press, New York, 1954, 243.
8. Swift, J. A. and Bews, B., The chemistry of human hair cuticle. II. The isolation and amino acid analysis of the cell membranes and A-layer, *J. Soc. Cosmet. Chem.*, 25, 355, 1974.
9. Swift, J. A. and Bews, B., The chemistry of human hair cuticle. III. The isolation and amino acid analysis of various subfractions of the cuticle obtained by pronase and trypsin digestion, *J. Soc. Cosmet. Chem.*, 27, 289, 1976.
10. Matoltsy, A. G., What is Keratin?, in *Advances in Biology of Skin: Hair Growth*, IX, Montagna, W. and Dobson, R. L., Eds., Pergamon Press, Braunschweig, Germany, 1967, 559.
11. Matoltsy, A. G., The chemistry of keratinization, in *The Biology of Hair Growth*, Montagna, W. and Ellis, R. A., Eds., Academic Press, New York, 1958, 135.
12. Baden, H. P., Hair keratin, in *Hair and Hair Diseases*, Orfanos, C. E. and Happle, R., Eds., Springer-Verlag, Berlin, 1989, 45.
13. Baden, H. P., Characterization of hair keratins, in *Hair Research: Status and Future Aspects*, Orfanos, C. E., Montagna, W., and Stuttgen, G., Eds., Springer-Verlag, Berlin, 1981, 73.
14. Gillespie, J. M. and Marshall, R. C., The proteins of normal and aberrant hair keratins, in *Hair Research: Status and Future Aspects*, Orfanos, C. E., Montagna, W., and Stuttgen, G., Eds., Springer-Verlag, Berlin, 1981, 76.
15. Mercer, E. H., The electron microscopy of keratinized tissues, in *The Biology of Hair Growth*, Montagna, W. and Ellis, R. A., Eds., Academic Press, New York, 1958, 91.
16. Baden, H. P., McGilvray, N., Lee, L. D., Baden, L., and Kubilus, J., Comparison of stratum corneum and hair fibrous proteins, *J. Invest. Dermatol.*, 75, 311, 1980.
17. Baden, H. P., Goldsmith, L. A., and Fleming, B., A comparative study of the physicochemical properties of human keratinized tissues, *Bioch. Biophys. Acta*, 322, 269, 1973.

18. Swift, J. A., The hair surface, in *Hair Research: Status and Future Aspects*, Orfanos, C. E., Montagna, W., and Stuttgen, G., Eds., Springer-Verlag, Berlin, 1981, 65.

19. Fitzpatrick, T. B., Brunet, P., and Kukita, A., The nature of hair pigment, in *The Biology of Hair Growth*, Montagna, W. and Ellis, R. A., Eds., Academic Press, New York, 1958, 255.

20. Mason, H. S., Structure of melanins, in *Pigment Cell Biology*, Gordon, M., Ed., Academic Press, New York, 1959, 563.

21. Ortonne, J. P. and Prota, G., Hair melanins and hair color: ultrastructural and biochemical aspects, *J. Invest. Dermatol.*, 101, 82S, 1993.

22. Cesarini, J. P., Hair melanin and hair color, in *Hair and Hair Diseases*, Orfanos, C. E. and Happle, R., Eds., Springer-Verlag, Berlin, 1990, 165.

23. Ito, S. and Fujita, K., Microanalysis of eumelanin and pheomelanin in hair and melanomas by chemical degradation and liquid chromatography, *Anal. Biochem.*, 144, 527, 1985.

24. Birbeck, M. S. C. and Barnicot, N. A., Electron microscope studies on pigment formation in human hair follicles, in *Pigment Cell Biology*, Gordon, M., Ed., Academic Press, New York, 1959, 549.

25. Thody, A. J., Higgins, E. M., Wakamatsu, K., Ita, S., Burchill, S. A., and Marks, J. M., Pheomelanin as well as eumelanin is present in human epidermis, *J. Invest. Dermatol.*, 97, 340, 1991.

26. Barnicot, N. A., Birbeck, S. C., and Cuckow, F. W., The electron microscopy of human hair pigments, *Ann. Hum. Genet.*, 19, 231, 1955.

27. Jimbow, K., Ishida, O., Ito, S., Hori, Y., Witkop, C. J., and King, R. A., Combined chemical and electron microscopic studies of pheomelanosomes in human red hair, *J. Invest. Dermatol.*, 81, 506, 1983.

28. Rook, A., Racial and other genetic variations in hair form, *Br. J. Dermatol.*, 92, 599, 1975.

29. Steggerda, M. and Seibert, H. C., Size and shape of head hair from six racial groups, *J. Hered.*, 32, 315, 1941.

30. Steggerda, M., Cross sections of human hair from four racial groups, *J. Hered.*, 31, 475, 1940.

31. Hrdy, D., Quantitative hair form variation in seven populations, *Am. J. Phys. Anthropol.*, 39, 7, 1973.

32. Gold, R. J. M. and Scriver, C. R., The amino acid composition of hair from different racial origins, *Clin. Chim. Acta*, 33, 465, 1971.

33. Lindelöf, B., Forslind, B., Hedblad, M. A., and Kaveus, U., Human hair form, *Arch. Dermatol.*, 124, 1359, 1988.

34. Dekio, S. and Jidoi, J., Amounts of fibrous proteins and matrix substances in hairs of different races, *J. Dermatol.*, 17, 62, 1990.

35. Hausman, L. A., A comparative racial study of the structural elements of human head-hair, *Am. Nat.*, 59, 529, 1925.

36. Barnicot, N. A. and Birbeck, M. S. C., The electron microscopy of human melanocytes and melanin granules, in *The Biology of Hair Growth*, Montagna, W. and Ellis, R. A., Eds., Academic Press, New York, 1958, 239.

37. Jimbow, K. and Kukita, A., Fine structure of pigment granules in the human hair bulb: ultrastructure of pigment granules, in *Biology of Normal and Abnormal Melanocytes*, Kawamura, T., Fitzpatrick, T. B., and Seiji, M., Eds., University Park Press, Baltimore, 1971, 171.

38. Kinebuchi, S., Kobori, T., and Hori, Y., Behavior of melanosomes in melanocytes and keratinocytes of Japanese skin and black hair, in *Biology of Normal and Abnormal Melanocytes*, University Park Press, Baltimore, 1971, 195.

39. Trotter, M. and Duggins, O. H., Hair of Australian aborigines (Arnhem Land), *Am. J. Phys. Anthropol.*, 14, 649, 1956.

40. Wassermann, H. P., *Ethnic Pigmentation: Historical, Physiological, and Clinical Aspects*, Excerpta Medica, New York, 1974, 96.

41. Montagna, W., Prota, G., and Kenney, J. A., Jr., *Black Skin: Structure and Function*, Academic Press, San Diego, 1993, 21.

42. Breuer, M. M., The binding of small molecules and polymers to keratin and their effects on the physicochemical and surface properties of hair fibers, in *Hair Research: Status and Future Aspects*, Orfanos, C. E., Montagna, W., and Stuttgen, G., Eds., Springer-Verlag, Berlin, 1981, 96.

43. Wolfram, L. J., The reactivity of human hair. A review, in *Hair Research: Status and Future Aspects*, Orfanos, C. E., Montagna, W., and Stuttgen, G., Eds., Springer-Verlag, Berlin, 1981, 479.

44. Robbins, C. R. and Kelly, C., Amino acid analysis of cosmetically altered hair, *J. Soc. Cosmet. Chem.*, 20, 555, 1969.

45. Tolgyesi, E. and Fang, F., Action of nucleophilic reagents on hair keratin, in *Hair Research: Status and Future Aspects*, Orfanos, C. E., Montagna, W., and Stuttgen, G., Eds., Springer-Verlag, Berlin, 1981, 116.

46. Asquith, R. S. and Carthew, P., An investigation of the mechanism of alkaline degradation of cystine in intact protein, *Biochim. Biophys. Acta*, 278, 8, 1972.

47. Kidwell, D. A. and Blank, D. L., Hair analysis: techniques and potential problems, in *Recent Developments in Therapeutic Drug Monitoring and Clinical Toxicology*, Sunshine, I., Ed., Marcel Dekker, New York, 1992, 555.

48. Larsson, P., Larsson, B. S., and Tjalve, H., Binding of aflatoxin B_1 to melanin, *Fed. Chem. Toxicol.*, 26, 579, 1988.

49. Conlee, J. W., Gill, S. S., McCandless, P. T., and Creel, D. J., Differential susceptibility to gentamicin ototoxicity between albino and pigmented guinea pigs, *Hearing Res.*, 41, 43, 1989.

50. Lindquist, N. G., Accumulation *in vitro* of ^{35}S-chlorpromazine in the neuromelanin of human substantia nigra and locus coeruleus, *Arch. Int. Pharmacodyn.*, 200, 190, 1972.

51. Chen, K. K. and Poth, E. J., Racial differences as illustrated by the mydriatic action of cocaine, euphthalmine, and ephedrine, *J. Pharmacol. Exp. Ther.*, 36, 429, 1929.

52. Patil, P. N., Cocaine-binding by the pigmented and the nonpigmented iris and its relevance to the mydriatic effect, *Invest. Ophthalmol.*, 11, 739, 1972.

53. Patil, P. N., Shimada, K., Feller, D. R., and Malspeis, L., Accumulation of $(-)^{14}$C-ephedrine by the pigmented and the nonpigmented iris, *J. Pharmacol. Exp. Ther.*, 188, 342, 1974.

54. Patil, P. N. and Jacobowitz, D., Unequal accumulation of adrenergic drugs by pigmented and nonpigmented iris, *Am. J. Ophthamol.*, 78, 470, 1974.

55. Shimada, K., Baweja, R., Sokoloski, T., and Patil, P. N., Binding characteristics of drugs to synthetic levodopa melanin, *J. Pharm. Sci.*, 65, 1057, 1976.

56. Larsson, B. and Tjalve, H., Studies on the mechanism of drug-binding to melanin, *Biochem. Pharmacol.*, 28, 1181, 1979.

57. Larsson, B. and Tjalve, H., Studies on the melanin-affinity of metal ions, *Acta Physiol. Scand.*, 104, 479, 1978.

58. Montagna, W., Prota, G., and Kenney, J. A., Jr., *Black Skin: Structure and Function*, Academic Press, San Diego, 1993, 85.

59. Kidwell, D. A. and Blank, D. L., Mechanisms of incorporation of drugs into hair and the interpretation of hair analysis data, in *NIDA Research Monograph*, Cone, E. J. and Welch, M., Eds., 1994.

60. Su, T. P., Tsai, W. J., Joseph, R., Tsao, L. I., and Cone, E. J., Cocaine binds in a sterospecific, saturable manner to hair: a precaution on hair testing for forensic purposes, presented at College on Problems of Drug Dependence, Palm Beach, June 18–23, 1994.

61. Mohr, K. and Struve, M., Differential influence of anionic and cationic charge on the ability of amphiphilic drugs to interact with DPPC-liposomes, *Biochem. Pharmacol.*, 41, 961, 1991.

62. Schneider, V., The role in hair in forensic investigations, in *Hair Research: Status and Future Aspects*, Orfanos, C. E., Montagna, W., and Stuttgen, G., Eds., Springer-Verlag, Berlin, 1981, 459.

63. Kopito, L. E. and Shwachman, H., Alterations in the elemental composition of hair in some diseases, in *The First Human Hair Symposium*, Brown, A. C., Ed., Medcom Press, New York, 1974, 83.

64. Harkey, M. R. and Henderson, G. L., Hair analysis for drugs of abuse, in *Advances in Analytical Toxicology*, 2nd ed., Baselt, R. C., Ed., Biomedical Press, Foster City, CA, 1984, 298.

65. Deeming, S. B. and Weber, C. W., Hair analysis of trace minerals in human subjects as influenced by age, sex, and contraceptive drugs, *Am. J. Clin. Nutr.*, 31, 1175, 1978.

66. Chittleborough, G., A chemist's view of the analysis of human hair for trace elements, *Sci. Total Environ.*, 14, 53, 1980.

67. Sky-Peck, H. H., Distribution of trace elements in human hair, *Clin. Physiol. Biochem.*, 8, 70, 1990.

68. Petering, H. G., Yeager, D. W., and Witherup, S. O., Trace metal content of hair, *Arch. Environ. Health*, 27, 327, 1973.

69. Shrestha, K. P. and Schrauzer, G. N., Trace elements in hair: a study of residents in Darjeeling (India) and San Diego, California (U.S.A.), *Sci. Total Environ.*, 79, 171, 1989.

70. Breuer, M. M., Binding of phenols by hair. I, *J. Phys. Chem.*, 68, 2067, 1964.

71. Lotzsch, K. R., Reng, A. K., Gantz, D., and Quack, J. M., The radiometric technique. Explained by the example of absorption and desorption of ^{14}C-labelled distearyl-dimethylammonium chloride on human hair, in *Hair Research: Status and Future Aspects*, Orfanos, C. E., Montagna, W., and Stuttgen, G., Eds., Springer-Verlag, Berlin, 1981, 638.

72. Price, V. H., The role of hair care products, in *Hair Research: Status and Future Aspects*, Orfanos, C. E., Montagna, W., and Stuttgen, G., Eds., Springer-Verlag, Berlin, 1981, 501.

73. Bernstein, E., Dynamic experiments on hair in the scanning electron microscope, in *The First Human Hair Symposium*, Brown, A. C., Ed., Medcom Press, New York, 1974, 317.

74. Gerdes, R. J. and Brown, A. C., Analysis of hair surface structure for product evaluation, in *The First Human Hair Symposium*, Brown, A. C., Ed., Medcom Press, New York, 1974, 302.

75. Goldblum, R. W., Goldbaum, L. R., and Piper, W. N., Barbiturate concentrations in the skin and hair of guinea pigs, *J. Invest. Dermatol.*, 22, 121, 1954.

76. Uematsu, T., Miyazawa, N., Okazaki, O., and Nakashima, M., Possible effect of pigment on the pharmacokinetics of ofloxacin and its excretion in hair, *J. Pharm. Sci.*, 81, 45, 1992.

77. Uematsu, T., Sato, R., Fujimori, O., and Nakashima, M., Human scalp hair as evidence of individual dosage history of haloperidol: a possible linkage of haloperidol excretion into hair with hair pigment, *Arch. Dermatol. Res.*, 282, 120, 1990.

78. Faergemann, J., Zehender, H., Denouel, J., and Millerioux, L., Levels of terbinafine in plasma, stratum corneum, dermis-epidermis (without stratum corneum), sebum, hair and nails during and after 250 mg terbinafine orally once per day for four weeks, *Acta Derm. Venereol.*, 73, 305, 1993.

79. Henderson, G. L., Mechanisms of drug incorporation into hair, *Forensic Sci. Int.*, 63, 19, 1993.

80. Forrest, I. S., Otis, L. S., Serra, M. T., and Skinner, G. C., Passage of ^{3}H-chlorpromazine and ^{3}H-delta-tetrahydrocannabinol into the hair (fur) of various mammals, *Proc. West. Pharmacol. Sci.*, 15, 83, 1972.

81. Ishiyama, I., Nagai, T., and Toshida, S., Detection of basic drugs (methamphetamine, antidepressants, and nicotine) from human hair, *J. Forensic Sci.*, 28, 380, 1983.

82. Reuschel, S. A. and Smith, F. P., Benzoylecgonine (cocaine metabolite) detection in hair samples of jail detainees using radioimmunoassay (RIA) and gas chromatography/mass spectrometry, *J. Forensic Sci.*, 36, 1179, 1991.

83. Reid, R. W., O'Connor, F. L., and Crayton, J. W., The *in vitro* differential binding of benzoylecgonine to pigmented human hair samples, *Clin. Toxicol.*, 32, 405, 1994.

84. Cone, E. J., Darwin, W. D., and Wang, W. L., The occurrence of cocaine, heroin and metabolites in hair of drug abusers, *Forensic Sci. Int.*, 63, 55, 1993.

85. Mieczkowski, T. and Newel, R., An evaluation of patterns of racial bias in hair assays for cocaine: black and white arrestees compared, *Forensic Sci. Int.*, 63, 85, 1993.

86. Henderson, G. L., Harkey, M. R., and Jones, R. *Hair Analysis for Drugs of Abuse*, Grant number NIJ 90-NIJ-CX-0012, National Institute on Justice and National Institute on Drug Abuse, 1993.

87. Skopp, G., Potsch, L., and Aderjan, R., Diffusion of opiates in human hair, presented at *TIAFT/SOFT Joint Congress*, Tampa, Nov. 28–Oct. 3, 1994.

88. Baumgartner, W. A. and Hill, V. A., Hair analysis for drugs of abuse: decontamination issues, in *Recent Developments in Therapeutic Drug Monitoring and Clinical Toxicology*, Sunshine, I., Ed., Marcel Dekker, New York, 1992, 577.

89. Sellers, J. K., Smith, F. P., Gruszecki, A. C., and Clouette, R., Effect of shampoo on cocaine uptake in hair, presented at *TIAFT/SOFT Joint Congress*, Tampa, Nov. 28–Oct. 3, 1994.

90. Kalasinsky, K. S., Magluilo, J., and Schaefer, T., Study of drug disposition in hair by infrared microscopy visualization, *J. Anal. Toxicol.*, 18, 337, 1994.

Chapter 4

THE ANALYTICAL TOOLS FOR HAIR TESTING

Manfred R. Moeller and Hans P. Eser

CONTENTS

I. ABSTRACT

This article reviews the analysis of 48 drugs and drug metabolites in human hair by different analytical methods which are described in the current literature. Washing steps to exclude external contamination, extraction, derivatization, stationary phases, detection modes (DETMODEs) and detection limits of nonchromatographic and chromatographic procedures are presented in ten tables. Besides immunoassays (IA), which are used in most cases for screening purposes, the most important detection method after chromatographic separation of the components is mass spectrometry (MS) because of its sensitivity and specificity. However, new extraction techniques like supercritical fluid extraction (SFE), separation procedures like capillary electrophoreses, and DETMODEs like infrared microscopy are advancing. Additionally, methods for gas chromatographic/mass spectrometric (GC/MS) screening procedures are presented.

II. INTRODUCTION

Hair as a marker of exposure to toxicants has been of considerable interest to toxicologists for more than a hundred years. Casper[1] reported in 1857 the analysis of hair for the detection of poisons in his famous *Practisches Handbuch der gerichtlichen Medicin*. He referred to Hoppe-Seyler, who had found arsenic in the hair of an 11-year-buried body and had discussed the possibility that the metal was incorporated before death. About 100 years later, the first report about the detection of an organic drug in hair was published. Goldblum et al.[2] described an ultraviolet method to detect barbiturates in guinea pig hair.

Baumgartner et al.[3] published a radioimmunological (RIA) method for the detection of opiates in hair of addicts. The proof of morphine in head hair by thin-layer chromatography (TLC) was reported by Klug.[4] Suzuki et al.[5,6] published a method to detect methamphetamine and amphetamine in a single human hair by mass fragmentography (MS). The detection of morphine in hair by high-performance liquid chromatography (HPLC) was reported by Marigo et al.[7]

With growing attention to hair analysis for drugs of abuse and pharmaceutical drugs, a number of pharmacological, analytical, and technical problems became of interest, which have been focused in a Consensus Report[8] and revised at the SOFT Conference 1992 in Cromwell, Hartford, CT.[9] Main problems arose with the questions of external contamination, the routes of incorporation of drugs into hair, possible dose-response relationships, interpretation of results, and their application to forensic, clinical, and occupational cases.

One of the most debated issues in hair analysis is the question of external contamination. The discussion is very controversial. Some authors have proved that external contamination cannot be excluded,[10-16] whereas, on the other hand, data have been presented to show that washing procedures allow distinguishing between drugs entering into hair through absorption from the capillary bloodstream or by environmental contamination.[17] However, the latter procedures are only partly published. Therefore, a scientific assessment is not possible.

The data presented here only focus on methods for the detection and quantitation of drugs and their metabolites in hair. The issue of incorporation of drugs into hair or external contamination cannot be discussed in detail.

III. ANALYTICAL PROCEDURES FOR THE DETECTION OF DRUGS IN HAIR

A. Extraction Methods

Due to the low amount of specimen used for hair analysis and the low concentration of drugs and their metabolites in hair, extraction procedures are mostly necessary to concentrate the analytes in the tested solution. Various methods have been used for this step.[18] The variety of described procedures show that there is no universal solution. The choice of an extraction procedure may be influenced by the kind of drug and drug metabolite(s) to be analyzed and the method used for the subsequent analysis. Preceding the real extraction, there are four treatments with more or less intermethodical differences to prepare the hair matrix for the extraction:

acid hydrolysis (HY)

alkaline hydrolysis (AL)

enzymatic digestion (ED)

treatment with buffer or organic solvents

The variations concern pH, temperature, time of processing, and, in the last case additionally, possible ultrasonic treatment. The following step consists in a liquid-phase or solid-phase extraction (SPE). However, other extraction methods like SFE[19,20] also have been used prior to the detection procedure. The final choice of the workup procedure is mainly dependent on the chemical stability of the analyte; e.g., benzodiazepines are highly unstable under alkaline conditions, but they are also affected under acid conditions.[22] Alkaline hydrolysis should be avoided for cocaine;[18,23] however, morphine can be extracted after acid or alkaline hydrolysis.[18]

In some cases, the resulting solution from the treatment with buffer is used directly for IA (Tables 1.1 to 1.3).

Further considerations should include the following analytical method(s) (e.g., denaturation of antibodies in IA by the digest must be avoided), the possible necessity of derivatization (e.g., methylation, acetylation, propionylation), pharmacokinetic and chemical reflections on formation of metabolites or degradation products, and/or the disposition pattern of parent drug or metabolite(s) to be detected.

B. Nonchromatographic Methods

1. Nonimmunological Methods

The majority of detection procedures are based on GC/MS,[20,24] followed by IA. However, new and promising methods for the detection of drugs and their metabolites have been investigated. Kalasinsky et al.[25] described a method using infrared microscopy and utilized it to differentiate passive contamination from drugs absorbed into the hair from ingestion. A three-dimensional infrared imaging of a single hair with Fourier transform infrared (FT-IR) shows the partition of the drug in the cross-sectioned or laterally microtomed hair. Perhaps this method allows, together with other approaches, a more detailed investigation of drug incorporation into hair. However, the specimen is always only a part of one single hair. Tagliaro et al.[26] compared the sensitivity, practicability, and reproducibility of capillary electrophoresis

TABLE 1.1

Immunological Techniques for the Determination of Opiates and Their Metabolites

Drug	Immunol. method	Decontamination procedure	Workup	LOD[a]/COL[b]/ LDV[c]/LPC[d]	2nd Method	Ref.
Opiates	RIA	n. i. a.*	HY	0.5 ng/ml[b]	GC/MS	42
	FPIA	EtOH	AL	0.1 ng/ml[b]	GC/MS	60
	RIA	DINT	AL	0.1 ng/mg[c]	None	64
Morphine/ opioids	RIA	ET-HCL	HY	1 ng/ml[d]	HPLC-FD CS	46
Morphine	RIA	DET-H_2O-DINT	MeOH	<1 ng/mg[c]	None	3
	RIA	ET-HCL	HY-EX	n. i. a.	HPLC-FD	7
	RIA	ET-HCL	AL-EX	n. i. a.	HPLC-FD	7
	RIA	EtOH-PB	n. i. a.	0.13 ng/mg[b]	None	30
	RIA	DINT-CH_2Cl_2	AL-EX	0.01 ng/mg[a]	GC/MS	33
	RIA	DINT-CH_2Cl_2	AL-EX	0.5 ng/mg[a]	GC/MS	33
	RIA	n. i. a.*	HY	4 ng/ml[b]	?	42
	RIA	ET-HCL	HY-EX	0.1 ng/mg[a]	HPLC-ECD	43
	RIA	ET	HY-EX	n. i. a.	HPLC-ECD GC/MS	45
	RIA	ET-HCL	HY-EX	0.07 ng/mg[b]	HPLC-ECD	47
	RIA	ET-HCL	HY	1 ng/ml[a]	HPLC-FD	48
	RIA	ET-HCL	HY	1 ng /ml[d]	HPLC-ECD	49
	RIA	H_2O-EtOH-DINT	HY	0.3 ng/mg[c]	GC/MS	53
	RIA	AC-H_2O-AC	AL	0.2 ng/mg[c]	GC/MS	59
	RIA	AC-H_2O-AC	AL	n. i. a.	None	65, 66
	RIA	DET-AC	EN	0.18 ng/mg[c]	None	67
	RIA	CH_2Cl_2	AL	1 ng/ml[a]	None	68
	RIA	CH_2Cl_2	EN	1 ng/ml[a]	None	68
	RIA	None	HY	0.4 ng/mg[b]	None	69
	EIA	DINT-CH_2Cl_2	AL-EX	3 ng/mg[a]	GC/MS	33
	EIA	CH_2Cl_2	AL	0.71 ng/mg[c]	GC/MS	57
	FPIA	CH_2Cl_2	AL	0.48 ng/mg[c]	GC/MS	57
Heroin/ morphine	RIA	EtOH-PB	n. i. a.	0.052 ng/mg[b]	GC/MS	27
Heroin	RIA	EtOH-PB	n. i. a.	n. i. a.	None	31
6-MAM	RIA	H_2O-AC-DINT	PB	0.2 ng/mg[a]	GC/MS	34
	RIA	n. i. a.*	HY	1 ng/ml[b]	?	42

Note: n.i.a. = no information available.

* Author just remarks "washed hair."

(CE) and a HPLC method for the detection of cocaine in hair. They found that CE was three to five times faster, but less sensitive (0.15 to 0.3 ng/mg) compared to HPLC with fluorimetric detection (0.015 ng/mg). In the CE procedure, tetracaine was used as the IS. CE was also used to detect morphine.

2. Immunological Methods

In hair analysis, three methodically different IA with different kinds of labeling are used: radioimmunoassay (RIA), which is the most common IA in hair analysis, enzymeimmunoassay (EIA), and fluorescence polarisation immunoassay (FPIA). In hair analysis, IA have the same advantages and disadvantages as compared to their use in urinalysis. They are fast, easy to handle, and can be automated. Their use can save time and expenses when a great number of negative samples has to be expected.

On the other hand, only classes of substances can be detected with most IA. Cross reactivity of interfering substances cannot be excluded. They have been developed for urinalysis, therefore the usefulness for the matrix "hair" has to be investigated. Additionally, IA can only be used as screening procedures in cases where the

TABLE 1.2

Immunological Techniques for the Determination of Cocaine and Its Metabolites

Drug	Immunol. method	Decontamination procedure	Workup	LOD[a]/COL[b]/ LDV[c]/LPC[d]	2nd Method	Ref.
Cocaine	RIA	EtOH-PB	n. i. a.	0.085 ng/mg[b]	GC/MS	27
	RIA	ET-PB	n. i. a.	0.2 ng/mg[b]	None	28
	RIA	EtOH-PB	n. i. a.	0.085 ng/mg[b]	None	29
	RIA	EtOH-PB	n. i. a.	0.12 ng/mg[b]	None	30
	RIA	EtOH-PB	n. i. a.	n. i. a.	None	31
	RIA	n. i. a.	n. i. a.	n. i. a.	None	32
	RIA	H_2O-EtOH	HY	0.3 ng/mg[a]	None	35
	RIA	n. i. a.*	HY	0.3 ng/mg[a]	GC/MS	36
	RIA	EtOH	MeOH	0.025 ng/mg[a]	GC/MS	37
	RIA	EtOH	MeOH	0.025 ng/mg[a]	None	38
	RIA	H_2O-EtOH-DINT	HY	0.1 ng/mg[c]	GC/MS	53
	RIA	EtOH?-PB	HY	2 ng/mg[a]	GC/MS	54
	RIA	DET-AC	EN	1.1 ng/mg[c]	None	67
	RIA	DINT-DET-H_2O-DINT	EtOH	0.0069 ng/mg[c]	None	70
	RIA	DET-H_2O-AC	HY-EX	n. i. a.	None	71
	RIA	DET-H_2O-AC	MeOH	n. i. a.	None	71
	RIA	DET-H_2O-AC	EN	n. i. a.	None	71
	FIA	EtOH	AL	0.3 ng/ml[b]	GC/MS	60
Benzoylecgonine	RIA	EtOH	MeOH	0.25 ng/mg[a]	GC/MS	37
	RIA	EtOH	MeOH	0.25 ng/mg[a]	None	38
	RIA	DET-H_2O	HY	3 ng/ml[a]	GC/MS	55
	RIA	DET-DINT	EtOH	0.16 ng/mg[c]	GC/MS	58
	RIA	CH_2Cl_2	AL	6 ng/ml[a]	None	68
	RIA	CH_2Cl_2	EN	6 ng/ml[a]	None	68
	RIA	None	HY	0.4 ng/mg[b]	None	69
	RIA	DET-H_2O	HY-PB/ saline	5 ng/ml[a]	None	72
	RIA	DET-DINT	EtOH	62 ng/ml[d]	None	73

Note: n.i.a. = no information available.

* Author just remarks "washed hair."

sensitivity and specificity of the test is sufficient. This must be proved in every case. However, this is often not clearly demonstrated in the cited papers.[27-32]

Moreover, there is no consensus about the definition of sensitivity limits (Tables 1.1 to 1.3): some authors indicate limits of detection (LOD), others claim only the limits of their calibration curves (lowest point of calibration curve [LPC]) or the lowest values detected (LDV), some refer to mg hair, others to mL of their extract. However, sensitivity is usually in the range of ng drug/mg hair. Some scientists even claim detection limits in the pg range (morphine and 6-monoacetylmorphine [6-MAM];[33,34,43] cocaine and benzoylecgonine [BZE];[35-37,38] nicotine, and cotinine,[38] and haloperidol).[39-41] In analogy to urinalysis, IA are screening methods which have to be confirmed by more specific methods.[24,42-44] Combinations with HPLC,[7,43,45-49] collisional spectroscopy (CS),[46] or GC/MS have been published. However, confirmation of immunological results with GC/MS is state of the art.[27,33,34,36,37,42,45,50-61]

Because it is obvious that primarily the parent drug and lipophilic metabolites are found in hair, some IA are rather specific for single substances of a drug class, e.g., RIA for cocaine (Table 2) or 6-MAM. In both cases, the substance with the highest cross reactivity to the antibody is identical with the main compound excreted into hair after drug consumption. Quantitative immunological results correlate with GC/MS.[34,42] In the case of the DPC-Kit for free morphine (Table 3), there are good correlations with sensitivity and results of GC/MS and/or HPLC-ECD confirmation.[43,59,62,63]

TABLE 1.3

Immunological Techniques for the Determination of Methadone, THC, and Miscellaneous Other Drugs and Their Metabolites

Drug	Immunol. method	Decontamination procedure	Workup	LOD[a]/COL[b]/ LDV[c]/LPC[d]	2nd Method	Ref.
Methadone	RIA	EtOH-PB	n. i. a.	0.065 ng/mg[b]	GC/MS	27
	RIA	n. i. a.*	HY	2.5 ng/mg[a]	GC/MS	50
	RIA	H_2O-EtOH-DINT	HY	0.1 ng/mg[c]	GC/MS	53
	RIA	H_2O-EtOH-DINT	HY	2.5 ng/ml[d]	None	74
	RIA	H_2O-EtOH-DINT	HY	0.5 ng/mg[c]	None	75
Marijuana	RIA	EtOH-PB	n. i. a.	0.008 ng/mg[b]	GC/MS	27
	RIA	EtOH-PB	n. i. a.	n. i. a.	None	31
THC	RIA	n. i. a.*	HY	0.4 ng/mg[c]	GC/MS	51
	RIA	H_2O-EtOH-DINT	HY	0.4 ng/mg[c]	GC/MS	53
Cannabis	FPIA	EtOH	AL	0.025 ng/ml[b]	GC/MS	60
Nicotine	RIA	EtOH-MeOH	AL	0.25 ng/mg[a]	None	38
	RIA	H_2O-EtOH-DINT	HY	0.4 ng/mg[c]	GC/MS	52
	RIA	H_2O-EtOH-DINT	HY	0.2 ng/mg[c]	GC/MS	53
	RIA	H_2O-EtOH-DINT	HY-EX	20 ng/ml[b]	GC/MS	61
	RIA	HEX	AC	1 ng/ml[a]	None	76
Cotinine	RIA	EtOH-MeOH	AL	0.1 ng/mg[a]	None	38
	RIA	HEX	AC	1 ng/ml[a]	None	76
Haloperidol	RIA	DET-H_2O-DINT	AL-EX	<0.625 ng/ml[a]	None	39
	RIA	DET-H_2O-DINT	AL-EX	<0.625 ng/ml[a]	None	40
	RIA	DET-H_2O-DINT	AL-EX	<0.625 ng/ml[a]	None	41
Phencyclidine	RIA	EtOH-PB	n. i. a.	0.043 ng/mg[b]	GC/MS	27
	RIA	DET-H_2O-DINT	MeOH	0.3 ng/mg[c]	None	77
Benzodiazepines	RIA	DET-AC	EN	0.36 ng/mg[c]	None	67
	FPIA	EtOH	AL	0.2 ng/ml[b]	GC/MS	60
Barbiturates	RIA	H_2O-EtOH-DINT	HY	2.1 ng/mg[c]	GC/MS	53
	FPIA	EtOH	AL	0.2 ng/ml[b]	GC/MS	60
Pholcodine	RIA	AC-DINT	SB	0.5 ng/mg[c]	GC/MS	56
Amphetamine	FPIA	EtOH	AL	0.15 ng/ml[b]	GC/MS	60
	EIA	H_2O	AL-EX	2000 ng/ml[a]	None	78
Antidepressants	FPIA	EtOH	AL	0.075 ng/ml[b]	GC/MS	60

Note: n.i.a. = no information available.

* Author just remarks "washed hair."

TABLE 2.

Examples for the Cross Reactivities (%) of Immunoassays

Drug/metabolite	CAC® cocaine (RIA, Diagnostic Products Corporation)	Abuscreen cocaine (RIA, Roche Diagnostics)	TDx (FPIA, Abbott Laboratories)
Cocaine	100	100	100
Benzoylecgonine	0.8	10000	8333
Cocaethylene	42	—	—
Ecgonine	0.0006	200	92
Ecgonine methyl ester	0.4	7	8
Norcocaine	0.5	500	8

From Cassani, M., Spiehler, V., *Forensic Sci. Int.*, 63, 175, 1993. With kind permission of Elsevier Science, The Netherlands.

In conclusion, specific IA for hair are not commercially available. However, it is possible to adopt kits designed for serum or urine for the purpose of hair analysis. These meet the requirements as analytical tools for detecting drugs in ppm or even ppb ranges. Moreover, they are rapid and easy to use. Therefore, if the limitations

TABLE 3.

Cross Reactivity of the DPC-Morphine-RIA to Some Selected Drugs
of Abuse (Concentration: c = 2 μg/mL)

Drug/metabolite	Cross reactivity in blood (%)	Cross reactivity in urine (%)
Morphine	100	100.00
Codeine	0.14	0
Methadone	0.08	0.05
Diacetylmorphine	2.16	0.36
Cocaine	0.05	0

From Matejczik, R. J., Kosinski, J. A., *J. Forensic Sci.*, 30, 677, 1985. With
permission.

of these methods are kept in mind, they are able to serve as sensitive pretests for
further analysis. In daily practice, they can be a facilitation and a time- and cost-
economizing factor when a great number of negative samples are expected. However,
the disadvantages have to be considerd, e.g., the division of the samples when several
groups of drugs have to be detected, or the need for confirmatory analysis by
chromatographic methods. Therefore, if only small amounts of hair are available, it
makes more sense to use a GC/MS screening procedure with subsequent quantitation.

In Tables 1.1 to 1.3, the immunological methods for the detection of drugs in hair
are listed by substances. Additional information is given about decontamination and
workup procedures, sensitivity, and whether or not a confirmation method was used.

C. Chromatographic Procedures for Individual Drugs/Drug Metabolites

Chromatographic techniques are the most powerful tools for the identification
and quantitation of drugs in hair due to their separation ability and their detection
sensitivity. TLC, HPLC, and GC were used in several cases for the detection of various
drugs and/or drug metabolites. However, the majority of detection procedures for
individual compounds are based on GC/MS.[20,24]

1. Thin-Layer Chromatography

In 1980, Klug[4] reported a method to detect morphine in head hair of drug abusers
by TLC. He dissolved the hair in sodium hydroxide and hydrolyzed the solution
with HCL. He extracted the solution with amyl alcohol and separated the components by TLC. Detection and quantitation were made by fluorimetry. The findings
were between traces and 4 ng/mg. A HPTLC method was used to determine morphine in human hair.[79] From 20 to 200 mg of hair were washed four times with water
to remove surface contamination. Then the hair was incubated in sodium hydroxide
and subsequently in HCL at 80°C in both cases. After SPE (type "Extrelut® 3"/Merck)
the solution was dansylated, analyzed by HPTLC (methanol (MeOH)/ammonia
99:1), and quantitated by densitometry.

2. High-Performance Liquid Chromatography

A HPLC method to detect morphine in hair samples was reported by Marigo
et al.[7] They tested several washing and dissolving procedures for the hair and
decided to use a single-step acid washing with dilute HCL. After incubation of the
hair sample with dilute HCL, they made a SPE and dansylated the extract. They
used silica as a stationary phase with a hexane/isopropanol/ammonia mixture as a

mobile phase and fluorescence detection to identify and quantitate the morphine. They found a good correlation of the amounts detected in the hair of heroin addicts by HPLC and by RIA (r = 0.997; n = 15). Staub and Robyr[43] also used a screening by RIA and confirmation by HPLC with electrochemical detection to determine morphine in hair of drug addicts.

The separation of the optical isomers of methamphetamine and amphetamine from hair by HPLC was reported.[81,82] A rather high amount of hair (200 to 250 mg) from stimulant abusers was used as the specimen. After acetylation, the samples were chromatographed on a column which contained an optical active carrier (Chiracel OB) and n-hexane (HEX)/isopropanol as a mobile phase to separate the d- and l-isomers of methamphetamine and amphetamine. Only d-methamphetamine and d-amphetamine could be found in the hair of the abusers.

Haloperidol and its major active metabolite, reduced haloperidol, were detected in hair samples by isocratic HPLC.[83] The hair samples were dissolved in different manners (sonificated in 0.1% SDS solution or MeOH resp. or in 2 n NaOH/80°C). The NaOH dissolution gave the best results. The samples were chromatographed on a TSK Gel-TM ODS column with potassium phosphate buffer/acetonitrile/MeOH (3:2:1, v/v) as a mobile phase and detected coulometrically. The concentration of haloperidol was determined by RIA in parallel. Both values showed good correlation. The concentrations of haloperidol and reduced haloperidol measured in hair correlated better than serum with the individual dosage history. The detection of a number of antidepressants (amitryptilin, doxepin, dothiepin, imipramine, mianserin, and trimipramine) and antipsychotic drugs (haloperidol, chlorpromazine, and thioridazine) and several of their metabolites has been described in post-mortem hair.[84] The drugs were quantitated by HPLC. Ofloxacin, norfloxacin, and ciprofloxacin were determined in human hair by HPLC with fluorescence detection[85] using a reversed-phase C 18 column together with a SPE. The hair was dissolved in 1 M sodium hydroxide. A new antimicrobial (Quinolone; AM-1155) was used as a "time marker" for analyzing other drugs in hair. The hair samples were dissolved in 1 M sodium hydroxide and analyzed by HPLC.[86] Buprenorphine and its dealkylated metabolite were detected in hair.[136]

3. Gas Chromatography

GC procedures are less useful for the analysis of drugs in hair. The enormous number of possible exogenous and endogenous compounds which can partly be found in hair make the interpretation of chromatograms with flame ionization detector (FID), nitrogen-phosphor flame ionization detector (NP-FID), or even with electron capture detector (ECD) very difficult. Therefore, the number of papers is very limited.

Viala et al.[87] published a method for GC detection of chloroquine and monodesmethylchloroquine in hair of patients who had been treated with chloroquine for several months. The samples were washed with detergent (DET) and then dissolved in hot potassium hydroxide. After the extraction with ether, a thin-layer procedure was used for cleaning up. The quantitation was made by GC with a NP-FID. Chloroquine and its metabolite were identified by GC/MS. The GC detection of chloroquine and with NP-FID detector was reported by Ochsendorf et al.[88] They quantitated the drug in different hair sections and found a correlation to the pharmacological behavior of the substance. Couper et al.[84] described a GC screening with NP-FID for antidepressants and antipsychotic drugs in hair from post-mortem cases.

Takahashi[89] reported the detection of methamphetamine and amphetamine in hairs of monkeys after administration of methamphetamine. He used trifluoroacetic anhydride (TFAA) for derivatization and OV-101 as the GC column. He found the highest amount of both compounds in the fourth week. Then they decreased gradually. Nagai et al.[90] presented a method for the detection of methamphetamine and amphetamine in hair of addicts, as well as in hair, bones, and teeth after animal experiments. They used a packed column (PC) with 5% PEG 6000 + 5%-KOH and N-ethylbenzylamine as the internal standard (IS).

Ishiama et al.[91] found amitriptyline, imipramine, and their metabolites in human hair of patients with long-term treatment with antidepressants. He also found methamphetamine and amphetamine in hair of addicts. In the hair of smokers, nicotine was detected. For the detection of the amphetamines, the hair was dissolved by treating with NaOH and then with HCL, extracted with chloroform, trifluoroacetylated, and analyzed on an OV-17 column. The antidepressants were extracted with heptane and also analyzed on an OV-17 PC.

4. Gas Chromatography-Mass Spectrometry

GC/MS is the most powerful tool for the detection of drugs in hair. The most important conditions of the investigations are overviewed in Tables 4 through 8, ordered by substance groups. Listed are the kind of removal of external contamination (DECON); the workup procedure, containing a possible acid (HY), alkaline (AL), or enzymatic (EN) hydrolysis, and the method of extraction: solid phase (SPE) or liquid/liquid extraction (EX); conditions of derivatization; the instrument requirements, e.g., gas chromatographic column (GC Col.); detection mode (DETMODE); the kind of quantitation: with internal standard (IS), deuterated internal standard (DIS), or external standard (ES); and the detection limits (DETLIMIT) of the methods, published for the different drugs.

a. Amphetamines

The GC/MS procedures for methamphetamine are described in Table 4. The papers published in Japanese[5,81,92,93] have corresponding reports in English.[6,82,94,95] Methamphetamine was detected and determined by mass fragmentography in rat hair after administration of the substance.[92,94] Nine methods also detected the metabolite amphetamine or amphetamine alone. Suzuki et al.[96] determined methamphetamine also in nail, sweat and saliva. The workup (EX after acid or alkaline hydrolysis) and derivatization technique (methanol-trifluoroacetic acid [TFA]) is rather uniform in most procedures. Nakahara et al.[80] used methoxyphenamine excretion into beard hair to discuss several washing procedures. Alkaline or methanolic extraction are used with one exception. Derivatization is mainly made by fluorinated anhydrides. A review[107] gives details on analytical procedures, incorporation rates of amphetamines from blood to hair, and relationship between drug history and drug distribution in hair.

b. Cannabinoids

The detection of cannabinoids in hair by GC/MS seems to be more difficult because there are only seven reports (Table 5) in the reviewed literature, although it is probably the most common drug of abuse in the U.S. and Europe. Balabanova et al.[51] published a method with RIA detection of cannabinoids and GC/MS confirmation of Δ9-tetrahydrocannabinol (THC). However, the selected ion monitoring (SIM) chromatograms shown in the publication are very poor.[97,98] For the detection

TABLE 4.

GC/MS Analysis for Amphetamine and Derivatives in Hair

Drug/drug metabolite	Decontam.	Workup	Derivat.	GC col.	DetMode	Quant.	DetLimit	Ref.
Methamphetamine	DET-HCL	HY-EX	TFA	OV-17 PC	EI-SIM	—	0.1 ng	92
Methamphetamine	MeOH-H_2O	AL-EX	TFA	OV-17 PC	EI-SIM	IS	—	91
Methamphetamine, amphetamine	—	AL-EX	TFA	Thermon PC-3000	CI-SCAN EI-SCAN	IS	0.01 ng	6
Methamphetamine, amphetamine	—	AL-EX	TFA	Thermon PC-3000	CI-SCAN EI-SCAN	IS	0.01 ng	5
Methamphetamine	HCL-MeOH	HY-EX	TFA	OV-17 PC	EI-SIM	—	1 ng	94
Methamphetamine, amphetamine	MeOH-H_2O	HY-EX	TFA	—	—	—	0.02 ng 0.1 ng	93
Methamphetamine, amphetamine	—	AL-EX	TFA	OV-17 PC	EI-SCAN	—	—	90
Methamphetamine, amphetamine *p*-OH-methamphetamine	MeOH-H_2O	HY-EX	TFA	OV-17 PC	EI-SIM	ES	0.02 ng 0.02 ng 0.1 ng	96
Methamphetamine, amphetamine	SDS-H_2O	HY-EX	TFA	WCOT-CC	EI-SIM	DIS	0.5 ng/mg	118
Amphetamine	EtOH	AL-EX	TFA	—	EI-SIM	—	—	60
Amphetamine	H_2O-AC	EN-SPE	PFPA	HP-5-CC	EI-SIM	DIS	0.1 ng/mg	111
Amphetamine	CH_2Cl_2	HY-EX	TFA	BP-5-CC	EI-SIM	IS	0.1 ng/mg	114
Methamphetamine, amphetamine, methoxyphenamine, desmethylmethoxyphenamine	SDS-H_2O	HY-EX	TFA	WCOT-CC	EI-SIM	DIS	1 ng/mg	80
Methylene-dimethoxy-methamphetamine (MDMA)	H_2O-AC	EN-SPE	PFPA-PFPOH	HP-1-CC	EI-SIM	DIS	0.1 ng/mg	132

TABLE 5.

GC/MS Analysis for THC and Derivatives in Hair

Drug/drug metabolite	Decontam.	Workup	Derivat.	GC col.	DetMode	Quant.	DetLimit	Ref.
Δ9-THC	—	HY	—	DB-1	EI-SIM	—	—	51
THC-COOH	H_2O-AC	AL-SPE	MET	HP-1	EI-SIM	—	—	99
THC-COOH	H_2O-AC	AL-SPE	PFPA-PFPOH	HP-1	EI-SIM	IS	—	100
Δ9-THC, THC-COOH	CH_2Cl_2	AL-EX	PFPA-PFPOH	HP-5-CC	EI-SIM	DIS	0.1 ng/mg 0.1 ng/mg	115
Δ9-THC, THC-COOH	CH_2Cl_2	AL-EX	HFBA-HFPOH	HP-1-CC	EI-SIM	DIS	0.01 ng/mg 0.01 ng/mg	113
THC-COOH	H_2O	AL-EX	HFPA-HFPOH	DB-5-CC	CI-MS-MS	DIS	0.05 pg/mg	119
Δ9-THC	H_2O-AC-PE	MeOH-US	PSA	DB-5-CC	EI-SIM	DIS	0.005 ng/mg	120

of THC-COOH[99] the hair was hydrolyzed in alkaline solution, extracted with Baker C_{18} columns from acid solution, and derivatized with methyliodide. The quantitation of THC-COOH was made after nearly similar workup and extraction conditions and derivatization with pentafluoropropionic anhydride (PFPA) and pentafluoropropionyl alcohol (PFPOH) using levallorphan as an IS.[100] The most sensitive method is a CI/MS/MS procedure, which detects and reports concentrations as low as fg/mg in hair of chronic users.[119] However, these low concentrations are not in accordance with other reports with concentrations in the ng/mg area.[113,115,120]

c. Cocaine and Metabolites

Procedures for the MS detection of cocaine and/or metabolites have been published in more than 25 papers (Table 6). A review[101] gives details on subjects investigated, amount of hair analyzed, decontamination procedures, workup, analytical methods, targeted analyte(s), quantitative range(s) for targeted analyte(s), and usefulness of the citation in providing specific information related to the issue of external contamination. There is a large variety in the workup and derivatization conditions. Especially the decontamination seems to be a problem, because cocaine could not be washed out from hair soaked with the drug.[10-15] Therefore, a passive contamination cannot be excluded. The determination of the metabolites may help to solve the problem.[10,102] It seems that the substances are incorporated according to their lipophilicity, because cocaine is found in most cases in higher concentration compared to BZE and methylecgonine.[10,103,104] Some papers describe detection by MS or MS/MS after direct inlet of the probe.[103,105,106] Alkaline extraction must be avoided.[18,23] Mild acid hydrolysis with silylation or perfluoracylation are used in most procedures. There are several advantages of direct insertion probe (DIP) and chemical ionization-mass spectrometry (CI/MS) over electron impact ionization (EI) with SIM (EI-SIM),[103] although the preferred routine method for the quantitation is GC/MS.[105] Kidwell[14] analyzed hair of addicts with tandem MS for cocaine, phencyclidine, and their metabolites without extraction to avoid degradation of the parent drugs during the workup procedure. He found cocaine and its metabolites BZE and EC, but no PCP metabolites.

d. Opiates

The papers dealing with the GC/MS detection of 10 different opiates in hair are listed in Table 7. The first report was published in 1984[108] about codeine detection in animal hair after administration of the drug. One paper is included with direct insertion of the probe and CI.[109] As heroin samples always contain codeine as an impurity, this substance also can be detected in cases of heroin abuse. Morphine is a metabolite of codeine and can be detected when codeine is abused. The quantitation of both drugs allows differentiation between codeine and heroin abuse.[99] The detection of heroin or 6-acetylmorphine opens the possibility to prove directly the abuse of heroin.[21,34,112,117,134] Here also, the more lipophilic 6-acetylmorphine exceeds the morphine in most samples.

e. Miscellaneous

In Table 8 the drugs are listed which have been found in hair, not belonging to one of the groups above. Ishiyama et al.[91] detected tricyclic antidepressants in hair and discussed the possibility to determine definitely if patients are under a long-term treatment of medicines. Pentazocine was abused by a medical doctor and could be detected by hair analysis.[110] In patients under methadone treatment, hair analysis perhaps may show a dose-related concentration.[111] Kintz et al.[60] detected several barbiturates (seco-, amo-, phenobarbital), benzodiazepines (diazepam, desmethyl-diazepam, flunitrazepam, nitrazepam), and antidepressants by IA (FPIA) and confirmed them by GC/MS. Couper et al.[84] analyzed antidepressants and antipsychotic drugs in post-mortem human scalp hair (amitriptyline, doxepin, dothiepin, imipramine, mianserin, trimipramine, haloperidol, chlorpromazine, thioridazine, and six of their metabolites) by GC, confirmed positive findings by GC/MS, and quantitated by HPLC.

D. Screening Procedures and Reviews

Screening procedures by GC/MS were published by Moeller and Fey.[100] They analyzed hair samples for morphine, codeine, THC-COOH, and amphetamine in one procedure. After pulverization of the hair in a ball mill and SPE, they derivatized with PFPA/PFPOH and analyzed by GC/MS using the EI-SIM technique. They extended the procedure to 6-MAM, DHC, methadone, and its metabolite EDDP, but excluded THC-COOH.[111] Cone et al.[112] described a GC/MS screening procedure for opiates (heroin, 6-MAM, morphine, normorphine, codeine, acetylcodeine, norcodeine) and cocainics (cocaine, norcocaine, cocaethylene, norcocaethylene, benzoylecgonine, ecgonine methyl ester) after incubation of the washed hair samples overnight with methanol at 40°C, followed by SPE, derivatization with BSTFA, and GC/MS analysis using DIS. Jurado et al.[113] described a procedure for the simultaneous quantitation of opiates (morphine, codeine, 6-MAM), cocainics (cocaine and BZE), and cannabinoids (THC and THC-COOH). They extracted the opiate and cocaine compounds by means of a soft acidic hydrolysis with 0.1 N HCL at 50°C overnight and organic solvent extraction at pH 9.2. Stronger basic hydrolysis with 11.8 N KOH for 10 min was used for cannabinoids before organic solvent extraction. Kintz and Mangin[114] described a method where they analyzed morphine, nicotine, benzoylecgonine, diazepam, oxazepam, and amphetamine simultaneously in the hair of neonates.

Hair analysis by chromatographic procedures was reviewed in 1992 by Moeller.[24] Analytical methods for the detection of drugs in hair were reviewed for amphetamines,[107] cannabinoids,[115] cocaine,[101] and opiates.[20]

IV. CONCLUSIONS

The concentrations of drugs in hair are in the ng/mg range at least in cases of chronic abuse. No information is available about the minimum dose of drug intake, which can be detected by hair analysis. Calculation by Moeller et al.[116] assumes that for heroin, one to two injections per week lead to a concentration of morphine of 0.5 to 1 ng/mg hair. This is approximately about ten times above the detection limit of the presented routine GC/MS methods. However, other drugs with different chemical and physical properties have other minimum dose limitations. Kintz and Mangin[134] proposed to use cutoffs of 1 ng/mg for cocaine and to lower the cutoff to 0.5 ng/mg when other evidence of drug intake is available (analysis of blood or urine). For heroin consumption, they propose a cutoff of 0.5 ng/mg 6-MAM. Depending on the amount of specimen used and the sensitivity of the test, IA are good methods to preanalyze hair samples. In positive cases, the results must be confirmed by a more specific method. Here the GC/MS procedure is state of the art. The GC/MS methods exceed by far all other chromatographic methods used. Sachs and Raff[19] compared different extraction and GC/MS methods for quantitations of drugs in

TABLE 6

GC/MS Analysis for Cocaine and Derivatives in Hair

Drug/drug metabolite	Decontam.	Workup	Derivat.	GC col.	DetMode	Quant.	DetLimit	Ref.
Cocaine	H₂O-EtOH	HY-EX	—	OV-1-CC	EI-SIM	—	—	36
Benzoylecgonine	Soap-AC	HY-EX	PFB-TBA	FSOT-CC	EI-SIM	ES	0.05 ng	121
Cocaine	MeOH	SPE	—	—	CI-MS-MS	DIS	—	105
Benzoylecgonine								
Cocaine	MeOH	SPE	—	—	EI-SCAN	DIS	—	105
Benzoylecgonine								
Cocaine	—	—	—	—	CI-MS-MS	ES	—	103
Benzoylecgonine								
Ecgonine								
Benzoylecgonine	H₂O-AC	EX	BSTFA	DB-5 CC	EI-SIM	DIS	—	58
Cocaine	H₂O-AC	EN-SPE	PFPA-PFPOH	HP-2 CC	EI-SIM	DIS	0.1 ng/mg	104
Benzoylecgonine							0.1 ng/mg	
Ecgonine methyl ester							1 ng/mg	
Cocaine	SDS-MeOH	EN-SPE	MTBSTFA	DB-5 CC	CI-SIM	IS	0.1 ng/mg	102
Benzoylecgonine								
Ecgonine methyl ester								

Analyte	Solvent	Cleanup	Derivatization	Column	Detection	IS	LOD	Ref.
Cocaine	MeOH-37°C	HY-SPE	BSTFA	HP-1 CC	EI-SIM	DIS	0.1 ng/mg	10
Cocaethylene								
Norcocaine								
Cocaine	MeOH	HY-SPE	—	—	CI-MS-MS	DIS	—	106
Benzoylecgonine								
Cocaine	MeOH	MeOH	BSTFA	HP-1-CC	EI-SIM	DIS	0.05 ng/mg	112
Norcocaine							0.05 ng/mg	
Benzoylecgonine							0.05 ng/mg	
Benzoylnorecgonine							0.5 ng/mg	
Cocaethylene							0.05 ng/mg	
Norcocaethylene							0.05 ng/mg	
Ecgonine methyl ester							0.05 ng/mg	
Benzoylecgonine	CH2Cl2	HY-EX	BSTFA	BP-5-CC	EI-SIM	IS	0.6ng/mg	114
Cocaine	MeOH	HY-SPE	BSTFA	DB-5-CC	EI-SIM	DIS	—	124
Benzoylecgonine								
Ecgonine methyl ester								
Cocaine	ET-HCL	HY-SPE	—	—	MS-MS	—	5pg/mg	125
Cocaine	CH2Cl2	HY-EX	HFBA-HFPOH	HP-1-CC	EI-SIM	DIS	0.03ng/mg	113
Benzoylecgonine							0.03ng/mg	
Cocaine	H2O-AC-PE	MeOH-US	PSA	DB-5-CC	EI-SIM	IS	0.005ng/mg	120

TABLE 7

GC/MS Analysis for Opiates and Derivatives in Hair

Drug/drug metabolite	Decontam.	Workup	Derivat.	GC col.	DetMode	Quant.	DetLimit	Ref.
Codeine	—	—	—	OV-17-PC	—	—	—	108
Morphine	Soap-AC	HY-EX	HFBA	Ultra-2-CC	EI-SIM	IS	—	126
Codeine								
Morphine	ET-HCL	HY-SPE	—	—	EI-CS	—	1–10 fg	109
Morphine	CH2Cl2	AL-EX	AC	OV-1-CC	EI-SIM	—	0.05 ng/mg	99
Morphine	AC	HY-SPE	HFBA	Ultra-2-CC	EI-SIM	IS	—	99
Codeine								
Dihydromorphine								
Dihydrocodeine								
Morphine	MeOH	AL-SPE	PFPA	HP-1-CC	EI-SIM	IS	5 ng	127
Codeine								
Pholcodine	AC	HY-SPE	AA	HP-1-CC	EI-SCAN	—	0.3 ng/g	56
Dihydrocodeine	AC	AL-SPE	HFBA	Ultra-2-CC	EI-SIM	IS	0.03 ng	122
Heroin								
6-Acetylmorphine	MeOH-37°C	EX	MBTFA	HP-5-CC	EI-SIM	DIS	0.1 ng/inj.	21
Morphine								
Codeine								
6-Acetylmorphine	MeOH-AC	EN-SPE	HFBA	Ultra-2-CC	EI-SCAN	IS	—	117
Morphine								
Codeine								
Ethylmorphine	CH2Cl2	HY-EX	BSTFA	BP-5-CC	EI-SIM	IS	—	128
Heroin	MeOH	MeOH	BSTFA	HP-1	EI-SIM	DIS	—	112
6-Acetylmorphine							0.05 ng/mg	
Morphine							0.05 ng/mg	
Normorphine							0.05 ng/mg	
Codeine							0.5 ng/mg	
Acetylcodeine							0.05 ng/mg	
Norcodeine							0.05 ng/mg	
							0.5 ng/mg	

Analyte	Solvent	Method	Derivatization	Column	Detection	IS/DIS	LOD	Ref
Morphine	CH2Cl2	HY-EX	BSTFA	BP-5 CC	EI-SIM	IS	0.1 ng/mg	129
6-Acetylmorphine	H2O-AC	EN-SPE	PFPA	HP-5	EI-SIM	DIS	0.16 ng/mg	111
Morphine							0.04 ng/mg	
Codeine							0.04 ng/mg	
Dihydrocodeine							0.04 ng/mg	
6-Acetylmorphine	H2O-SDS	MeOH-US	BSTFA	NB-1-CC	EI-SIM	IS	—	130
Morphine	H2O-SDS	HY-EX	BSTFA	NB-1-CC	EI-SIM	IS	—	130
Heroin	H2O-AC-PE	SFE	PFPA	Ultra-2-CC	EI-SIM	DIS	—	19
6-Acetylmorphine								
Morphine								
6-Acetylmorphine	ET	MeOH	BSTFA	BP-5-CC	EI-SIM	—	—	45
Morphine	ET	HY-EX	AC	BP-5-CC	EI-SIM	—	—	45
Heroin	MeOH	SIL-130° (1h)	MSTFA	DB-5-CC	MS-MS	IS	—	131
6-Acetylmorphine								
Morphine								
Acetylcodeine								
Codeine								
6-Acetylmorphine	CH2Cl2	HY-EX	HFBA-HFPOH	HP-1-CC	EI-SIM	DIS	0.02 ng/mg	113
Morphine							0.06 ng/mg	
Codeine							0.07 ng/mg	
Dihydrocodeine	H2O-AC-PE	AL-SPE	HFBA	Ultra-2-CC	EI-SIM	IS	30 pg	122
Heroin		MeOH-US	PSA	DB-5-CC	EI-SIM	IS	—	108
6-Acetylmorphine							0.005 ng/mg	
Morphine							0.01 ng/mg	
Codeine							0.01 ng/mg	
Dihydrocodeine							0.01 ng/mg	

TABLE 8.

GC/MS Analysis for Miscellaneous Drugs/Drug Metabolites in Hair

Drug/drug metabolite	Decontam.	Workup	Derivat.	GC col.	DetMode	Quant.	DetLimit	Ref.
Amitriptyline	DET-SDS	AL-EX	—	OV-17 PC	EI-SIM	IS	—	91
Nortriptyline								
Imipramine	DET-SDS	AL-EX	—	OV-17 PC	EI-SIM	IS	—	91
Desmethylimipramine								
Nicotine	DET-SDS	AL-EX	—	OV-17 PC	EI-SIM	IS	—	91
Chloroquine	DET-H2O	AL-EX	—	OV-1CC	NCI	—	—	87
Monodesmethylchloroquine								
Methadone	—	HY-EX	—	DB-1 CC	EI-SIM	—	6 pg/inj.	50
Phencyclidine	—		—	—	CI-MS-MS	—	—	103
Bromazepam	H2O-AC	HY-SPE	BP	HP-1 CC	NCI-SCAN	IS	—	123
Oxazepam	H2O-AC	HY-SPE	BP	HP-1 CC	EI-SIM	IS	—	123
Nicotine	H2O-AC	HY-SPE	—	HP-1 CC	EI-SIM	—	—	123
Caffeine	H2O-AC	HY-SPE	—	HP-1 CC	EI-SIM	—	—	123
Pentazocine	H2O-AC	EN-AL-SPE	PFPA-PFPOH	HP-1 CC	EI-SIM	IS	—	110
Oxazepam	CH2Cl2	HY-EX	BSTFA	BP-5-CC	EI-SIM	IS	0.1ng/mg	129
Diazepam								
Nicotine	CH2Cl2	EX-ET	—	BP-5-CC	EI-SIM	IS	—	114
Methadone	H2O-AC	EN-SPE	PFPA	HP-5-CC	EI-SIM	DIS	0.22 ng/mg	111
EDDP							0.22 ng/mg	
Phenobarbital	CH2Cl2	DINT-SPE	—	CC	EI-SIM	DIS	0.2 ng/mg	135
Zipeprol	CH2Cl2	HY-EX	BSTFA	BP-5-CC	EI-SIM	IS	0.03 ng/mg	133

human hair. They concluded that two different GC/MS methods would be best to meet the requirements of analysis and confirmation.

Welch et al.[13] evaluated analytical procedures and environmental issues. They investigated methods for the extraction and detection of cocaine, some of its metabolites, morphine, and codeine from hair. They concluded from a round-robin study that extractions with 0.1 N HCL are as efficient to remove the target compounds from hair as enzymatic digestion that dissolve the hair. GC/MS with either EI or CI provides accurate determinations of the targeted compounds. External contamination by powdered or vapor-deposited cocaine was incompletely removed by all approaches tested, making it difficult to differentiate incorporated drug from external contamination. This was also proved in extensive laboratory experiments by Wang and Cone.[15]

Independent from this controversy, hair analysis for drugs offers a way to uncover chronic use in cases where blood or urine analyses fail.[116] The detection of the parent drug and the metabolites seem to be a possibility to decide whether the drug has been incorporated or not. In cases of incorporation, the concentration of the more lipophilic parent drug seems to be nearly always higher than that of the metabolites,[21,103-105,117] contrary to urinalysis.

ACKNOWLEDGMENT

We thank the Elsevier Science B.V., Amsterdam, The Netherlands for permission to include material from our paper, Drug Detection in Hair by Chromatographic Procedures, published in *International Journal of Chromatography*, 580, 125–134, 1992, with kind permission from Elsevier Science, The Netherlands.

LIST OF ABBREVIATIONS

AA	Acetic anhydride
AC	Acetone
AL	Alkaline hydrolyzed
BP	Benzophenon
BSTFA	N,O-bis(trimethylsilyl)trifluoroacetamide
BS	Buffer solution
BZE	Benzoylecgonine
CC	Capillary column
CI	Chemical ionization
COL	Cutoff level
CS	Collisional spectroscopy
DECON	Removal of external contamination
DER	Derivatization reagent
DET	Detergent
DETMODE	Detection mode
DHC	Dihydrocodeine
DINT	Disintegration of hair samples by ball mills or scissors
DIP	Direct insertion probe

DIS	Deuterated internal standard
DPC	Diagnostic Products Corporation, Los Angeles, U.S.
EC	Ecgonine
ECD	Electron capture detector
EDDP	2-Ethyl-1,5-dimethyl-3,3-diphenylpyrrolinium perchlorate
EI	Electron-impact ionization
EIA	Enzyme immunoassay
EN	Enzymatic cleavage of conjugates by proteolytic enzymes
ES	External standard
ET	Ether
EtOH	Ethanol
EX	Liquid-liquid extraction
FD	Fluorescence detector
FID	Flame ionization detection
FPIA	Fluorescence polarization immunoassay
GC	Gas chromatography
GC/MS	Gas chromatography/mass spectrometry
HCL	Hydrochloric acid
HEX	n-Hexane
HFBA	Heptafluorobutyric acid
HFPOH	Hexafluoropropanol
HPLC	High-performance liquid chromatography
HPTLC	High-performance thin-layer chromatography
HY	Acid hydrolyzed
IA	Immunoassay
IS	Internal standard
LDV	Lowest detected value
LPC	Lowest point of calibration curve
LOD	Limit of detection
6-MAM	6-Monoacetylmorphine
MBTFA	N-Methyl-bis-trifluoroacetamide
ME	Methylated/methylation
MeOH	Methanol
MS	Mass spectrometry
MTBSTFA	N-Methyl-N-(tert-butyldimethylsilyl)-trifluoroacetamide
m/z	Mass-to-charge ratio
NB	Neutrabond
n. i. a.	No information available
NP-FID	Nitrogen phosphor flame ionization detector
PB	Phosphate buffer
PC	Packed column
PCP	Phencyclidine
PE	Petrolether
PEG	Polyethyleneglycol

PFB	Pentafluorobenzylbromide
PFPA	Pentafluoropropionic anhydride
PFPOH	Pentafluoropropionyl alcohol
PSA	Propionic acid anhydride
QUANT	Quantitation
RIA	Radioimmunoassay
RIAH	Radioimmunoassay for hair (Psychemedics Inc.)
SB	Soerensen buffer
SDS	Sodium dodecyl sulfate
SIM	Selected ion monitoring
SPE	Solid-phase extraction
TBA	Tetrabutylammoniumhydrogensulfate
TFA	Trifluoroacetic acid
TFAA	Trifluoroacetic anhydride
THC	Δ9-Tetrahydrocannabinol
THC-COOH	11-Nor-9-carboxy-Δ9-tetrahydrocannabinol
TLC	Thin-layer chromatography
US	Ultra sonication

REFERENCES

1. Casper, J. L., Practisches Handbuch der gerichtlichen Medicin, Hirschwald, 2 vols, Berlin, 1857/58.
2. Goldblum, R. W., Goldbaum, R. L., Piper, W. N., Barbiturate Concentrations in the Skin and Hair of Guinea Pigs, *J. Invest. Dermatol.*, 22, 121, 1954.
3. Baumgartner, A. M., Jones, P. J., Baumgartner, W. A., Black, T. C., Radioimmunoassay of Hair for Determining Opiate-Abuse Histories, *J. Nucl. Med.*, 20, 748, 1979.
4. Klug, E., Determination of Morphine in Human Hair, *Z. Rechtsmed.*, 84, 189 1980.
5. Suzuki, O., Hattori, H., Detection of Methamphetamine and Amphetamine in a Single Human Hair and Nail Clipping by GC/CIMS, *Koenshu-Iyo Masu Kenkyukai*, 8, 201, 1983.
6. Suzuki, O., Hattori, H., Asano, M., Detection of Methamphetamine and Amphetamine in a Single Human Hair by Gas Chromatography/Chemical Ionization Mass Spectrometry, *J. Forensic Sci.*, 29, 611, 1984.
7. Marigo, M., Tagliaro, F., Poiesi, C., Lafisca, S., Neri, C., Determination of Morphine in the Hair of Heroin Addicts by High Performance Liquid Chromatography with Fluorimetric Detection, *J. Anal. Toxicol.*, 10, 158, 1986.
8. Consensus Report, NIDA/SOFT-Conference on Hair Analysis for Drugs of Abuse, Washington, D.C., May 27–29 1990.
9. SOFT Advisory Committee, Revised Consensus Opinion on Applicability of Hair Analysis, *Tox Talk*, 16, 3, 1992.
10. Cone, E. J., Yousefnejad, D., Darwin, W. D., Maguire, T., Testing Human Hair for Drugs of Abuse. II. Identification of Unique Cocaine Metabolites in Hair of Drug Abusers and Evaluation of Decontamination procedures, *J. Anal. Toxicol.*, 15, 250, 1991.
11. Blank, D. L., Kidwell, D. A., External Contamination of Hair by Cocaine: An Issue in Forensic Interpretation, *Forensic Sci. Int.*, 63, 145, 1993.
12. Henderson, G. L., Mechanisms of Drug Incorporation into Hair, *Forensic Sci. Int.*, 63, 19, 1993.
13. Welch, M. J., Sniegoski, L. T., Allgood, C. C., Habram, M., Hair Analysis for Drugs of Abuse: Evaluation of Analytical Methods, Environmental Issues, and Development of Reference Materials, *J. Anal. Toxicol.*, 17, 389, 1993.
14. Kidwell, D. A., Analysis of Phencyclidine and Cocaine in Human Hair by Tandem Mass Spectrometry, *J. Forensic Sci.*, 38, 272, 1993.

15. Wang, W. L., Cone, E. J., Testing Human Hair for Drugs of Abuse. IV. Environmental Cocaine Contamination and Washing Effects, *Forensic Sci. Int.*, 70, 39, 1995.

16. Blank, D. L., Kidwell, D. A., Decontamination Procedures for Drugs of Abuse in Hair: Are They Sufficient?, *Forensic Sci. Int.*, 70, 13, 1995.

17. Baumgartner, W. A., Hill, V. A., Sample Preparation Techniques, *Forensic Sci. Int.*, 63, 121, 1993.

18. Chiarotti, M., Overview on Extraction Procedures, *Forensic Sci. Int.*, 63, 161, 1993.

19. Sachs, H., Raff, I., Comparison of Quantitative Results of Drugs in Human Hair by GC/MS, *Forensic Sci. Int.*, 63, 207, 1993.

20. Staub, C., Analytical Procedures for Determination of Opiates in Hair: A Review, *Forensic Sci. Int.*, 70, 111, 1995.

21. Goldberger, B. A., Caplan, Y. H., Maguire, T., Cone, E. J., Testing Human Hair for Drugs of Abuse. III. Identification of Heroin and 6-Acetylmorphine as Indicators of Heroin Use, *J. Anal. Toxicol.*, 15, 226, 1991.

22. Couper, F. J., McIntyre, I. M., Drummer, O. H., Extraction of Psychotropic Drugs from Human Scalp Hair, *J. Forensic Sci.*, 40, 83, 1995.

23. Sachs, H., Moeller, M. R., Interlaboratory Comparison of Quantitative GC/MS-Methods for Drugs in Hair, in *Proceedings of the 30th International Meeting, October 19–23, 1992, Fukuoka, Japan*, Nagata, T., Ed., Organizing Commitee of the 30th TIAFT Meeting and Department of Forensic Medicine, Kyushu University, Fukuoka, Japan, 1992, 33.

24. Moeller, M. R., Drug Detection in Hair by Chromatographic Procedures, *J. Chromatogr.*, 580, 125, 1992.

25. Kalasinsky, K. S., Magluilo, J., Schaefer, T., Hair Analysis by Infrared Microscopy for Drugs of Abuse, *Forensic Sci. Int.*, 63, 253, 1993.

26. Tagliaro, F., Poiesi, C., Aiello, R., Dorizzi, R., Ghielmi, S., Marigo, M., Capillary Electrophoresis for the Investigation of Illicit Drugs in Hair: Determination of Cocaine and Morphine, *J. Chromatogr.*, 638, 303, 1993.

27. Baumgartner, W. A., Hill, V. A., Blahd, W. H., Hair Analysis for Drugs of Abuse, *J. Forensic Sci.*, 34, 1433, 1989.

28. Mieczkowski, T., Newel, R., An Evaluation of Patterns of Racial Bias in Hair Assays for Cocaine: Black and White Arrestees Compared, *Forensic Sci. Int.*, 63, 85, 1993.

29. Callahan, C. M., Grant, T. M., Phipps, P., Clark, G., Novack, A. H., Streissguth, A. P., Raisys, V. A., Measurement of Gestational Cocaine Exposure: Sensitivity of Infants Hair, Meconium, and Urine, *J. Pediatr.*, 102, 163, 1992.

30. Magura, S., Freeman, R. C., Siddiqi, Q., Lipton, D. S., The Validity of Hair Analysis for Detecting Cocaine and Heroin Use among Addicts, *Int. J. Addict.*, 27, 51, 1992.

31. Hindin, R., McCusker J., Vickers-Lahti, M., Bigelow, C., Garfield, F., Lewis, B., Radioimmunoassay of Hair for Determination of Cocaine, Heroin, and Marijuana Exposure: Comparison with Self-Report, *Int. J. Addict.*, 29, 771, 1994.

32. Marques, P. R., Tippetts, A. S., Branch, D. G., Cocaine in the Hair of Mother-Infant-Pairs: Quantitative Analysis and Correlations with Urine Measures and Self-Report, *Am. J. Drug Alcohol Abuse*, 19, 159, 1993.

33. Centini, F., Offidani, C., Carnevale, A., Chiarotti, M., Barni Comparini, I., Determination of Morphine in Hair by Immunochemical and Gas Chromatographic Mass Spectrometric Techniques, *Development in Analytical Methods in Phurmaceutical, Biomedical, and Forensic Sciences*, 1st ed., Piemonte, G., Tagliaro, F., Marigo, M., Frigerio, A., Eds., Plenum Press, New York, 1987, 107.

34. Moeller, M. R., Mueller, C., The Detection of 6-Monoacetylmorphine in Urine, Serum and Hair by GC/MS and RIA, *Forensic Sci. Int.*, 70, 125, 1995.

35. Balabanova, S., Brunner, H., Nowak, R., Radioimmunological Determination of Cocaine in Human Hair, *Z. Rechtsmed.*, 98, 229, 1987.

36. Balabanova, S., Homoki, S., Determination of Cocaine in Human Hair by Gas Chromatography/Mass Spectrometry, *Z. Rechtsmed.*, 98, 235, 1987.

37. Koren, G., Klein, J., Forman, R., Graham, K., Hair Analysis of Cocaine: Differentiation Between Systemic Exposure and External Contamination, *J. Clin. Pharmacol.*, 32, 671, 1992.

38. Klein, J., Forman, R., Eliopoulos, C., Koren, G., A Method for Simultaneous Measurement of Cocaine and Nicotine in Neonatal Hair, *Ther. Drug Monit.*, 16, 67, 1994.

39. Sato, R., Uematsu, T., Sato, R., Yamaguchi, S., Nakashima, M., Human Scalp Hair as Evidence of Individual Dosage History of Haloperidol: Prospective Study, *Ther. Drug Monit.*, 11, 686, 1989.

40. Uematsu, T., Sato, R., Fujimori, O., Nakashima, M., Human Scalp Hair as Evidence of Individual Dosage History of Haloperidol: A Possible Linkage of Haloperidol Excretion into Hair with Hair Pigment, *Arch. Dermatol. Res.*, 282, 120, 1990.

41. Uematsu, T., Sato, R., Suzuki, K., Yamaguchi, S., Nakashima, M., Human Scalp Hair as Evidence of Individual Dosage History of Haloperidol: Method and Retrospective Study, *Eur. J. Clin. Pharmacol.*, 37, 239, 1989.

42. Cassani, M., Spiehler, V., Analytical Requirements, Perspectives and Limits of Immunological Methods for Drugs in Hair, *Forensic Sci. Int.*, 63, 175, 1993.
43. Staub, C., Robyr, C., Analysis of Morphine in Hair by Radio-Immunoassay and HPLC with Electrochemical Detection, in *Forensic Toxicology: Proceedings of the 25th International Meeting, June 27–30, 1988, Groningen, The Netherlands*, Uges, D. R. A., de Zeeuw, R. A., Eds., University Press, Groningen, 1988, 230.
44. Baselt, R. C., Inappropriate Use of Immunoassays as a Quantitative Tool, *J. Anal. Toxicol.*, 13, 1, 1989.
45. Staub, C., Hair Analysis: Its Importance for the Diagnosis of Poisoning Associated with Opiate Addiction, *Forensic Sci. Int.*, 63, 69, 1993.
46. Tagliaro, F., Traldi, P., Pelli, B., Maschio, S., Neri, C., Marigo, M., Determination of Morphine and Other Opioids in the Hair of Heroin Addicts by RIA, HPLC and Collisional Spectroscopy, *Development in Analytical Methods in Pharmaceutical, Biomedical, and Forensic Sciences*, 1st ed., Piemonte, G., Tagliaro, F., Marigo, M., Frigerio, A., Eds., Planum Press, New York, 1987, 115.
47. Gamaleya, N., Tagliaro, F., Parshin, A., Vrublevskii, A., Bugari, G., Dorizzi, R., Ghielmi, S., Marigo, M., Immune Response to Opiates: New Findings in Heroin Addicts Investigated by Means of an Original Enzyme Immunoassay and Morphine Determination in Hair, *Life Sci.*, 53, 99, 1993.
48. Tagliaro, F., Lafisca, S., Maschio, S., Parolin, A., Lubli, G., Marigo, M., Quantitative Determination of Morphine in Hair: A Comparison between RIA and HPLC Methods, *Acta Med. Leg. Soc. (Liege)*, 35, 181, 1985.
49. Marigo, M., Tagliaro, F., Trabetti, E., Rapid and Reliable Determination of Morphine in Hair by Means of RIA Screening and HPLC Confirmation to Investigate Opiate Abuse Histories, in *Forensic Toxicology: Proceedings of the 25th International Meeting, June 27–30, 1988, Groningen, The Netherlands*, Uges, D. R. A., de Zeeuw, R. A., Eds., University Press, Groningen, 1988, 230.
50. Balabanova, S., Arnold, P. J., Brunner, H., Luckow, H., Wolf, H. U., Detection of Methadone in Human Hair by Gas Chromatography/Mass Spectrometry, *Z. Rechtsmed.*, 102, 495, 1989.
51. Balabanova, S., Arnold, P. J., Luckow, V., Brunner, H., Wolf, H. U., Tetrahydrocannabinole in the Hair of Hashish Smokers, *Z. Rechtsmed.*, 102, 503, 1989.
52. Balabanova, S., Schneider, E., Buehler, G., Nachweis von Nikotin in Haaren, *Dtsch. Apoth. Ztg.*, 130, 2200, 1990.
53. Balabanova, S., Albert, W., Untersuchung zum Transport und zu Stabilitaet von Drogen im Haar, *Arch. Kriminol.*, 193, 100, 1994.
54. Fritch, D., Groce, Y., Rieders, F., Cocaine and Some of Its Products in Hair by RIA and GC/MS, *J. Anal. Toxicol.*, 16, 12, 1992.
55. Martinez, F., Poet, T. S., Pillai, R., Erickson, J., Estrada, A. L., Watson, R. R., Cocaine Metabolite (Benzoylecgonine) in Hair and Urine of Drug Users, *J. Anal. Toxicol.*, 17, 138, 1993.
56. Maurer, H. H., Fritz, C. F., Toxicological Detection of Pholcodine and Its Metabolites in Urine and Hair Using Radio Immunoassay, Fluorescence Polarisation Immunoassay, Enzyme Immunoassay and Gas Chromatography-Mass Spectrometry, *Int. J. Leg. Med.*, 104, 43, 1990.
57. Kintz, P., Mangin, P., Analysis of Opiates in Human Hair with FPIA, EMIT, and GC/MS, *Adli Tip Derg.*, 7, 129, 1991.
58. Reuschel, S. A., Smith, F. P., Benzoylecgonine (Cocaine Metabolite) Detection in Hair Samples of Jail Detainees Using Radioimmunoassay (RIA) and Gas Chromatography/Mass Spectrometry (GC/MS), *J. Forensic Sci.*, 36, 1179, 1991.
59. Sachs, H., Arnold, W., Results of Comparative Determination of Morphine in Human Hair Using RIA and GC/MS, *J. Clin. Chem. Clin. Biochem.*, 27, 873, 1989.
60. Kintz, P., Ludes, B., Mangin, P., Detection of Drugs in Human Hair Using Abott ADx, with Confirmation by Gas Chromatography/Mass Spectrometry (GC/MS), *J. Forensic Sci.*, 37, 328, 1992.
61. Spiehler, V., Analytical Requirements of Immunological Methods for Drugs in Hair, *SOFT Conference on Drug Testing in Hair, October 29–30, 1994*, Tampa, FL, Society of Forensic Toxicologists, Mesa, AZ, 1994.
62. Spiehler, V., Brown, R., Unconjugated Morphine in Blood by radioimmunoassay and gas chromatography/mass spectrometry, *J. Forensic Sci.*, 32, 906, 1987.
63. Matejczik, R. J., Kosinski, J. A., Determination of Cross-Reactant Drugs with a New Morphine Radioimmunoassay Procedure, *J. Forensic Sci.*, 30, 677, 1985.
64. Pueschel, K., Thomasch, P., Arnold, W., Opiate Levels in Hair, *Forensic Sci. Int.*, 21, 181, 1983.
65. Arnold, W., The Determination of Drugs and Their Substitutes in Head Hairs, *GIT Labor-Medizin*, 3, 82, 1987.
66. Arnold, W., Radioimmunological Hair Analysis for Narcotics and Substitutes, *J. Clin. Chem. Clin. Biochem.*, 25, 753, 1987.
67. Offidani, C., Strano Rossi, S., Chiarotti, M., Drug Distribution in the Head, Axillary and Pubic Hair of Chronic Addicts, *Forensic Sci. Int.*, 63, 105, 1993.

68. Offidani, C., Carnevale, A., Chiarotti, M., Drugs in Hair: A New Extraction Procedure, *Forensic Sci. Int.*, 41, 35, 1989.

69. Valente, D., Cassani, M., Pigliapochi, M., Vansetti, G., Hair as the Sample in Assessing Morphine and Cocaine Addiction, *J. Clin. Chem.*, 27, 1952, 1981.

70. Baumgartner, W. A., Black, C. T., Jones, P. F., Blahd, W. H., Radioimmunoassay of Cocaine in Hair: Concise Communication, *J. Nucl. Med.*, 23, 790, 1982.

71. Offidani, C., Strano Rossi, S., Chiarotti, M., Improved Enzymatic Hydrolysis of Hair, *Forensic Sci. Int.*, 63, 171, 1993.

72. Graham, K., Koren, G., Klein, J., Schneiderman, J., Greenwald, M., Determination of Gestational Cocaine Exposure by Hair Analysis, *J. Am. Med. Assoc.*, 262, 3328, 1989.

73. Smith, F. P., Liu, R. H., Detection of Cocaine Metabolite in Perspiration Stain, Menstrual Bloodstain, and Hair, *J. Forensic Sci.*, 31, 1269, 1986.

74. Balabanova, S., Wolf, H. U., Determination of Methadone in Human Hair by Radioimmunoassay, *Z. Rechtsmed.*, 102, 1, 1989.

75. Balabanova, S., Wolf, H. U., Methadone Concentrations in Human Hair of the Head, Axillary and Pubic Hair, *Z. Rechtsmed.*, 102, 293, 1989.

76. Haley, N. J., Hoffmann, D., Analysis for Nicotine and Cotinine in Hair to Determine Cigarette Smoker Status, *Clin. Chem.*, 31, 1598, 1985.

77. Baumgartner, A. M., Jones, P. J., Black, T. C., Detection of Phencyclidine in Hair, *J. Forensic Sci.*, 26, 576, 1981.

78. Yamamoto, Y., Yamamoto, K., Comparison of Enzyme Multiplied Immunoassay Technique and High Performance Liquid Chromatography in Determination of Amphetamine in Urine, *Nippon Hoigaku Zasshi*, 34, 158, 1980.

79. Jeger, A. N., Raas, R. E., Hamberg, C., Briellmann, T., Morphine Determination in Human Hair by Instrumental HPTLC, *Camac Bibliogr. Serv.*, 68, 7, 1992.

80. Nakahara, Y., Takahashi, K., Konuma, K., Hair Analysis for Drugs of Abuse. VI. The Excretion of Methoxyphenamine and Methamphetamine into Beards of Human Subjects, *Forensic Sci. Int.*, 63, 109, 1993.

81. Nagai, T., Takahashi, M., Saito, K., Kamiyama, S., Nagai, T., A New Analytical Method for Methamphetamine Optical Isomers and Its Habitual Users' Hair by HPLC, *Igaku To Seibutsugaku*, 115, 147, 1987.

82. Nagai, T., Kamiyama, S., Nagai, T., Forensic Toxicologic Analysis of Methamphetamine Optical Isomers by High Performance Liquid Chromatography, *Z. Rechtsmed.*, 101, 151, 1988.

83. Matsuno, H., Uematsu, T., Nakashima, M., The Measurement of Haloperidol and Reduced Haloperidol in Hair as an Index of Dosage History, *Br. J. Clin. Pharmacol.*, 29, 187, 1990.

84. Couper, F. J., McIntyre, I. M., Drummer, O. H., Detection of Antidepressant and Antipsychotic Drugs in Post-Mortem Human Scalp Hair, *J. Forensic Sci.*, 40, 87, 1995.

85. Mizuni, A., Uematsu, T., Nakashima, M., Simultaneous Detection of Ofloxacin, Norfloxacin and Ciprofloxacin in Human Hair by High-Performance Liquid Chromatography and Fluorescence Detection, *J. Chromatogr.*, 653, 187, 1994.

86. Uematsu, T., Kusajima, H., Umemura, K., Ishida, R., Ohkubo, H., Nakashima, M., A New Antimicrobial (Quinolone) Analysed in Hair as an Index of Drug Exposure and as a Time Marker, *J. Pharm. Pharmacol.*, 45, 1012, 1993.

87. Viala, A., Deturmeny, E., Aubert, C., Estadieu, M., Durand, A., Cano, J. P., Delmont, J., Determination of Chloroquine and Monodesethylchloroquine in Hair, *J. Forensic Sci.*, 28, 922, 1983.

88. Ochsendorf, F. R., Schöfer, H., Runne, U., Schmidt, K., Raudonat, H. W., Sequentielle Chloroquin-Bestimmung im menschlichen Haar bei toxischer/therapeutischer Dosierung: Korrelation zur Dosis und Therapiedauer, *Zbl. Rechtsmed.*, 31, 866, 1989.

89. Takahashi, K., Determination of Methamphetamine and Amphetamine in Biological Fluids and Hair by Gas Chromatography, *Nippon Hoigaku Zasshi*, 38, 319, 1984.

90. Nagai, T., Nagai, T., Ikeda, T., Study on the Drug Abuse in Forensic Science — Detection of Methamphetamine and Amphetamine from Hair, Bone and Teeth, in *TIAFT Proceedings of the 21st International Meeting, Brighton (1984)*, Dunnett, N., Kimber, K. J., Eds., McMillan, New York, 1984, 89.

91. Ishiyama, I., Nagai, T., Toshida, S., Detection of Basic Drugs (Methamphetamine, Antidepressants, and Nicotine) from Human Hair, *J. Forensic Sci.*, 28, 380, 1983.

92. Suzuki, S., Inoue, T., Niwaguchi, T., Determination of Trace Amounts of Methamphetamine in Biological Materials by Mass Fragmentography, *Koenshu-Iyo Masu Kenkyukai*, 6, 129, 1981.

93. Suzuki, S., Inoue, T., Yasuda, T., Niwaguchi, T., Hori, H., Inayama, S., Analysis of Methamphetamine in Human Hair by Mass Fragmentography, *Eisei Kagaku*, 30, 23, 1984.

94. Niwaguchi, T., Suzuki, S., Inoue, T., Determination of Methamphetamine in Hair After Single and Repeated Administration to Rat, *Arch. Toxicol.*, 52, 157, 1983.

95. Suzuki, S., Inoue, T., Yasuda, T., Analysis of Methamphetamine in Hair, Nail, Saliva and Sweat by Mass Fragmentography, in *TIAFT Proceedings of the 21st International Meeting, Brighton (1984)*, Dunnett, N., Kimber, K. J., Eds., McMillan, New York, 1984, 95.

96. Suzuki, S., Inoue, T., Hori, H., Inayama, S., Analysis of Methamphetamine in Hair, Nail, Sweat, and Saliva by Mass Fragmentography, *J. Anal. Toxicol.*, 13, 176, 1989.

97. Käferstein, H., Sticht, G., Comments on: "Detection of Methadone in Human Hair by Gas Chromatography/Mass Spectrometry" and "Tetrahydrocannabinols in the Hair of Hashish Smokers", *Int. J. Leg. Med.*, 103, 393, 1990.

98. Bogusz, M., Comments on: "Tetrahydrocannabinols in the Hair of Hashish Smokers", S. Balabanova et al. and on the response by Dr. Balabanova to comments by H. Kaeferstein and G. Sticht, *Int. J. Leg. Med.*, 103, 621, 1990.

99. Sachs, H., Moeller, M. R., Detection of Drugs in Hair by GC/MS, *Fres. J. Anal. Chem.*, 334, 713, 1989.

100. Moeller, M. R., Fey, P., Detection of Drugs in Hair by GC/MS, *Bull. Soc. Sci. Med. Grand-Duchè Luxemb.*, 172, 460, 1990.

101. Selavka, C., Rieders, F., The Determination of Cocaine in Hair: A Review, *Forensic Sci. Int.*, 70, 155, 1995.

102. Harkey, M. R., Henderson, G. L., Zhou, C., Simultaneous Determination of Cocaine and Its Major Metabolites in Human Hair by Gas Chromatography/Chemical Ionization Mass Spectrometry, *J. Anal. Toxicol.*, 15, 260, 1991.

103. Kidwell, D. A., Analysis of Drugs of Abuse in Hair by Tandem-Mass-Spectrometry, in *Proceedings of the 36th American Society of Mass Spectrometry Conference on Mass Spectrometry and Allied Topics, San Francisco, June 6–10, 1988*, 1364, 1988.

104. Moeller, M. R., Fey, P., Rimbach, S., Identification and Quantitation of Cocaine and Its Metabolites, Benzoylecgonine and Ecgonine Methyl Ester, in Hair of Bolivian Coca Chewers by Gas Chromatography-Mass Spectrometry, *J. Anal. Toxicol.*, 16, 291, 1992.

105. Martz, R. M., The Identification of Cocaine in Hair by GC/MS and MS/MS, *Crime Lab. Dig.*, 15, 67, 1988.

106. Martz, R., Donnelly, B., Fetterolf, D., Lasswell, L., Hime, G. W., Hearn, W. L., The Use of Hair Analysis to Document a Cocaine Overdose Following a Sustained Survival Period Before Death, *J. Anal. Toxicol.*, 15, 279, 1991.

107. Nakahara, Y., Detection and Diagnostic Interpretation of Amphetamines in Hair, *Forensic Sci. Int.*, 70, 135, 1995.

108. Nagai, T., Nagai, T., Detection of Codeine from Animal Hair and the Lapse of Time of Its Accumulation, *Igaku To Seibutsugaku*, 109, 145, 1984.

109. Pelli, B., Traldi, P., Tagliaro, F., Lubli, G., Marigo, M., Collisional Spectroscopy for Unequivocal and Rapid Determination of Morphine at ppb Level in the Hair of Heroin Addicts, *Biomed. Environ. Mass Specrom.*, 14, 63, 1987.

110. Moeller, M. R., Fey, P., Screening Procedure for Drugs in Hair, *43rd Meeting of the American Academy of Forensic Sciences, Anaheim, CA, Feb. 18–23 (1991) Abstracts*, K 45 182.

111. Moeller, M. R., Fey, P., Wennig, R., Simultaneous Determination of Drugs of Abuse (Opiates, Cocaine and Amphetamine) in Human Hair by GC/MS and Its Application to a Methadone Treatment Program, *Forensic Sci. Int.*, 63, 185, 1993.

112. Cone, E. J., Darwin, W. D., Wang, W. L., The Occurrence of Cocaine, Heroin and Metabolites in Hair of Drug Abusers, *Forensic Sci. Int.* 63, 55, 1993.

113. Jurado, C., Gimenez, M. P., Menendez, M., Repetto, M., Simultaneous Quantification of Opiates, Cocaine and Cannabinoids in Hair, *Forensic Sci. Int.*, 70, 165, 1995.

114. Kintz, P., Mangin, P., Determination of Gestational Opiate, Nicotine, Benzodiazepine, Cocaine and Amphetamine Exposure by Hair Analysis, *J. Forensic Sci. Soc.*, 33, 139, 1993.

115. Cirimile, V., Kintz, P., Mangin, P., Testing Human Hair for Cannabis, *Forensic Sci. Int.*, 70, 175, 1995.

116. Moeller, M. R., Fey, P., Sachs H., Hair Analysis as Evidence in Forensic Cases, *Forensic Sci. Int.*, 63, 43, 1993.

117. Raff, I., Denk, R., Sachs, H., Monoacetylmorphin in Haaren, *Zbl. Rechtsmed.*, 36, 479, 1991.

118. Nakahara, Y., Takahashi, K., Shimamine, M., Takeda, Y., Hair Analysis for Drugs of Abuse. I. Determination of Methamphetamine and Amphetamine in Hair by Stable Isotope Dilution Gas Chromatography/Mass Spectrometry Method, *J. Forensic Sci.*, 36, 70, 1991.

119. Kippenberger, D. J., Hayes, G., Scholtz, H., Donahue, T., Matsui, P., Baumgartner, W., Determination of Carboxy-THC in Hair by Tandem Mass Spectrometry, *2nd International Meeting on Clinical and Forensic Aspects of Hair Analysis, Genua, June 6–8, 1994*.

120. Kauert, G., Meyer, L. v., Herrle, I., Drogen- und Medikamentennachweis im Kopfhaar ohne Extraktion des Haaraufschlusses mittels GC/MS, *Zbl. Rechtsmed.*, 38, 33, 1992.

121. Brunner, H., Balabanova, S., Homoki, J., Wolf, H. U., Determination of Benzoyl Ecgonine in Human Scalp Hair Following Cocaine Abuse by Gas Chromatography-Mass Spectrometry (GC/MS), *Beitr. Gerichtl. Med.*, 46, 127, 1988.

122. Sachs, H., Denk, R., Raff, I., Determination of Dihydrocodeine in Hair of Opiate Addicts by GC/MS, *Int. J. Leg. Med.*, 105, 247, 1993.

123. Moeller, M. R., Fey, P., Sachs, H., Kettenbaum, F., Drug Analysis in Hair by GC/MS, *S.O.F.T. Conference on Hair Analysis for Drugs of Abuse, Washington, D.C., May 27–29 (1990).*

124. Springfield, A. C., Cartmell, L. W., Aufderheide, A. C., Buikstra, J., Ho, J., Cocaine and Metabolites in the Hair of Ancient Peruvian Coca Leaf Chewers, *Forensic Sci. Int.*, 63, 269, 1993.

125. Traldi, P., Favretto, D., Tagliaro, F., Ion Trap Mass Spectrometry, a New Tool in the Investigation of Drugs of Abuse in Hair, *Forensic Sci. Int.*, 63, 239, 1993.

126. Sachs, H., Brunner, H., Gas Chromatography-Mass Spectrometry Findings of Morphine and Codeine in the Vitreous Humor and Hair, *Beitr. Gerichtl. Med.*, 44, 281, 1986.

127. Cone, E. J., Testing Human Hair for Drugs of Abuse. I. Individual Dose and Time Profiles of Morphine and Codeine in Plasma, Saliva, Urine, and Beard Compared to Drug-Induced Effects on Pupils and Behaviour, *J. Anal. Toxicol.*, 14, 1, 1990.

128. Kintz, P., Jamey, C., Mangin, P., Ethylmorphine Concentrations in Human Samples in an Overdose Case, *Arch. Toxicol.*, 68, 210, 1994.

129. Kintz, P., Mangin, P., Evidence of Gestational Heroin or Nicotine Exposure by Analysis of Fetal Hair, *Forensic Sci. Int.*, 63, 99, 1993.

130. Nakahara, Y, et al., Hair Analysis for Drugs of Abuse. IV. Determination of Total Morphine and Confirmation of 6-Acetylmorphine in Monkey and Human Hair by GC/MS, *Arch. Toxicol.*, 66, 449, 1992.

131. Polettini, A., Groppi, A., Montagna, M., Rapid and Highly Selective GC/MS/MS Detection of Heroin and Its Metabolites in Hair, *Forensic Sci. Int.*, 63, 217, 1993.

132. Moeller, M. R., Maurer, H. H., Roesler, M., MDMA in Blood, Urine and Hair: A Forensic Case, in *Proceedings of the 30th International Meeting, October 19–23, 1992, Fukuoka, Japan*, Nagata, T., Ed., Organizing Commitee of the 30th TIAFT Meeting and Department of Forensic Medicine, Kyushu University, 1992, 347.

133. Kintz, P., Traqui, A., Potard, D., Petit, G., Mangin, P., An Unusual Death by Zipeprol Overdose, *Forensic Sci. Int.*, 64, 159, 1994.

134. Kintz, P., Mangin, P., What Constitutes a Positive Result in Hair Analysis: Proposal for the Establishment of Cut-Off Values, *Forensic Sci. Int.*, 70, 3, 1995.

135. Goullé, J. P., Noyon, J., Layet, A., Rapoport, N. F., Vaschalde, Y., Pignier, Y., Bouige, D., Jouen, F., Phenobarbital in Hair and Drug Monitoring, *Forensic Sci. Int.*, 70, 191, 1995.

136. Kintz, P., Mangin, P., personal communication.

Chapter 5

IMPORTANCE OF SUPERCRITICAL FLUID EXTRACTION (SFE) IN HAIR ANALYSIS

Christian Staub, Patrick Edder, and Jean-Luc Veuthey

CONTENTS

I. INTRODUCTION

Preparing the sample is a key step in all biological analyses, and hair analysis is no exception to this rule. Over the last ten years, there has been an ever increasing interest in the use of supercritical fluid extraction (SFE) as an alternative to traditional methods of preparing samples. The driving force[1] behind this development is, without doubt, the need for a simple, rapid, automated, and selective method which should also be environmentally friendly. In this context, the use of supercritical fluids fulfills these conditions, due to their unique physicochemical properties.[2] The following is a list of advantages:

- The desired selectability can be obtained by adjusting parameters such as pressure, temperature, and the degree of modifier.
- The advantageous diffusion properties of supercritical fluids permits rapid penetration and extraction from the samples.
- The use of low critical temperatures allows for good stability of thermically labile components.

This list of advantages is, without doubt, incomplete. In the field of solid matrix analysis with respect to hair samples, where classical methods are both long and tedious, it has become important to devote a chapter to this new and promising technique.

II. SUPERCRITICAL FLUIDS

A. Thermodynamic Properties and Definitions

The thermodynamic state of a pure component[3] is determined by three variables: the pressure P, the volume V, and the temperature T. The relationship between these three variables is known as the state equation and is represented by a surface in the three-dimensional plotting of P, V, and T. Any pure component, following the value of these three parameters, will be either a solid (S), a liquid (L), or a gas (G). A plot of P against T, or P against V is generally preferred because of its easier application.

Diagram P,T (Figure 1) shows the three states of the matter: S, L, G. The links between these different domains correspond to phase transitions, accompanied by changes in physical properties (in particular P), and to the equilibria liquid-gas, solid-liquid, and solid-gas. The three phases solid-liquid-gas co-exist at the triple point T. It can be seen therefore that it is impossible to move from a solid to a liquid state without crossing a segment, in other words without a discontinuity in the densities of the phases. It is, however, possible to move from a liquid to a gas without crossing the segment (L-G), by going round the point C.

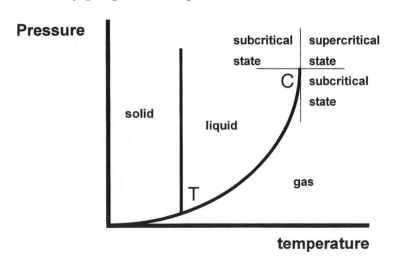

Figure 1
Schematic phase diagram of a single substance.

The point C is known as the critical point, and corresponds to a critical temperature Tc, a critical pressure Pc, and a critical density ρc.

At the point C exists an intermediate state between a liquid and a gas called a critical state, or, if sufficiently removed from C, a **supercritical** state. This state is obtained at temperatures and pressures superior to those of the critical temperature and critical pressure, respectively. If one of these parameters falls under its critical value while the other remains superior to its critical value, we are talking about a **subcritical** state.

B. Physicochemical Properties

Table 1 groups together differing orders of magnitude of physical parameters for the three states of the same fluid. It should be noted that in spite of high densities (similar to liquids), supercritical fluids are only slightly viscous and, from this point of view, have similar properties to gases.

TABLE 1.

Typical Physical Properties for Gases, Supercritical Fluids, and Liquids

	Density (g/cm³)	Viscosity (Pa · s)	Diffusion (m² · s⁻¹)
Gas 15–30°C, 1 bar	6×10^{-4} to 2×10^{-3}	1×10^{-5} to 3×10^{-5}	1×10^{-5} to 4×10^{-5}
Supercritical fluid			
at T_c, P_c	0.2 to 0.5	1×10^{-5} to 3×10^{-5}	7×10^{-8}
at T_c, $4P_c$	0.4 to 0.9	3×10^{-5} to 9×10^{-5}	1×10^{-8}
Liquid 15–30°C, 1 bar	0.6 to 1.6	2×10^{-4} to 3×10^{-3}	2×10^{-10} to 2×10^{-9}

Data from Reference 3.

1. Density

If liquids can be considered to be uncompressible over 300 bar, for a fluid in a supercritical state, this compressibility K_T is important around the critical point.

$$K_T \text{ is defined by the relationship } K_T = \frac{1}{\rho}\left(\frac{d\rho}{dP}\right)$$

where ρ is the density in the state under consideration.

A slight increase in pressure leads to an increase in the density (Figure 2). In the zone surrounding the critical point, the compressibility is about 500 times greater than that of a liquid. Consequentially, contrary to liquid extraction, it is possible to vary the density of the extraction phase by acting upon the pressure and/or the temperature. As a result, SFE has a supplementary degree of freedom compared to liquid extraction and it is therefore possible to modulate the solvent power of the fluid, thus obtaining more selective extractions.

2. Viscosity

The viscosity of supercritical fluids is greater than that of gases. This is important because a pronounced viscosity is responsible for a noticeable drop of pressure in an extraction cell.

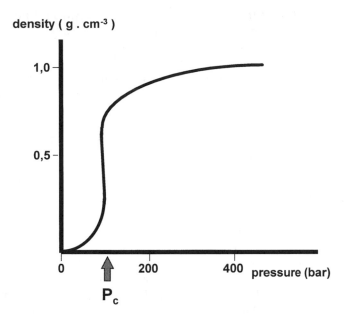

Figure 2
Variation of the density near the critical pressure for CO_2. (Adapted from Reference 3.)

The viscosity of supercrital fluids increases with pressure, but approaches that of a liquid less rapidly than the density. Even if very high pressures are used, the viscosity of a fluid in a supercritical state is inferior to that of a liquid. This is followed by improved penetration in porous materials coupled with a rapid kinetic.

3. Coefficient of Diffusion

Although noticeably weaker than that of a gas, the coefficient of diffusion of a solute in a supercritical fluid is about ten times greater than that of a liquid.

This advantage permits an important decrease in the duration of extraction as opposed to that of other liquid extraction.

4. Polarity of Supercritical Fluids

The polarity of a solvent can be measured by the solubility parameter of Hildebrand δ, where E^v is the coherent energy of the solvent and Vm the molal volume.

$$\delta = \sqrt{\frac{E^v}{Vm}}$$

For a supercritical fluid, δ increases with the density while remaining independent of the temperature.

$$\delta = 1,25\, P_c^{1/2}\left(\frac{\rho}{\rho_{liq}}\right)$$

where $\delta = (cal\ cm^{-3})^{1/2}$ solubility parameter; Pc = critical pressure of the fluid (atm); ρ = density of the fluid in the considered state; and ρ_{liq} = density of the corresponding liquid.

This equation indicates that the polarity of the supercritical fluid depends on its state by the relationship ρ/ρ_{liq} and on its nature (chemical effect from the term $1{,}25$ $P_c^{1/2}$).

Figure 3 compares values of δ of supercritical fluids for $\rho = \rho_{liq}$ (where only the chemical nature of the fluid is taken into consideration). From this figure it can be seen that CO_2 is a slightly polar fluid, and that at first sight the extraction of solutes strongly linked to a matrix will be difficult.

C. Principal Fluids Used in SFE

Table 2 shows critical parameters of the fluids most used for SFE. When it comes to choosing a supercritical fluid, the critical pressure and the critical temperature are two important parameters. The critical pressure determines, from a first approximation, the importance of the solvent power of the fluid. Ethane, for example, which has a lower critical pressure than carbon dioxide, will not dissolve a moderately polar soluble in the same way as carbon dioxide. Similarly, fluids with a higher critical pressure are more able to dissolve polar compounds. The critical temperature has practical implications. Indeed, one should always consider the influence of the extraction temperature on the stability of the component to extract.

D. Carbon Dioxide

Among the fluids shown in Table 2, carbon dioxide is, by far, the most used for the following reasons:

- ease of manipulation, given its temperature and critical pressure ($31\,°C$ and $73.8\,bar$),
- good solvent strength and miscibility with most solvents,
- compatibility of its temperature with the thermal stability of the solutes,
- lack of toxicity,
- nonflammable,
- noncorrosive,
- odourless,
- and, finally, economical (moderate costs).

For temperatures superior to the critical temperature, it can be seen that the density increases with the pressure. The solvent strength can therefore be increased by increasing the pressure.

Adding an organic solvent known as a modifier to a supercritical fluid can act significantly on the strength. For example, to extract more polar compounds, it can be interesting to add an alcohol to CO_2.

Figure 4 shows a diagram indicating the pressure-fraction molal of the modifier for a mixture of CO_2-MeOH at a temperature of $50\,°C$. The diagram calls for the following comments:

- At a pressure higher than 95 bar, whatever the composition of the mixture, only one phase exists.
- At intermediate pressures, the two phases (gas and liquid) can exist. At high proportions of CO_2 the mixture is a gas (G). At high proportions of MeOH the mixture is, on the contrary, a liquid (L). At intermediate levels the two phases co-exist (G + L).

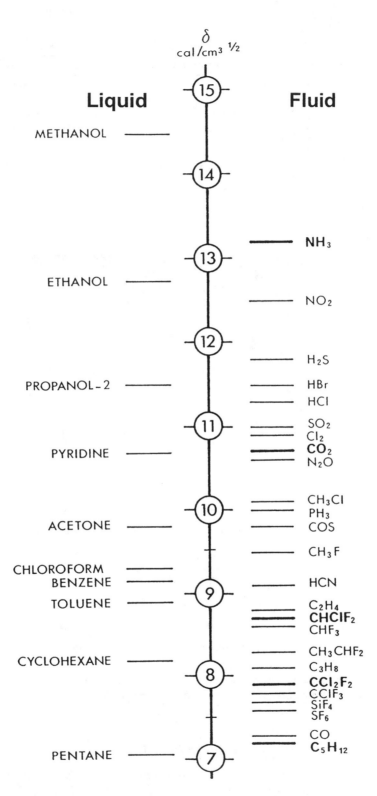

Figure 3
Comparison of the Hildebrand solubility parameters δ for different useful solvents. (Adapted from Rosset, R., Claude, M., Jardy, A., *Chromatographies en Phases Liquides et Supercritiques*, Masson, Paris, 1991, 830. With permission.)

TABLE 2.

Characteristics of some Supercritical Fluids Used for SFE

	Critical pressure (bar)	Critical temperature (°C)	Critical density (g·cm⁻³)
CO_2	73.8	31.0	0.448
N_2O	71.4	36.5	0.457
NH_3	111.3	132.3	0.240
SF_6	37.1	45.6	0.752
Isopropanol	47.0	253.3	0.273
n-Propane	41.9	96.7	0.217
Ethane	48.3	32.4	0.203
Butane	37.5	152.0	0.228
Xenon	58.0	16.6	1.105

Data from Reference 3.

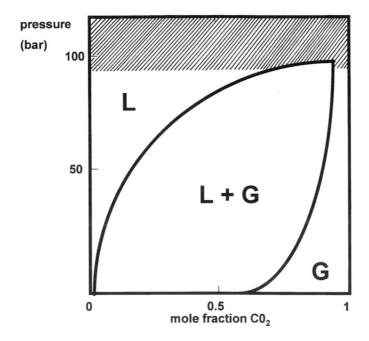

Figure 4
Schematic phase diagram for methanol-CO_2 at 50 °C. (Adapted from Reference 2.)

- The pressure is at its maximum at the critical point for a temperature of 50 °C.
- For a critical composition of the mixture, a lowering of the pressure leads to a two-phase system.
- As fluids in a subcritical state retain most of the advantages of supercritical fluids, some extractions can be undertaken under these conditions.

III. SAMPLE MATRIX EFFECTS

The nature of the matrix[4] can have an important effect on the results obtained by SFE. Unfortunately, incomplete knowledge on the solubility of the analyte in supercritical fluids does not always allow the prediction of the efficiency of the supercritical extraction for a given component. Consequently, no theoretical extrapolation can ever replace laboratory tests. In this section we will review different influencing matrix effects in supercritical phase extraction.

A. Physical Matrix Effects

The morphology of the matrix on which we wish to make a SFE can have an enormous influence on the efficiency of the extraction rate. Generally, a rapid and complete extraction depends upon the relative size of the matrix particles, the smaller being the better. This is due principally to the short internal distance that the solute must cover in order to attain the core of the supercritical fluid solution. Some studies[5,6] have shown that the geometrical form can also have an influence on the rate and efficacity of the extraction. As in the case of an extraction solid-liquid, an increase in the porosity of the matrix will lead to an efficient and rapid extraction.

B. Chemical Changes in the Sample Matrix

The chemical composition of the matrix can have either a positive or a negative effect on the results obtained by SFE. One of the most important parameters influencing the extraction is the degree of humidity of the sample. For example, it is generally admitted that a partial dehydration of the sample allows for a faster extraction. Indeed, hydrophylic matrices have a tendency of preventing contact between the supercritical fluid and the analyte. This is particulary true for the extraction of drugs from skin tissue.[7] In specific cases, however, the presence of water can, on the contrary, improve the rate of extraction by acting as an "internal cosolvent."

C. Matrix Effects and Extraction Kinetic

The rate of extraction is due, firstly, to the solubility of the analyte in the supercritical fluid and, secondly, to the transfer of the mass of the solute outside of the matrix. As indicated in Figure 5, we can therefore conveniently consider four mechanisms of mass transfer:

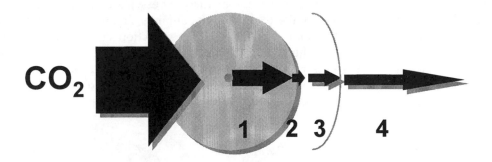

Figure 5
Simplified mass transport steps in the SFE of an analyte from a matrix particle. (Adapted from Reference 4.)

1. the diffusion of the analyte across the internal volume of the sample,
2. the desorption from the surface of the sample,
3. the diffusion of the analyte across the boundary layer on the surface of the sample,
4. the transfer in the core of the supercritical phase.

If the most important determining step of the rate is the diffusion across the internal volume of the sample, the rate of extraction will be dependent upon the size of the particles contained in the sample. A pulverization will consequently increase the rate of extraction. The desorption from the surface of the matrix, as well as the diffusion across the boundary layer on the surface of the sample, can be significantly improved by adding a polar modifier such as MeOH.[8,9]

IV. INSTRUMENTATION

As seen in Figure 6, the instrumentation needed to perform a SFE is relatively simple.

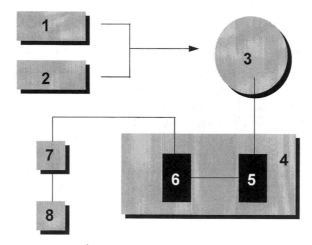

Figure 6
Schematic of a supercritical fluid extractor. 1 = CO_2; 2 = modifier; 3 = pump; 4 = oven; 5 = mixing column; 6 = extraction cell; 7 = pressure regulator; 8 = collection of the extract.

The instrumentation should allow for two extraction modes: one dynamic and the other static.[10] In the dynamic mode, the extraction vessel is constantly supplied with fresh supercritical fluid, while the recovery container is continually supplied with the solute. In the static mode, the outlet of the extraction vessel is closed and extraction takes place without the regeneration of supercritical fluid. On completion of the extraction, the vessel is rapidly rinsed by the supercritical fluid in order to permit the recuperation of the solute. At this point we should return to our discussion of the four components of the instrumentation.

A. Pumping

The high-pressure pump should provide a reproducible and constant flow. Either a syringe pump or a piston pump can be used. In one case, it will be necessary to cool the pumps in order to avoid phenomena of cavitation when filling the head of the pump.

As seen previously, during a SFE it is sometimes necessary to add an organic modifier. This adjunction can be made in different ways. If simple impregnation of the matrix before SFE is used, important concentration gradients could occur if working in dynamic mode. It is also possible to directly use a cylinder containing the mixture, but in spite of the elegance of the method it has been shown that the mixture of polar modifier becomes richer during the emptying of the cylinder.[11] It is therefore recommended to use a second pump and a mixing chamber in order to obtain a homogenous mixture.

B. Extraction vessels

At first, scientists used capillary tubes or small HPLC columns as extraction vessels. Now, however, it is possible to buy vessels dedicated to these applications directly on the market. Because of the necessity of working at high pressures, the resistance to pressure and the absence of leaks are the principal characteristics that an extraction vessel should have.

While most SFE instrumentation requires the use of vessels resistant to high pressures, Isco (Lincoln, NE) has found a way around this problem. The vessel is placed in a chamber under high pressure, and thus both the inside and the outside of the vessels are under pressure simultaneously. It is no longer necessary, therefore, to use expensive stainless steel vessels. As the temperature of the vessel needs to be controlled, an oven is required.

Lastly, the shape of the vessel does not appear to have an influence on the efficiency of the extraction, although some practical considerations such as the ease of filling the vessels needs to be taken into account. The SFE of liquid matrices, on the other hand, is technically problematic. A vessel permitting the extraction of liquid matrices is today commercialized by Jasco (Hachioji, Japan).

C. Flow Control Device

As previously mentioned, the SFE pump should produce a constant pressure of supercritical fluid with a rate controlled by a flow restrictor after the extraction vessel. There are a number of types of flow control devices, including a capillary made from fused silica, a pinched stainless steel tube, or a variable orifice allowing for electronic control of the pressure.

D. Analyte Collection

There are three ways of recuperating the analyte after depressurization of the supercritical fluid:

1. Thermal trapping
2. Sorbent trapping
3. Solvent trapping

Thermal trapping is a simple technique because the supercritical fluid is simply depressurized in a cooled recovery container. Unfortunately, this technique is limited to nonvolatile components as high gas flow can lead to the loss of relatively volatile compounds. Even slightly volatile compounds can be led by aerosol formation.

The other two techniques offer increased and interesting potential for the quantitative extraction of highly or moderately volatile compounds. Sorbent trapping provides favourable results and is relatively easy to apply. The supercritical fluid is depressurized and then absorbed on a solid support. Once the supercritical phase extraction has terminated, the analyte is recovered by elution with a small volume (a few milliliters) of solvent liquid. This method has the advantage of being able to increase the selectivity of the extraction by way of the elution solvent.

The simplest technique, however, remains that of trapping in a liquid solvent. While this procedure does not offer the selectivity seen in sorbent trapping, it is undoubtedly simpler and quicker. The choice of the solvent can be difficult; this is shown in Table 3 which describes the collection efficiencies for different models of organic components.

TABLE 3.

SFE Collection Efficiencies (%) in Different Organic Solvents

	Hexane[a]	Methanol[a]	CH_2Cl_2[a]	CH_2Cl_2 (5°C)[b]
Phenol	43	55	77	98
Nitrobenzene	60	58	82	100
2-Nitrophenol	57	61	80	99
2-Nitroaniline	72	57	86	98

[a] Collection efficiencies were determined by spiking 1-g samples of clean sand with about 18 µg of each component, and extracted using supercritical CO_2 at 400 atm and 50°C at a flow rate of 0.6 mL/min.
[b] Collection efficiencies when the collection vial was placed in a heating block set to turn on at 5°C.
Results adapted from Reference 10.

V. SFE OF OPIATES IN HAIR

A. Current State of Research

It was at the end of the 1970s that Baumgartner et al.[12] published the first method of analysing opiates in hair.[12] Since this time many techniques, which include four stages, have been proposed:[13]

- A washing stage, which allows for the elimination of external contamination,
- A pretreatment stage, which varies between the simple elution by an aqueous solvent and the complete dissolution of the proteinic structure by a basic hydrolysis,
- An extraction stage, where techniques such as the extraction liquid-liquid or the extraction on a solid phase are used, and finally,
- An analytical stage, where simple methods such as radioimmunological tests and more sophisticated methods such as GC/MS or MS/MS are used.

At first sight, SFE would appear to be the preferred method as it allows for the execution of all four stages with the same apparatus.

It was Sachs and Uhl[14] and Sachs and Raff[15] in 1992–3 that demonstrated for the first time the use of supercritical fluids in the extraction of drugs in hair. They illustrated the possibility of extracting opiates and cocaine in hair by means of a mixture of CO_2-ethyl acetate. The recovery of the extraction as well as the reproducibility of the method, however, remain inferior to other conventional techniques. Two years later, Edder et al.,[16] demonstrated the quantitative extraction of opiates in hair, and Morrison et al.[17] the quantitative extraction of cocaine. The results presented in this chapter are therefore a synthesis of these two works.

B. SFE of Opiates from Standard Hair and Hair of Drug Addicts

For ethical reasons, it is sometimes difficult to obtain, from the same person, a large quantity of hair. The elaboration of a method calls for numerous tests and therefore a large quantity of hair. The preparation of reference material, reproducing as closely as possible real hair of drug abusers, is of real interest. The simple adsorption of drugs on the surface of hair is insufficient. It is imperative that the drugs be allowed to penetrate into the matrix and preferably be strongly bonded. Recently, Welch et al.[18] have successfully produced a reference material of this type by simple immersion of hair in a mixture of drugs, dissolved in dimethylsulfoxide (DMSO). In our laboratories[19] we have also developed a quicker procedure that gives comparable results.

1. Procedure Used to Soak Hair

In this procedure, 120 to 200 mg of "white" hair, meaning hair not containing any opiates, pre-washed and reduced to a powder, are immersed into 20 mL of an aqueous solution containing 0.1 to 5 mg/L of codeine, ethylmorphine, 6-MAM, and morphine. This mixture is magnetically stirred for 5 h.

The hair is then recovered by filtration and washed with 20 mL of H_2O, followed by 20 mL of MeOH in order to eliminate the proportion of the drugs adsorbed by the hair surface. The concentrations of opiates in hair have been determined by GC/MS using the extraction method as described by Ahrens et al.[19] The average soaking efficiency and standard deviations of soaking of 5 successive samples of soaked hair for codeine, ethylmorphine, 6-MAM, and morphine can be seen in Table 4. In addition, the levels of opiates in hair are directly proportional to the concentration of the soaking aqueous solution. Indeed, the graphical representation of the concentrations of opiates in the solution coupled with the same concentration in the hair follows a linear distribution (with correlation coefficients of 0.995 for codeine, 0.996 for ethylmorphine, 0.992 for 6-MAM, and 0.995 for morphine).[20]

TABLE 4.

Efficiencies and Standard Deviations of Soaking Procedures for Hair

	Codeine	Ethylmorphine	6-MAM	Morphine
Soaking efficiencies (%)	13.5	14.5	12.0	14.6
Standard deviations (%)	1.4	2.3	1.7	1.4

Note: Each value is the result of five replicate soaking and extraction procedures.

2. Preparing Hair Before Extraction

a. Washing Procedures

Washing the hair should permit not only the elimination of inhibiting substances, but also prevent eventual outside contamination by transpiration and from sebum.[21,22] In addition, these external contaminations can equally come from the environment to which the subject is exposed. This is particularly important in the case where drugs can be smoked (cocaine, marijuana). Heroin is principally consumed by intravenous injection in Europe. However, as it can sometimes be smoked, this obliges us to wash all hair when looking for opiates. Some laboratories use shampooings,[12] solutions containing surfactants,[23] organic solvents such as ethylether,[24] methanol,[25] or hexane.[26]

It is generally admitted that these procedures are efficient and eliminate the majority of drugs fixed to the external surface of hair. Cone et al.[27] and Welch et al.[18] have shown that when the hair is very contaminated, washing procedures eliminate only a portion of the drugs retained (see Chapter 2 for further information on this problem). The washing procedure that we have retained is the following: a tuft of hair (about 200 to 500 mg) is rapidly rinsed by percolation with 10 mL of MeOH, 10 mL HCl 0.1 M, 10 mL H_2O, and, lastly, 10 mL of MeOH. The hair is then dried.

b. Pulverization of Hair

As previously mentioned, the quantity extracted and the reproducibility will be higher if the matrix is finely pulverized. The size of the particles has therefore a significant influence and should be controlled. Hair powder is obtained be means of a ball mill pulverizer (Retsch Schieritz, Hauenstein, Switzerland). About 0.5 g of hair is placed in a tungsten cell with a tungsten ball. The cell is then vigorously agitated. In order to determine the optimum pulverization time, the powder is examined by electronic microscopy (microscope scanning JSM 6400) after pulverization times of 3, 5, and 10 min. Some photos of hair powders obtained can be seen in Figure 7. It can be seen that 3 min are insufficient (Figure 7a), that large particles of around 500 μm remain and that the powder is not homogenous (the size of the particles varies between 50 and 500 μm). After 5 min (Figure 7b) the powder is already more homogenous, but there remains some fibres of about 50 μm. Finally, a pulverization time of 10 min produces a fine homogenous powder (Figure 7c) with residual fibres not exceeding 30 μm. The comparison between three hair samples obtained from three different individuals obtained after 10 min of pulverization show no significant differences.

A pulverization time of 10 min has therefore been retained.

3. Supercritical Phase Extraction

The supercritical phase extractor used for this application is described in Figure 8. The supercritical fluid is delivered by means of a double syringe pump (SFC 300, Carlo Erba, purchased from Brechbühler, Geneva, Switzerland); the first contains CO_2 while the other contains the modifier. A mixing column placed in an oven heated to 40°C is placed before the extraction cell in order that the mixture of CO_2 and the modifier should be as homogenous as possible.

The sizes of the extraction cell are the following:

- length 45 mm
- width 3.5 mm
- volume 430 μL

A Rheodyne six-port valve is added in order to purge the system without crossing the extraction cell.

In order to avoid all possible problems caused by small extraction yields, a new type of restrictor has been designed. It consists of a capillary of polyether-ether-ketone (PEEK, 0.016-in internal diameter), compressed by two disks by means of a screw. The PEEK is a polymer which has the advantage of being sufficiently supple to retain its original shape even after being strongly compressed. If the restrictor is blocked, it is possible to unblock it by simple decompression using the screw without removing the recovery vial. A perfect control of the pressure and of the flow is obtained with this economic system. In addition, the PEEK resists to a temperature

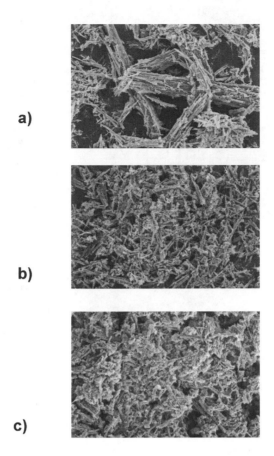

Figure 7
Scanning electronic micrographs of pulverized hair: (a) pulverization time of 3 min; (b) pulverization time of 5 min; (c) pulverization time of 10 min.

of 150°C, thus the heating of the valve, needed in order to avoid the formation of carbonic ice, is possible. The extracts are recuperated by trapping the decompressed fluid in a solvent (MeOH).

4.   GC/MS Analysis of Extracts Obtained by SFE

The extracts obtained by supercritical phase extraction are evaporated until dried under nitrogen flow. The opiates are then derived by propionylation.[21] After evaporation of the solvent, 100 μL of propionic anhydride (99%, Aldrich) and 100 μL of pyridine (99.5%, Merck) are added to the residue obtained and heated at 60°C for 30 min. After evaporation of the derivatization reagents under nitrogen, the residue is dissolved in 50 μL of ethyl acetate. The nalorphine, added after the SFE, is used as a chromatographic standard.

The characterisation and quantification of opiates (morphine, 6-MAM, codeine, and ethylmorphine) are obtained with the aid of a GC/MS. The apparatus and the conditions used are the following:

GC/MS: HP 5988

Injection: splitless 2 μL at T = 270°C

Column: 15 m × 25 mm i.d. DB-5ms (J & W Scientific, Folsom, CA)

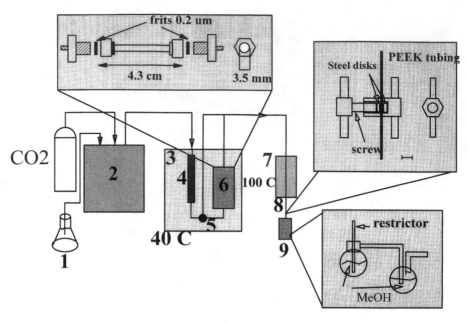

Figure 8
Diagram of the SFE system and details of the PEEK restrictor. 1 = polar modifier; 2 = syringe pumps; 3 = oven; 4 = mixing column; 5 = switching valve; 6 = extraction cell; 7 = block heater; 8 = PEEK restrictor; 9 = collection vessel. (Data from Reference 16.)

Chromatography:

$T_0 = 170\,°C$, 1 min

$T_1 = 250\,°C$, 20 °C/min

$T_2 = 285\,°C$, 5°C/min

Transfer line: $T = 280\,°C$

Source: $T = 200\,°C$

Detection: SIM in electronic impact mode with an energy of 70 eV.

Codeine $m/z = 355,282$, 6-MAM $m/z = 383,327$, morphine $m/z = 397,341$, ethylmorphine $m/z = 369,296$, nalorphine $m/z = 423,367$ (chromatographic standard).

5. Results

For this preliminary study, morphine has been chosen as a compound model. The polarity of the extractant phase is varied by increasing the percentage of polar modifier (composed of MeOH/TEA/H_2O 40/40/20 v/v) in the CO_2. The elute is recovered in the MeOH after 30 min of extraction with 0, 2, 5, 8, 12, 15, and 20% of polar modifier, where the SFE conditions retained are the following: $P = 25$ MPa, $T = 40\,°C$, flow rate = 0.7 mL/min. After the evaporation of the MeOH, the remainder is derived and analyzed by GC/MS following the preceding conditions. Table 5 shows that the morphine is poorly extracted with mixtures containing 0 and 2% of polar modifier. This observation shows that morphine is probably strongly fixed to the structure of the hair and not simply adsorbed on the surface. By using a mixture containing 2% of polar modifier (mixture of CO_2/MeOH/TEA/H_2O 98/0.8/0.8/0.4 v/v), it is therefore possible to add a preliminary washing stage to the extraction itself. It is interesting to note that these preliminary results are in perfect agreement with those obtained by Morrison et al.[17] for cocaine. This preliminary stage should render useless the initial washing with $MeOH/HCl/H_2O/MeOH$ and allow for the elimination of that fraction of drugs weakly linked to the matrix, as well as most slightly polar endogenic composites capable of disturbing the analysis. The quantitative extraction of morphine is obtained in eight of the nine cases with a mixture containing 15% of polar modifier, therefore with a composite eluent phase of $CO_2/MeOH/TEA/H_2O$ 85/6/6/3 v/v. In addition the soaked hair strongly resembles veritable drug abusers' hair from the point of view of the extractability of morphine (see Figure 9).

TABLE 5.

Extracted Morphine in Hair as a Function of the Polarity of the Supercritical Fluid by Varying the Percentage of the Modifier

Modifier (%)	0	2	5	8	12	15	20
Case 1	0	0	25.4	46.3	20.9	7.4	0
Case 2	0	0	84.0	16.0	0	0	0
Case 3	0	0	13.1	15.8	71.1	0	0
Case 4	0	0	100	0	0	0	0
Case 5	0	6.5	21.2	67.4	4.9	0	0
Case 6	0	0	3.3	28.4	40.3	28	0
Case 7	1.5	5.3	21.3	68.9	0	0	0
Case 8	0	0	0	14.0	22.5	30.3	33.2
Soaked hair	0	0	68.5	29.8	1.7	0	0

Note: SFE conditions: ($MeOH - TEA - H_2O$; 2:2:1 v/v) in CO_2; P = 25 Mpa; T = 40°C; flow rate = 0.7 mL/min; t = 30 min.

C. Optimisation of the Method

This is made using standard hair where the amount of opiates is in the order of 2 ng/mg.

1. *Influence of the Polar Modifier*

The composition of the extractant mixture, particularly the proportions of H_2O and of TEA, is an important parameter determining the efficiency of the extraction.

a. *Influence of the Amount of Water in the Polar Modifier*

The influence of water is studied by varying the percentage of water from 0 to 3% while maintaining the other parameters, MeOH (6%) and TEA (6%), constant. The remaining supercritical phase extraction conditions are identical: P = 25 MPa, T = 40°C, flow rate = 0.7 mL/min, and t = 30 min.

Figure 10a shows that the amounts of opiates are at a peak using 3% of water, while only 30% of the maximum amounts are recuperated, without water. However, because of the risks of hydrolysis of the 6-MAM, it is necessary to limit as much as possible the proportion of water in the extractant mixture.

b. *Influence of the Amount of TEA*

As for water, the influence of TEA is examined by varying the percentage of TEA (from 1 to 6%) while maintaining the other parameters constant. Figure 10b

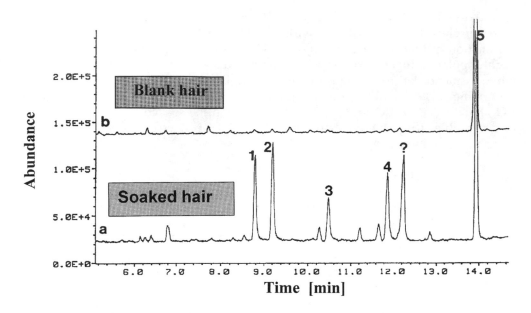

Figure 9

Total ion current chromatograms of opiates in soaked hair and blank hair. SFE conditions: P = 25 MPa, T = 40°C, flow rate = 0.7 mL/min, t = 30 min, $CO_2/MeOH/TEA/H_2O$ (85/6/6/3 v/v). 1 = codeine; 2 = ethylmorphine; 3 = 6-MAM; 4 = morphine; 5 = nalorphine.

shows that amounts of opiates are at a peak in the presence of 6% TEA in the extractant mixture, while with 1% of TEA, only 25% of this maximum is recovered.

2. Influence of the Duration of Extraction

Under standard conditions (P = 25 MPa, T = 40°C, flow rate = 0.7 mL/min) the eluent is collected after 2, 5, 10, 15, 20, 25, 30, and 40 min of SFE; following this, morphine is determined in each fraction. Figure 11 shows that maximum recovery (normalised at 100%) is obtained after 20 min of extraction. However, to be sure of extracting the totality of the opiates, 30 min have been retained.

The optimum conditions retained therefore for the extraction of opiates in hair are the following:

P = 25MPa, T = 40°C, flow rate = 0.7 mL/min, t = 30 min

Eluent phase: $CO_2/MeOH/TEA/H_2O$ (85/6/6/3 v/v)

D. Statistical Data of the Method

1. Extraction Yield

Even if the preceding results show that one can arrive at optimum levels of extraction by SFE, the absolute yield of extraction is unknown.

In order to estimate this yield, we have doped hair by means of morphine marked with [125]Iode (coming from a radioimmunoassay kit, Coat-a-Count™ Morphine, Diagnostic Product Corporation, Los Angeles, CA).[20] Six replicate SFE steps give an average yield of 93.5% with a variation coefficient of 2.7%. It is therefore permitted to consider that the SFE is complete for morphine as well as for most other opiates, as morphine, because of its strong polarity, is the most difficult opiate to extract.

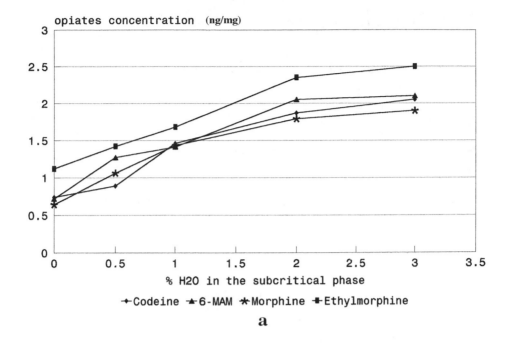

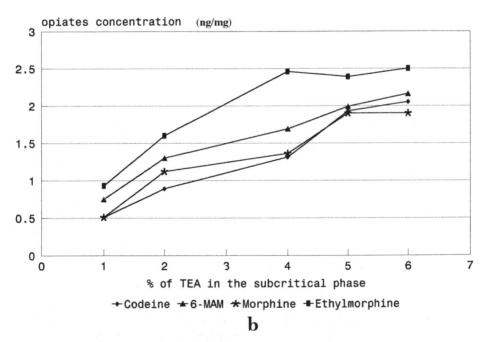

Figure 10
Opiates concentrations (ng/mg) in soaked hair as a function of water content (a) and the triethylamine (b).

2. *Repeatability*

The repeatability of the overall method proposed (SFE extraction and GC analysis) is determined by using optimum conditions previously established, on soaked hair.

Average concentrations in opiates and coefficients of variation obtained for 5 different extractions are presented in Table 6 below. The variation coefficients are less than 10%, apart from codeine, which is perfectly satisfactory for this type of analysis.

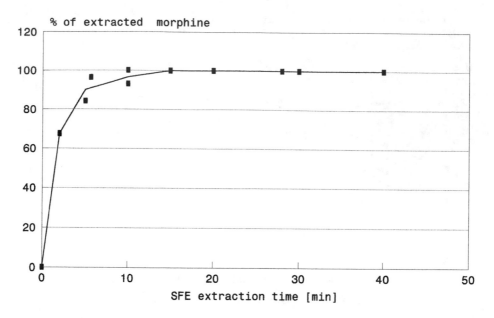

Figure 11

SFE recovery profile of morphine in soaked hair. SFE conditions: P = 25 MPa, T = 40°C, flow rate = 0.7 mL/min, t = 30 min, $CO_2/MeOH/TEA/H_2O$ (85/6/6/3 v/v).

TABLE 6.

Comparison of SFE with Other Extraction Techniques for Opiates in Hair: Concentrations (ng/mg) in Hair and Coefficients of Variation (CV; %)

	Methanolic extraction		Enzymatic digestion		Acid hydrolysis		Basic hydrolysis		SFE	
	(n = 5) (ng/mg)	CV (%)	(n = 5) (ng/mg)	CV (%)	(n = 3) (ng/mg)	CV (%)	(n = 3) (ng/mg)	CV (%)	(n = 5) (ng/mg)	CV (%)
Codeine	1.98	18	1.92	17	1.22	15	1.47	10	1.99	12
Ethylmorphine	2.41	9.8	2.19	3.3	1.95	10	2.15	14	2.56	5.6
6-MAM	2.18	7.9	2.49	10	0	—	0	—	2.18	10
Morphine	1.21	7.5	1.73	13	2.39	14	2.46	13	1.93	3.0

3. Limits of Detection (LOD) and of Quantification (LOQ)

Six 50-mg blank hair samples were extracted by SFE and the noise was integrated for the ion used for quantification (m/z = 355 for codeine, 369 for ethylmorphine, 383 for 6-MAM, and 397 for morphine) in a retention time window of $t_r \pm 0.5$ min. The LOD and LOQ were determined (n = 6) using IUPAC methods.[28] For each substance the standard deviation of the blank value (S_B) was determined. The mean area converted from the noise was calculated as concentration equivalent based on a calibration graph. The LOD is defined as $3S_B$ and the LOQ as $10S_B$.

In these conditions, the detection limits are 30 pg/mg for codeine, ethylmorphine, and morphine, and 50 ppb for 6-MAM, while the quantification limits are 100 pg/mg for the first three opiates and 200 pg/mg for 6-MAM. These values are comparable to those obtained by Moeller et al.,[28] who proposed LODs of 100 pg/mg for opiates.

E. Comparison with Classical Methods

In order to undertake a comparative study of various extraction methods of opiates from hair, standard hair was extracted employing some methods currently used in this context. The classical methods studied were the following.

1. Basic Hydrolysis

Basic hydrolysis was performed in a closed vial for 1 h at 100 °C by adding 2 mL of 1 M NaOH to 50 mg of hair. The mixture was neutralised with 1 mL of 2 M HCl and buffered with 1 mL of 1/15 M phosphate buffer (pH 6.0). This solution was then extracted by SPE on Bond Elut® cartridges, activated beforehand with 3 mL of methanol and water. For the washing step 2 mL of water, 1 mL of 0.1 M acetate buffer (pH 4.0), and 2 mL of methanol were used. Elution of opiates was performed using 3 mL of dichloromethane-isopropanol (80:20, v/v) containing 2% of ammonia.

2. Acid Hydrolysis

Acid hydrolysis was performed in a closed vial at 120 °C for 30 min by adding 2 mL of 2 M HCl to 50 mg of hair. The mixture was neutralised with 10 M NaOH and buffered with 1 mL of 1/15 M phosphate buffer (pH 6.0) prior to extraction on Bond Elut cartridges with the same procedure as above.

3. Methanolic Extraction

Methanolic extraction was performed in a closed vial at 37 °C for 18 h by adding 2 mL of methanol to 50 mg of washed and pulverized hair. The supernatant was collected and hair was rinsed with 2 mL of methanol. After evaporation of the solvent, the residue was dissolved in 2 mL of 1/15 M phosphate buffer (pH 6.0). Finally, the solution was extracted on Bond Elut cartridges with the same procedure as above.

4. Enzymatic Digestion

Enzymatic digestion was performed in a closed vial by adding 50 mg of guanidine hydrochloride, 50 μL of 2-mercaptoethanol, 25 μL of glucuronidase, and 1 mL of 1/15 M phosphate buffer (pH 7.0) to 50 mg of hair. The mixture was incubated for 4 h at 45 °C. The supernatant was collected and the hair was rinsed with 2 mL of 0.1 M phosphate buffer (pH 7.0). The solution was extracted on Bond Elut cartridges as described above.

Table 6 shows a comparison of opiate concentrations in hair, together with coefficient variations obtained by SFE and by other extraction techniques. It can be seen that methanolic extraction gives good results with the exception of morphine, where the extraction yield appears poor (~60% of the amounts obtained by SFE). This could be a result of the polar nature of morphine, which must be more solidly fixed on the matrix. Acid hydrolysis and basic hydrolysis are of little interest, because 6-MAM, which is effectively found in high quantities in hair, is destroyed during the procedure and is partially transformed into morphine.

Enzymatic digestion gives equally excellent results for the four opiates. Comparing this method with SFE shows that the two techniques are comparable for codeine and ethylmorphine, but show differences when it comes to 6-MAM and morphine. In fact the higher extraction yield for morphine and the lower yield for

6-MAM, observed by SFE, compared to those obtained by enzymatic digestion, indicate the presence of slight hydrolysis of 6-MAM to morphine.

VI. SFE OF OTHER DRUGS IN HAIR

SFE has also been successfully applied to cocaine and to methadone.

A. Cocaine

Cocaine and its principal metabolite, benzoylecgonine, is also found in significant concentrations in hair.[17,20] Whereas in urine, the levels of benzoylecgonine are superior to those of cocaine, in hair, cocaine is often the major component.[29,30]

1. Soaking Procedures for Hair

We have applied the same procedure that we developed for opiates. Morrison and colleagues[17] used the same method presented recently by Welch et al.[18]

2. Supercritical Phase Extraction

While we use the same apparatus as described for opiates, Morrison et al.[17] use an apparatus which can briefly be described in the following manner:

The supercritical fluid is delivered by a syringe pump (model 100D, Isco, Lincoln, NE). The modifier is added by means of a loop of 100 µL and a Rheodyne valve (model 7125) directly into the CO_2. The extraction cell is heated to 110°C and the flow of supercritical fluid is adjusted to 1.1 to 1.2 mL/min. The extract is then recovered by trapping the decompressed fluid in a vial containing approximately 3.5 mL of methanol.

3. GC/MS Analysis of Extracts Obtained by SFE

The extracts obtained by supercritical phase extraction are evaporated under a current of nitrogen. Benzoylecgonine, which possesses a carboxylic function, is derivatized, either by the addition of 100 µL pentafluoropropionic anhydride (99%, Aldrich, Milwaukee, WI) and of 100 µL of pentafluoropropanol (97%, Aldrich, Milwaukee, WI), or by the addition of 50 µL of N,O-bis(trimethylsilyl)-acetamide (BSA, Pierce, Rockford, IL). The final extracts are then injected into the GC/MS.

The apparatus and conditions used are the following.

a. Edder and colleagues[16]

GC/MS: HP 5988

Injection: splitless 2 µL at T = 270°C

Column DB-5ms 15m× 0.25 mm i.d.

with a film thickness of 0.25 µm (J & W Scientific, Folsom, CA).

Chromatography:

T = 100°C, 1 min

T = 280°C, 20°C/min, 1 min

Interface: T = 280 °C

Source: T = 200 °C

Detection: SIM in electronic impact mode

Cocaine: m/z = 303,182

Benzoylecgonine: m/z = 421,300

b. *Morrison and colleagues*[17]

GC/MS: HP 5970

Injection: on column 2 µL

Column DB-5ms 60 × 0.25 mm i.d.

with a film thickness of 0.25 µm (J & W Scientific, Folsom, CA)

Chromatography:

T = 100 °C, 1 min

T = 250 °C, 25 °C/min

T = 290 °C, 10 °C/min

Detection: SIM in electronic impact mode

Cocaine: m/z = 303,272,182

Benzoylecgonine: m/z = 361,346,240

4. Results and Discussion

Table 7 gives levels of cocaine and benzoylecgonine for 5 hair samples (obtained from cocaine abusers). The supercritical phase extraction has been made under the same conditions as that for opiates (P = 25 MPa, T = 4 °C, flow rate = 0.7 mL/min; eluent phase: CO_2/MeOH/TEA/H_2O 85/6/6/3 v:v).

TABLE 7.

Concentrations of Cocaine
and Benzoylecgonine in Hair

Case	Cocaine (ng/mg)	Benzoylecgonine (ng/mg)
1	1.5	1.5
2	0.3	0
3	0.4	0
4	0.3	0.2
5	0.2	0

It is interesting to note that all hair samples contain cocaine, which once again proves the potential of analysing hair for the detection of drug abuse. Figure 12 shows the chromatograms obtained after SFE and GC/MS, where we see that the selectivity of the method proposed is entirely satisfactory. Morrison and colleagues[17] have studied the influence of the matrix on the extraction of cocaine. For this comparative study, three matrices with different adsorption properties were chosen:

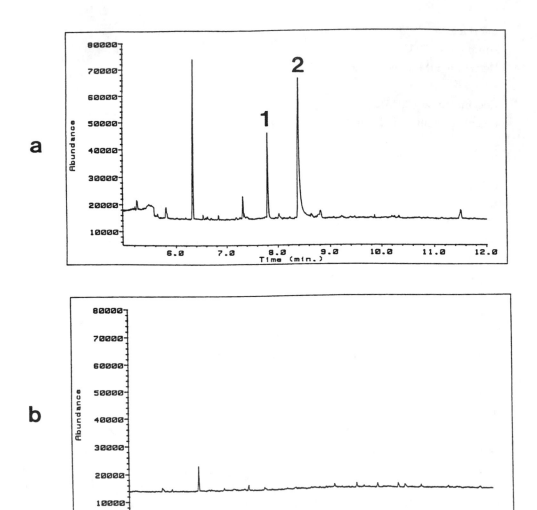

Figure 12
Total ion current chromatograms of cocaine (2) and benzoylecgonine (1) in soaked hair (a) and blank hair
(b). SFE conditions: P = 25 MPa, T = 40°C, flow rate = 0.7 mL/min, t = 30 min, CO_2/MeOH/TEA/H_2O
(85/6/6/3 v/v).

- A piece of teflon was chosen as a model of a matrix slightly or nonadsorbant.
- Hair on which cocaine had been adsorbed, as a model of a sample contaminated on its surface.
- Lastly, hair soaked with cocaine, as a model of a matrix where there is a strong interaction with the analyte (since cocaine has had the time to penetrate the hair in an incubation period of about 1 month).

The extractability of these three matrices was studied using supercritical fluids with increasing solvent powers (CO_2, CO_2 + MeOH, CO_2 + H_2O/TEA).
Figure 13 shows varying results, each depending on the matrix studied. Cocaine is partially extracted from the teflon and hair adsorbed with cocaine, with CO_2 (70 and 54%, respectively). On the other hand, pure CO_2 is totally ineffective for soaked hair. The addition of MeOH results in a slight increase of the extraction efficiency, while only the addition of TEA and of H_2O allows for the quantitative extraction (extraction

efficiency close to 90%) of cocaine. These results show that the way that the analyte is incorporated into the matrix determines the required conditions for supercritical phase extraction. Therefore, it should be possible to exploit this effect in order to distinguish between the drugs remaining on the hair surface and those incorporated in the hair.

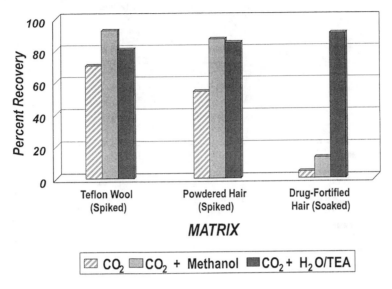

Figure 13

Influence of matrix on the supercritical fluid extractability of cocaine using CO_2, CO_2 + MeOH and CO_2 + H_2O/TEA (85:15 v/v). SFE conditions: 400 atm, 110°C, 10 min static, 15 min dynamic mode; 100 μL modifier. (From Morrison, J. F., MacCrehan, W. A., Selavka, C. M., submitted. With permission from the National Institute of Standards and Technology.)

Finally, note that with a mixture of CO_2 and H_2O/TEA (85:15 v/v), Morrison and colleagues[17] were unable to extract benzoylecgonine from hair.

B. Methadone

Methadone is frequently used as a drug substitute in the detoxification of heroin addicts.

In this context, hair analysis is particularly interesting because it allows the practitioner to control the methadone intake and to determine a simultaneous consumption of other drugs.

We have therefore applied the procedure, developed for opiates, to three samples coming from subjects susceptible to having consumed methadone and/or opiates.

The MS detection is then applied for the following ions:

Methadone: m/z = 72

Morphine: m/z = 397 and 341

6-MAM: m/z = 383 and 327

We chose only one ion for methadone because the molecular ion m/z = 294 is weak and represents only 1 to 2% of the base peak (m/z = 72).

Of the three cases tested (see Table 8), two are positive for methadone. The chromatograms of Figure 14 show that the technique remains sufficiently selective and that, in the extraction conditions optimised for opiates, analysis of methadone is also possible.

TABLE 8.

Morphine, 6-MAM, and Methadone
Concentrations (ng/mg) in Hair
of Three Drug Addicts

	Morphine	6-MAM	Methadone
Case 1	0.3	0	0.6
Case 2	1.2	0	1.2
Case 3	2.4	4.9	0

Note: SFE conditions: P = 25 Mpa; T = 40°C;
flow rate = 0.7 mL/min; t = 30 min;
$CO_2/MeOH/TEA/H_2O$ (85/6/6/3 v/v).

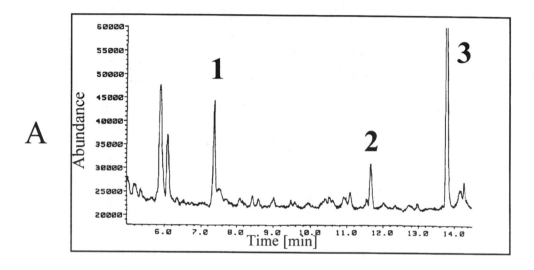

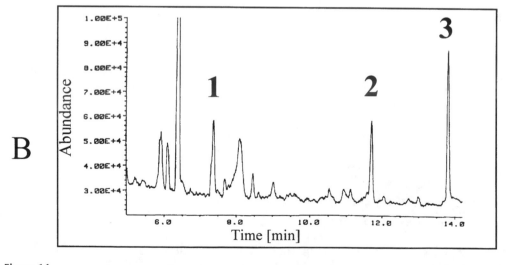

Figure 14

Total ion current chromatograms of hair obtained from drug addicts. 1 = methadone; 2 = morphine; 3 = nalorphine. SFE conditions: P = 25 MPa, T = 40°C, flow rate = 0.7 mL/min, t = 30 min, $CO_2/MeOH/TEA/H_2O$
(85/6/6/3 v/v).

VII. CONCLUSIONS

Supercritical phase extraction is the chosen method for extracting drugs, including opiates, cocaine, and methadone from a solid matrix such as hair. In its present state, supercritical phase extraction can be used with these three groups of drugs. In the near future, however, it is probable that this method will be applicable to other drugs and even to other medical substances. At that time, SFE will probably become the reference method for hair analysis.

The results presented in this chapter demonstrate that a polar modifier (MeOH, TEA, H_2O) should be added to the CO_2 in order to extract drugs accumulated in hair. On the other hand, CO_2 by itself permits the extraction of drugs slightly adsorbed on the surface of hair. It then becomes possible by SFE to distinguish between drugs remaining on the surface, or left by outside contamination, and those incorporated in the hair.

We have demonstrated in particular that SFE is an appropriate method for analysing opiates in hair. It is rapid (30 min), efficient, and reproducible. The procedure applied to standard hair, or to hair coming from drug abusers, has given excellent results. In addition, the quantification limits are comparable to those generally observed in hair analysis.

Comparing SFE with other classical extraction methods, we have observed that, either the method is more efficient (basic/acid hydrolysis, extraction with methanol), or that it gives comparable results (enzymatic hydrolysis). The results obtained for cocaine and for methadone, although necessitating complementary studies, remain very encouraging. Lastly, if we take into account the ease of the method and its economic implication, this technique should be adopted without delay in laboratories involved in this type of analysis.

VIII. ACKNOWLEDGMENTS

The authors thank gratefully Emily Chapalay and Graham Lawrence for their assistance in typing and translating this paper.

LIST OF ABBREVIATIONS

P	Pressure
T	Temperature
V	Volume
S	Solid
L	Liquid
G	Gas
ρ	Density
SFE	Supercritical fluid extraction
HPLC	High-performance liquid chromatography
GC	Gas chromatography
MS	Mass spectrometry
CO_2	Carbon dioxide
DMSO	Dimethylsulfoxide

6-MAM	6-monoacetylmorphine, metabolite of heroin
MeOH	Methanol
TEA	Triethylamine
Pa	Pascal
SPE	Solid-phase extraction
t	Time
min	Minute
atm	Atmosphere
LOD	Limit of detection
LOQ	Limit of quantification
t_r	Retention time
h	Hour(s)
mL	Milliliter
μL	Microliter
i.d.	Internal diameter
Vm	Molal volume
δ	Solubility parameter of Hildebrand
E^v	Coherent energy
K_T	Compressibility

REFERENCES

1. Chester, T.L., Pinkston, J.D., Raynie, D.E., Supercritical fluid chromatography and extraction, *Anal. Chem.*, 66, 106R–130R, 1994.
2. Clifford, A.A., Introduction to supercritical fluid extraction in analytical science. In *Supercritical Fluid Extraction and Its Use in Chromatographic Sample Preparation*, S.A. Westwood (Ed.), Blackie Academic and Professional, Glasgow, 1–38, 1993.
3. Rosset, R., Claude, M., Jardy, A., *Chromatographies en Phases Liquides et Supercritiques*, Masson, Paris, 815–907, 1991.
4. King, J.W., France, J.E., Basic principles of analytical supercritical fluid extraction. In *Analysis with Supercritical Fluids: Extraction and Chromatography*, B. Wenclawiak (Ed.), Springer-Verlag, Berlin, 32–57, 1992.
5. Pawliszyn, J., Kinetic model of supercritical fluid extraction, *J. Chromatogr. Sci.*, 31, 31–37, 1993.
6. Hawthorne, S.B., Miller, D.J., Burford, M.D., Langenfeld, J.J., Eckert-Tilotta, S., Luie, P., Factors controlling quantitative supercritical fluid extraction of environmental samples, *J. Chromatogr.*, 642, 301–317, 1993.
7. Ramsey, E.D., Perkins, J.R., Games, D.E., Startin J.R., *J. Chromatogr.*, 464, 353, 1989.
8. Fahmy, T.M., Pulaitis, M.E., Johnson, D.M., McNally, M.E.P., Modifier effects in the supercritical fluid extraction of solutes from clay, soil, and plant materials, *Anal. Chem.*, 65 (10), 1462–1469, 1993.
9. Langenfeld, J.J., Hawthorne, S.B., Miller, D.J., Pawliszyn, J., Role of modifiers for analytical scale supercritical fluid extraction of environmental samples, *Anal. Chem.*, 66(6), 909–916, 1994.
10. Hawthorne, S.B., Methodology for off-line supercritical fluid extraction. In *Supercritical Fluid Extraction and Its Use in Chromatographic Sample Preparation*, Westwood S.A. (Ed.), Blackie Academic and Professional, 39–64, 1993.
11. Camel, W., Tambuté, A., Claude, M., L'Extraction en phase supercritique à l'échelle analytique: principe, mise en oeuvre et applications, *Analysis*, 20, 503–528, 1992.
12. Baumgartner, A.M., Jones, P.F., Baumgartner, W.A., Black, C.T., Radioimmunoassay of hair for determining opiate abuse histories, *J. Nucl. Med.*, 20, 748–752, 1979.
13. Staub, C., Analytical procedures for determination of opiates in hair: a review, *Forensic Sci. Int.*, 70, 111–123, 1995.

14. Sachs, H., Uhl, M., Opiat-Nachweis in Haar-Extrakten mit Hilfe von GC/MS/MS und Supercritical Fluid Extraction (SFE), *Toxichem. Krimtech.*, 59, 114–120, 1992.

15. Sachs, H., Raff, I., Comparison of quantitative results of drugs in human hair by GC/MS, *Forensic Sci. Int.*, 63, 207–216, 1993.

16. Edder, P., Staub, C., Veuthey, J.L., Pierroz, I., Haerdi, W., Subcritical fluid extraction of opiates in hair of drug addicts, *J. Chromatogr. B.*, 658, 75–86, 1994.

17. Morrison, J.F., MacCrehan, W.A., Selavka, C.M., Evaluation of supercritical fluid extraction for the selective recovery of drugs of abuse from hair, 2nd International Meeting on Clinical and Forensic Aspects of Hair Analysis, National Institute on Drug Abuse. Special Publication, submitted, 1995.

18. Welch, M.J., Sniegoski, L.T., Allgood, C.C., Habram M., Hair analysis for drugs of abuse: evaluation of analytical methods, environmental issues, and development of reference materials, *J. Anal. Toxicol.*, 17(6), 389–398, 1993.

19. Ahrens, B., Erdmann, F., Rochholz, G., Schütz, H., Detection of morphine and monoacetylmorphine (MAM) in human hair, *Fresenius J. Anal. Chem.*, 344, 559–560, 1992.

20. Edder, P., Utilisation des Fluides Supercritiques pour l'Extraction de Stupéfiants dans des Matrices Biologiques Liquides (Urine) et Solides (Cheveux), Doctoral thesis, University of Geneva, 1994.

21. Blank, D.L., Kidwell, D.A., External contamination of hair: an issue in forensic interpretation, *Forensic Sci. Int.*, 63, 145–156, 1993.

22. Henderson, G.L., Mechanisms of drug incorporation into hair, *Forensic Sci. Int.*, 63, 19–29, 1993.

23. Ishiyama, I., Nagai, T., Toshida, S., Detection of basic drugs (methamphetamine, antidepressants and nicotine) from human hair, *J. Forensic Sci.*, 28, 380–385, 1983.

24. Marigo, M., Tagliaro, F., Poiesi, C., Lafisca, S., Neri, C., Determination of morphine in the hair of heroin addicts by high performance liquid chromatography with fluorometric detection, *J. Anal. Toxicol.*, 10, 158–161, 1986.

25. Suzuki, S., Inoue, T., Hori, H., Inayama, S., Analysis of methamphetamine in hair, nail, sweat and saliva by mass fragmentography, *J. Anal. Toxicol.*, 13, 176–178, 1989.

26. Haley, N.J., Hoffman, D., Analysis for nicotine and cotinine in hair to determine cigarette smoker status, *Clin. Chem.*, 31, 1598–1600, 1985.

27. Cone, E.J., Yousefnejad, D., Darwin, W.D., Maguire, T., Testing human hair for drugs of abuse. II. Identification of unique cocaine metabolites in hair of drug abusers and evaluation of decontamination procedures, *J. Anal. Toxicol.*, 15, 250–255, 1991.

28. Moeller, M.R., Fey, P., Wennig, R., Simultaneous determination of drugs of abuse (opiates, cocaine and amphetamine) in human hair by GC/MS and its application to a methadone treatment program, *Forensic Sci. Int.*, 63, 185–206, 1993.

29. Möller, M.R., Fey, P., Rimbach, S., Identification and quantification of cocaine and its metabolites, benzoylecgonine and ecgonine methyl ester, in hair of Bolivian coca chewers by gas chromatography/mass spectrometry, *J. Anal. Toxicol.*, 16, 291–296, 1992.

30. Nakahara, Y., Ochiai, T., Kikura, R., Hair analysis for drugs of abuse. V. The facility in incorporation of cocaine into hair over its major metabolites, benzoylecgonine and ecgonine methyl ester, *Arch. Toxicol.*, 66, 446–449, 1992.

Chapter 6

DETERMINATION OF COCAINE AND OPIOIDS IN HAIR

Diana Garside and Bruce A. Goldberger

CONTENTS

I. INTRODUCTION

The growing concern regarding the widespread abuse of illicit drugs has inspired the development of new technologies for the analysis of therapeutic and abused drugs and their metabolites in unusual biological tissues, including hair. Analysis of hair for the presence of drugs readily provides valuable information regarding recent and past drug use. Results of hair tests have been utilized in clinical, forensic and epidemiologic studies, historical research, and even presented as evidence in civil, criminal, and military courts of law.

Initial studies performed during the late 1970s and early 1980s revealed the presence of benzoylecgonine (cocaine metabolite) and morphine (heroin metabolite) in hair. Since then, further studies of hair analysis have identified additional cocaine analytes and opioids. The aim of this chapter is to provide a review of the analytical techniques utilized in hair analysis, including preliminary wash/decontamination procedures, methods of analyte isolation and purification, and analysis of the hair extracts. In addition, this chapter will present research findings.

A. Pharmacology of Cocaine

Cocaine is the predominant alkaloid present in the leaves of *Erythroxylan coca*, a shrub of the coca family Erythroxylaceae. These plants, native to South America, are also cultivated in Africa, South Asia, and Australia for the narcotic value of cocaine.

Historically, cocaine has been applied topically as a local anesthetic, especially for the nose, throat, and cornea, but it has now been replaced by less toxic, synthetic anesthetics. Systemically, cocaine stimulates the central nervous system, producing feelings of euphoria, enhanced physical strength and mental capacity, a lessened sense of fatigue, decreased appetite, and sometimes increased sexual interest. Because of these stimulating effects, cocaine is an abused drug. However, it also causes increased heart rate, blood pressure, and temperature; in large doses it can cause death. The chronic user of cocaine develops paranoia, increased anxiety, hallucinations, and perceptual changes. Cocaine is psychologically addictive, and abrupt cessation of use leads to depression.

For over three thousand years, native South American people have chewed the leaves of the coca shrub to dull the sense of hunger and to lessen fatigue. As an illicit substance, cocaine is readily self-administered as the chlorine salt, cocaine hydrochloride, either intravenously as an aqueous solution or intranasally as a powder. Cocaine hydrochloride, prepared by purification of the crude form of cocaine known as coca-paste, is a high-melting point, white, crystalline powder. Recently, cocaine has been chemically reconstituted from the salt to give a highly concentrated form of the drug known as "crack." In this chemical state, cocaine is administered by smoking as it is relatively volatile, with a lower melting point than cocaine hydrochloride.

Blood concentrations of cocaine are highest and peak immediately when the drug is administered intravenously or is smoked. Intranasal administration leads to a slower absorption into the blood with ultimately lower concentrations. Metabolism of cocaine begins almost immediately after absorption with hydrolysis of the ester functionalities to give benzoylecgonine and ecgonine methyl ester; cholinesterases in serum and liver are responsible for the hydrolysis to ecgonine methyl ester, while benzoylecgonine is formed spontaneously in biological fluids. Further hydrolysis of both products leads to ecgonine. Cocaethylene, an analyte with pharmacological activity similar to cocaine, is produced by transesterification when cocaine and alcohol are consumed simultaneously. An analyte unique to users of crack is anhydroecgonine methyl ester and is the direct result of cocaine pyrolysis. Removal of the *N*-methyl group of cocaine, cocaethylene, and benzoylecgonine results in the respective nor-products. The cocaine analytes found in hair are illustrated in Figure 1. Cocaine has a half-life of approximately 0.7–1.5 h in humans, and along with many of its metabolites is found in urine up to 24 to 72 h following administration.

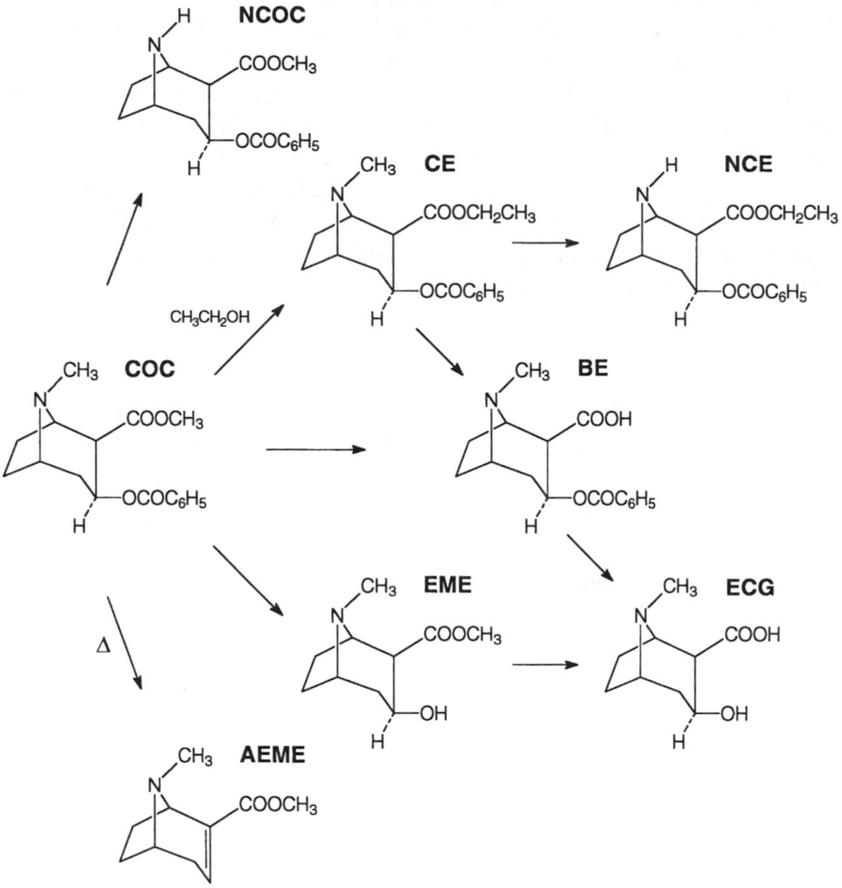

Figure 1
The metabolic scheme of cocaine analytes in hair. (Abbreviations: COC, cocaine; BE, benzoylecgonine; EME, ecgonine methyl ester; ECG, ecgonine; CE, cocaethylene; NCOC, norcocaine; NCE, norcocaethylene; and AEME, anhydroecgonine methyl ester).

B. Pharmacology of Opioids

Opium is an alkaloid obtained from the dried latex of incised unripe flower pods of the poppy plant *Papaver somniferum* and is comprised of as many as 25 different alkaloids, including pharmacologically active morphine, codeine, and papaverine. Opium is cultivated in Southeast Asia, Turkey, Pakistan, Afghanistan, India, Iran, Nigeria, and Mexico. Heroin is produced by acetylation of morphine.

Opioids are potent drugs that elicit their primary pharmacological effects on the central nervous, respiratory, cardiovascular, and gastrointestinal systems. Euphoria, anesthesia, analgesia and increased tolerance to pain, stimulation of chemotrigger zone, and production of miosis are common central nervous system effects. Effects on the respiratory system include depression in rate, volume, and exchange of respiration and decreased responsiveness. Opoids produce orthostatic hypotension. Decreased motility and delayed gastric emptying are common gastrointestinal effects. Opiods are abused primarily for production of euphoria and avoidance of withdrawal symptoms.

Heroin is usually self-administered parenterally, either by intravenous or intramuscular injection, but may also be snorted or smoked. Other less common routes of administration include oral, subcutaneous, sublingual, and rectal. Following administration, heroin is rapidly metabolized by deacetylation to 6-acetylmorphine (6-AM). 6-AM is rapidly hydrolyzed to morphine. The approximate half-lives of heroin and 6-AM are 5 and 45 min, respectively. Morphine is then metabolized by conjugation to morphine-glucuronide and by N-demethylation to normorphine. Normorphine is also conjugated to normorphine-glucuronide. Morphine and conjugated morphine are the primary heroin metabolites found in urine, but depending upon dose and time of collection, heroin and 6-AM may also be present. Since illicit heroin preparations also contain acetylcodeine and codeine, codeine is often present in urine following heroin use. The metabolic pathway of heroin is shown in Figure 2.

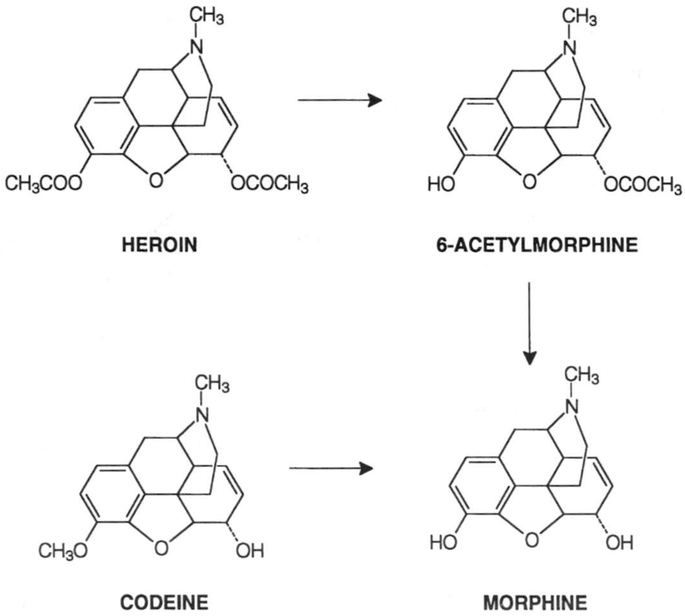

Figure 2
The metabolic scheme of heroin, morphine, and codeine in hair.

Morphine and codeine occur naturally in poppy seed, and as a result, urine specimens from poppy seed consumers resemble those obtained from heroin users. Thus, the presence of morphine alone is not a reliable indicator of heroin exposure, and a more specific marker, 6-AM, has been utilized. Unfortunately, detection of this metabolite in urine is limited due to its short half-life.

II. METHODS

A. Preliminary Wash/Decontamination

Prior to isolation and purification of analytes from hair, samples must be washed to remove externally applied drug, "dirt," and oils from the exterior of the hair to produce clean extracts and prevent false positive test results. Hair may be externally contaminated with drug following environmental exposure to drug through vapor or from contact with contaminated material such as hands, money, and clothing.

1. Cocaine

A wide variety of washing techniques have been utilized. Many reports describe the use of ultrasonication or heating to improve the efficiency of the wash procedure. It is doubtful, however, whether any of the washing procedures can completely remove all of the external contamination. Cone et al.[1] demonstrated this fact using two washing methods on environmentally contaminated hair samples. First, hair cut into about 1-mm segments was incubated with methanol for 15 min at 37°C. This was repeated twice, and the combined eluates were evaporated to dryness. Analysis of the residue after extraction and derivatization showed the procedure to be more than 90% effective in removing vaporized cocaine, but the procedure was inefficient in removing cocaine after the samples had been soaked in a concentrated aqueous solution of cocaine hydrochloride. The second method investigated by Cone et al.[1] was the decontamination procedure of Baumgartner et al.[2] Hair was soaked in 1 mL of ethanol for 15 min at 37°C, followed by 3 × 1 mL washes with phosphate buffer (pH 7) for 30 min, also at 37°C. Analysis showed only 81% of cocaine was removed from the hair that had been exposed to cocaine vapor, but again the sample contaminated with an aqueous cocaine solution remained extremely contaminated. Other investigators who report using ethanol as a wash solvent include Fritch et al.,[3] who used Baumgartner's method for their study of cocaine in hair, and Balabanova et al.,[4] who chose ethanol and distilled water. Methanol has also been utilized many times to wash hair, and although Martz et al.[5] report that no cocaine analytes were found in their washings (possibly because they did not heat their methanol), Cone et al.[1] and Springfield et al.[6] note that methanol also behaves as an extracting agent, removing analytes incorporated in the interior of the hair sample. Cocaine, benzoylecgonine, norcocaine and cocaethylene were all found in the wash performed by Cone et al.[1] Since the cocaine metabolites norcocaine and cocaethylene are not present in illicit cocaine, they had to have been extracted from the interior of the hair. Springfield et al.[6] attribute the lack of cocaine, relative to its metabolites benzoylecgonine and ecgonine methyl ester, found in the hair of ancient Peruvian coca-leaf chewers to the methanolic wash extracting some of the parent drug. These are not isolated reports of the extracting behavior of the methanol. Henderson et al.[7] noted that washing with 1% sodium dodecyl sulfate (SDS), methanol, and distilled water led to reduced concentrations of cocaine and its metabolites in the hair of South American coca chewers. The same procedure was used by the group in a previous report.[8]

SDS is a commonly utilized agent for washing hair. Welch et al.[9] washed the majority of their samples three times with 0.05% SDS, followed by rinsing three times with distilled water and three times with ethanol. In the same report, other wash solutions including methanol, ethanol, phosphate buffer, and SDS followed by methylene chloride were evaluated with environmentally contaminated hair samples. All of the wash procedures left detectable quantities of cocaine, although more than 90% was removed in each case. The combination of SDS and methylene chloride led to almost complete removal of cocaine, but when the method was applied to a drug user's hair sample, it was discovered that it also extracted cocaine from within the sample. Ethanol was shown to be more efficient than the phosphate buffer at washing hair, in two out of three samples. Both blond and black hair samples were used in the study. Smith and Liu,[10] in an early assay for detecting the cocaine metabolite benzoylecgonine in hair, also used a 0.05% SDS solution as a washing agent. Michalodimitrakis[11] collected hair from rats injected with cocaine and analyzed it after washing the hair in a 1% solution of SDS in saline, followed by water. Ultrasonication was utilized to facilitate the washing procedure used by Nakahara et al.,[12] who washed hair samples three times with 0.1% SDS and distilled water.

One instance where shampoo was chosen to wash the hair sample before analysis was reported by Baumgartner et al.[13] The hair had to be thoroughly rinsed to ensure no residual traces of shampoo were left, as this was found to interfere with the subsequent immunoassay. In a more recent hair analysis, performed by Martinez et al.,[14] a 10% soap solution was used to remove environmental contamination from samples supplied by male drug users. The hair was rinsed with distilled water after the wash and left to dry before extraction took place.

Organic solvents other than ethanol and methanol have been utilized for washing hair including acetone, by Moeller et al.,[15] and pentane, by Kidwell.[16] The latter was chosen to remove surface oils without extracting the drug. Welch et al.[9] conducted a study on some common hair treatments such as shampoo, conditioner, peroxide, bleach, dye, alkaline wave solution, simulated sweat (1% NaCl), and ethanol to assess their effect on cocaine content in drug abusers' hair. The study showed that dandruff shampoo had an intermediate effect on the drug levels found compared to the most drastic reduction, which occurred in hair treated with alkaline wave solution and 30% hydrogen peroxide. Slight decreases in drug were found following treatment with hair dye, 1% NaCl, and ethanol. After 30 treatments, however, the hair still retained 20 to 40% of the original cocaine content.

2. Opioids

A large number of wash procedures have been utilized for cleansing hair samples prior to opioid analysis by immunoassay and other techniques. These include: acetone, followed by water, followed by acetone; dichloromethane; methanol; ethanol; methylene chloride; and distilled water, followed by ethanol; 10% SDS in water.

Unfortunately, few comprehensive investigations of the efficacy of wash procedures with opioids in externally contaminated hair samples have been performed. A study of the effect of washing upon the morphine content in hair was carried out by Marsh et al.[17] Hair samples were subjected to the following five independent wash protocols prior to analysis by radioimmunoassay (RIA): (i) unwashed; (ii) ethanol, then distilled water; (iii) acetone-distilled water (50:50, v/v); (iv) shampoo (2% in warm distilled water), followed by warm distilled water rinse; and (v) SDS (1% in warm distilled water), followed by warm distilled water rinse. Compared to no wash at all, a mean reduction in morphine content among the subjects ranged from 19 to 27%. Although the results of all wash procedures produced a decrease in morphine content, the investigators were unable to determine statistically which procedure demonstrated the most efficient removal of drug. It was also noted, as with cocaine studies, that hair samples washed with shampoo must be thoroughly rinsed since detergents may interfere with immunoassay procedures.

A similar study with methadone in hair was performed by Marsh and Evans.[18] Drug-free hair samples were contaminated with methadone through soaking or by rubbing a solution of methadone along the outer surface of the hair. The hair was then washed with the three following independent protocols: (i) methanol; (ii) acetone, followed by distilled water; and (iii) dodecyl sulfate (1% in distilled water), followed by distilled water. The results of the wash procedures were variable; depending on the mode of contamination (soaking or surface exposure), the effectiveness of the wash procedure varied from completely ineffective to very effective.

B. Isolation of Analytes

Before cocaine and opioid analytes can be assayed, they must be released from the matrix where they are bound. Generally, isolation of the analytes is performed by direct solvent extraction with alcohol, or by acid, base, or enzymatic digestion of hair. The extreme basic conditions that are required for the complete dissolution of hair, however, will hydrolyze cocaine, heroin, and 6-AM.

1.   Cocaine

In the first report of cocaine analysis in hair, Valente et al.[19] compared the isolation of cocaine from hair with dilute acid (0.1 M HCl), dilute base (0.1 M NaOH), and methanol. Compared to methanol and dilute acid, isolation in base produced unsatisfactory results due to low recovery. By heating pulverized hair samples under reflux in methanol at 45 to 60°C, Valente et al.[19] recovered 84% of cocaine (and cocaine analytes) from hair. Baumgartner et al.[13] developed a similar method with ethanol rather than methanol.

Digestion of hair with acid is the most widely used procedure for liberating cocaine. This type of isolation, first described by Valente et al.[19] in their comparison with other isolating methods (sodium hydroxide, methanol, and direct RIA on hair), showed that overnight incubation with 0.1 M HCl at 45°C gave complete recovery of cocaine. Balabanova et al.[4] and Balabanova and Homoki[20] adopted Valente's procedure in their work; crushed hair samples were incubated overnight at 45°C with 0.1 M HCl. The acid extracts were then neutralized with 1 M NaOH and buffered with a pH 7.4 phosphate buffer. Despite Harkey et al.[8] reporting that hair samples were not completely dissolved by this procedure, even with 1 M HCl, and that enzymatic digestion is the only method that will dissolve the hair properly without degrading cocaine, Welch et al.[9] found the method to be as efficient as enzyme digestion. Nakahara and Kikura,[21] however, support the view of Harkey et al.[8] Other authors have utilized the hydrochloric acid procedure, but with different acid strengths ranging from 0.25 to 0.5 M, or, like Cone et al.,[1] have utilized 0.05 M H$_2$SO$_4$ at 37°C.

Enzymatic digestions utilize purified enzyme preparations such as proteinase K, β-glucuronidase-aryl-sulfatase, collagenase, and pronase to digest and release the target compounds. Buffered solutions of washed hair, which have either been cut into small pieces or pulverized, are usually incubated between 25 and 40°C overnight with the chosen enzyme preparation. Cassani and Spiehler[22] suggest that enzyme digestion is the best method when immunological-based assays are to be used for detection, since the hair-digesting method must not denature the antibody proteins of the immunoassay reagents. If acid digestion solutions are used, they must be neutralized before immunoassay.

There have been a few reports in the literature where hair has been analyzed directly for cocaine analytes, without analyte isolation. The earliest report was by Valente et al.,[19] who compared direct incubation of hair with the antiserum in an RIA test with acid, base, and methanol isolation. Although cocaine analytes were detected using this method, only 31% were recovered. In 1993, Kidwell[16] and Welch et al.[9] noted that cocaine analytes could be observed in hair using tandem mass spectrometry (MS/MS) with no prior isolation. The washed samples were rapidly heated and the vaporized material analyzed. However, Welch et al.[9] report that unambiguous results were only obtained when the hair samples were cryogenically powdered, which required more effort than preisolation of the analytes.

2.   Opioids

In the first published report of opioids in hair by Baumgartner et al.,[23] hair samples were pulverized and heated under reflux in methanol for 2 h. The methanol was removed, evaporated, and the extract was reconstituted in phosphate buffer. The extract was subjected to RIA analysis. Shortly thereafter, Valente et al.[19] compared this technique using other solutions including dilute acid (0.1 M HCl), dilute base (0.1 M NaOH), methanol, water, pH 7.4 buffer, and sodium chloride solution. Dilute acid and base and methanol produced the best results.

In studies performed by Arnold[24] and Sachs and Arnold,[25] hair samples were treated with a solution of 1 M NaOH and boiled until the hair disintegrated. The hydrolyzed sample was neutralized with 1 M HCl. The same approach was employed by Kintz and Mangin.[26] Hair digests were analyzed directly by the enzyme multiplied immunoassay technique (Emit®, Syva Corporation, Palo Alto, CA) and fluorescence polarization immunoassay (FPIA, Abbott Laboratories, Abbott Park, IL).

Similarly, Marsh et al.[17] and Marsh and Evans[27] incubated hair samples in 0.1 M HCl and neutralized with 1 M NaOH before diluting further with pH 7, 1 M phosphate buffer. The final preparation was assayed by the Diagnostic Products Corporation (DPC) morphine-specific and methadone RIAs. Franceschin et al.[28] also incubated samples overnight in a solution of 0.1 M HCl. After neutralizing with 1 M NaOH, analytes were extracted for FPIA analysis.

Offidani et al.[29] compared chemical hydrolysis with enzymatic hydrolysis utilizing hair samples obtained from heroin users. Hair samples were subjected to incubation in a solution of 1 M NaOH, then neutralized with hydrochloric acid and pH 7, 1 M phosphate buffer. Samples were also treated with a 1-mg/mL pronase solution in 0.05 M Tris buffer for 24 h at 39°C, followed by neutralization with pH 7, 1 M phosphate buffer. Both procedures yielded comparable results for morphine by RIA.

For ease of analysis, Goldberger et al.[30] and Cone et al.[31] utilized methanol to isolate cocaine and heroin analytes. Cut or pulverized hair samples were incubated at 37°C for 18 to 20 h in methanol. A magnetic stir bar was used to provide constant agitation. After incubation, the methanol was collected and evaporated. Although heroin is not generally stable in alcoholic solutions, the rate of hydrolysis to 6-AM was <10%.

In a report by Nakahara et al.,[32] the extraction efficiency of various methods for 6-AM and morphine was compared. Finely cut hair samples were placed into solutions of methanol, 0.1 M HCl, methanol-5 M HCl (20:1), helicase, or methanol-trifluoroacetic acid (TFA) (9:1). Extraction was performed overnight following ultrasonication for 1 h, except for the methanol extract, which was sonicated for 14 h. Methanol-TFA was found to be the best solvent for extracting 6-AM and morphine with minimal hydrolysis and maximum extraction efficiency. Nakahara et al.[32] noted that heroin was not detected in heroin users' hair by this method, possibly due to hydrolysis to 6-AM. The extraction rates of 6-AM and morphine from heroin users' hair with methanol-TFA reached a plateau after 8 to 10 h.

Hair samples obtained from surgery patients administered fentanyl during anesthesia were assayed by Wang et al.[33] Samples were extracted with methanol overnight at 40°C. The methanol extract was evaporated, reconstituted in 0.1 M citrate buffer, and assayed by DPC fentanyl RIA.

C. Extraction

Generally, after isolation and before analysis, cocaine analytes and opioids are extracted with either liquid-liquid extraction (LLE) or solid-phase extraction (SPE) methods. Immunoassay detection techniques, however, often use the crude isolation solution. Baumgartner et al.[2] noted that while isolating and extracting is the best procedure for forensic cases, it is not cost effective for routine medical or employee drug screening. He introduced an analytical procedure where the antibody subsequently used in an RIA assay acts as a specific analyte extracting agent.

Recent advances in copolymeric bonded-phase SPE cartridges has improved the efficiency and ease of analyte extraction. There are many other advantages to SPE over LLE, which have been reported by Krishnan and Ibraham.[34] Briefly, SPE is faster than LLE, requires less solvent, produces cleaner extracts, gives better reproducibility of results, and can be used with small samples. Harkey et al.[8] also report that SPE lends itself to higher recoveries for the combination of cocaine analytes, more than LLE yields. Commonly utilized extraction columns include Chromabond™ (Machery-Nagel, Dülmen, Germany); Bond Elut Certify™ (Varian, Harbor City, CA); Clean Screen® (United Chemical Technologies, Bristol, PA), and Extrelut® (Merck, Darmstadt, Germany). A variety of solvent mixtures including chloroform:isopropanol:n-heptane (50:17:33; v/v) and toluene:heptane:isoamyl alcohol (70:20:10; v/v/v) have been utilized for LLE. The commercially available LLE extract tube, Toxi-Tube A® (Ansys, Laguna Hills, CA), has also been utilized.[5,28]

Both Valente et al.[19] and Baumgartner et al.[13] note that the popular technique of pulverizing hair samples to a fine powder with a pestle and mortar does not increase the amount of isolated cocaine analyte. However, pulverizing is commonly used by many laboratories performing mass-production hair analyses.

Until recently, hair assays for morphine utilized analytical procedures which resulted in the chemical hydrolysis of heroin and 6-AM to morphine. Newer assays developed for opioids in hair utilizing conditions designed for the stabilization of heroin and 6-AM during analysis have been reported by Goldberger et al.[30,35] The first procedure developed for the purification and removal of heroin, 6-AM and morphine from hair extracts utilized LLE techniques.[30] Hair wash and incubation fractions were dissolved in water and mixed with saturated sodium bicarbonate buffer (pH 8.4) and a solution of toluene-heptane-isoamyl alcohol (70:20:10; v/v/v). After mixing and centrifugation, the final extract solvent was removed, evaporated, and derivatized. Improvements were made to the heroin assay by replacing the LLE procedure with SPE.[35] Hair wash and incubation fractions were added to the conditioned cartridges and the cartridges were washed with deionized water, pH 6 acetate buffer, and acetonitrile. The cartridges were aspirated to dryness and treated with ethyl acetate:diethylamine (98:2; v/v). The solvent extract was divided into two aliquots of equal volume and the individual portions were evaporated. Acetonitrile was added to one set of extracts and N-methyl-bis-trifluoroacetamide (MBTFA) was added to the other set.

The SPE assay was further modified by Cone et al.[31] by combining the assay for heroin and its metabolites with cocaine and its metabolites into a single procedure. Hair wash and incubation fractions were added to a conditioned Clean Screen SPE cartridge. The cartridges were washed with deionized water, pH 4 acetate buffer, and acetonitrile. The cartridges were aspirated to dryness and the analytes were eluted with a solution of methylene chloride:2-propanol (80:20) with 2% ammonium hydroxide. The solvent extract was evaporated and derivatized.

Sachs and Raff[36] described several methods of cocaine and heroin analyte extraction. Washed hair samples were pulverized and subjected to either enzymatic or base dissolution, or methanolic extraction. Following initial preparation, samples were extracted by Extrelut, Bond Elut Certify, Chromabond C-18, or supercritical fluid extraction.

A relatively new technique utilizing subcritical fluid extraction (SFE) was developed for the analysis of opioids in hair by Edder et al.[37] Hair samples were washed, pulverized, and extracted directly without further treatment by SFE with carbon dioxide-methanol-triethylamine-water (85:6:6:3, v/v/v/v). The procedure was compared to other techniques including acidic and basic hydrolysis, methanolic solvent, and enzyme digestion, followed by SPE. According to Edder et al.,[37] SFE was easy to perform and rapid (quantitative extraction within 30 min), whereas other techniques required several hours to complete. In addition, the SFE procedure resulted in minimal hydrolysis of 6-AM to morphine.

D. Analysis

1. Cocaine

RIA was the first technique utilized for the detection of cocaine in hair and was reported in 1981 by Valente et al.,[19] who used hair samples from known users. This technique was subsequently employed throughout the 1980s by other investigators including Balabanova et al.[4] and Baumgartner et al.,[13] who successfully detected cocaine analytes in 100% of their subjects, all of whom acknowledged cocaine use. The only commercially available RIA kit at the time was manufactured by Abuscreen™ (Hoffmann La Roche, Nutley, NJ). Smith and Liu[10] applied the RIA technique to an alleged sexual assault victim's hair, on which the results indicated cocaine use. This was the first time that results of hair drug testing were utilized in a U.S. court of law.

Because of its sensitivity and ease of use, RIA is generally the technique of choice for initial screening of hair extracts. However, immunoassays often cross react with other structurally related compounds. The first immunoassays employed to detect cocaine in hair used a benzoylecgonine antiserum, an immunoassay which not only reacts with benzoylecgonine, but displays inevitable cross reactivity to analogs and homologues of benzoylecgonine. These include cocaine, ecgonine, ecgonine methyl ester, norcocaine, benzoylnorcocaine, and norecgonine. The relative reactivity of the antiserum with each substance varied, but as a consequence, the concentration of cocaine measured by immunoassay techniques was the sum of the analytes present. The results of such analyses were expressed in benzoylecgonine equivalents or reported as the sum concentration of all cross-reactive compounds. When it was discovered, however, that hair generally contains more cocaine than benzoylecgonine, an immunoassay with an antiserum that exhibits much higher affinity for cocaine than for benzoylecgonine was used (Coat-A-Count™, DPC, Los Angeles, CA). This DPC RIA is an extremely sensitive assay for cocaine and exhibits very little cross reactivity to benzoylecgonine. For example, Klein et al.[38] utilized two different immunoassay kits to separately quantitate benzoylecgonine and cocaine. The kit used to detect benzoylecgonine demonstrated 4% cross reactivity with cocaine, while the one best suited to detect exclusively cocaine demonstrated only 0.5 and 6% cross reactivity to benzoylecgonine and ecgonine methyl ester, respectively.

RIA, although not sufficient by itself to fulfill the requirements of forensic drug testing including potential legal challenge, is suitable for preliminary testing (screening) of hair samples. It is simple, fast, convenient, inexpensive, sensitive, and requires less sample preparation than other methods. RIA is also ideal for screening large numbers of hair samples. A confirmatory test utilizing gas chromatography/mass spectrometry (GC/MS), and more recently, tandem mass spectrometry (MS/MS) is required for confirmation of positive RIA results. An example of RIA being used as a screening method for cocaine use is given by Martinez et al.[14] The investigators tested hair samples from men suspected of cocaine abuse. All of the hair samples that tested negative for cocaine (using benzoylecgonine as the target analyte) by RIA also tested negative by GC/MS. A good quantitative correlation was also demonstrated between the two detection techniques; samples that showed high quantities of cocaine and benzoylecgonine by RIA were also found to contain the highest quantities of analyte by GC/MS. Employing an RIA, Reuschel and Smith[39] tested for benzoylecgonine in hair samples collected anonymously from 48 jail detainees. Of the 22 samples that tested positive by RIA, all were confirmed positive by electron-impact (EI) GC/MS.

Balabanova and Homoki[20] were the first to utilize GC/MS as a qualitative tool for the determination of cocaine in human hair. The ionization mode was EI and quantitation was achieved by RIA. In 1991, Harkey et al.[8] described a more sensitive method for the simultaneous detection of cocaine, benzoylecgonine, and ecgonine methyl ester in human hair using chemical ionization (CI) GC/MS. Both EI and CI modes were evaluated in the report with the conclusion that CI produced greater sensitivity than EI. The protonated molecular ion generated with CI could also be used for quantitation. The same GC/MS (CI) procedure was used later by Henderson et al.[7] to analyze the hair of coca chewers. In addition to cocaine, benzoylecgonine, and ecgonine methyl ester, Cone et al.[1] were able to identify cocaethylene and nor-cocaine, two unique analytes of cocaine using GC/MS. The presence of these compounds, coupled with cocaine, produces convincing evidence of cocaine use. GC/MS was also used in a comprehensive assay, developed by Cone et al.,[31] for the detection of cocaine and opiates in a single procedure. A typical selective ion chromatogram of a hair extract obtained from a subject who self-administered both cocaine and heroin, and a standard is shown in Figure 3.

Depending upon the chromatographic system utilized, cocaine analytes (especially polar metabolites) may require derivatization prior to GC/MS analysis. Commonly utilized derivatization reagents include: N,O-bis(trimethylsilyl)trifluoroacetamide (BSTFA), pentafluoropropionic acid anhydride/hexafluoroisopropanol (PFPA)/(HFIP), N-methyl-N-trimethylsilyltrifluoroacetamide (MSTFA), pentafluoropropionic acid anhydride/pentafluoropropanol (PFPA)/(PFP), N,O-bis(trimethylsilyl)acetamide (BSA) and N-methyl-N-($tert$-butyldimethylsilyl)-trifluoroacetamide (MTBSTFA). For ease of analysis and stability of the derivatives, the authors prefer BSTFA with 1% trimethylchlorosilane (TMCS).

MS/MS has been utilized by a few investigators. It can be used for direct analysis of hair with minimal or no sample preparation. Direct vaporization of prewashed hair in a solids probe, followed by CI of the vaporized sample components, was reported by Kidwell.[16] In this manner, cocaine, benzoylecgonine, and ecgonine were detected and quantified using the peak areas of external standards. Welch et al.[9] concluded that the results of MS/MS from short hair segments vaporized directly were unsatisfactory and unreliable; hair from drug users could not always be distinguished from blank hair samples. Samples that had been cryogenically pulverized, however, produced acceptable results, and hair extracts analyzed by GC/MS/MS produced results comparable to GC/MS detection.

High-pressure liquid chromatography (HPLC) with fluorescence detection, ion-trap mass spectrometry (ITMS), and capillary electrophoresis (CE) with ultraviolet detection, three novel detection techniques for the analysis of cocaine in hair, have been evaluated by Tagliaro et al.[40,41] and Traldi et al.[42] The HPLC technique was highly sensitive, capable of detecting 0.015 ng/mg of cocaine in hair. The CE method was sensitive and highly selective. ITMS analysis of hair readily demonstrated the presence of cocaine in hair, but cocaine metabolites were more difficult to identify.

A)   Hair Extract

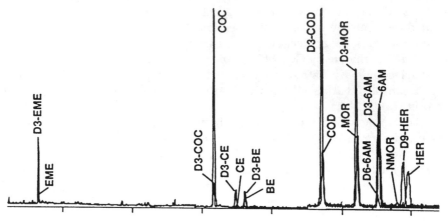

B)   Standards

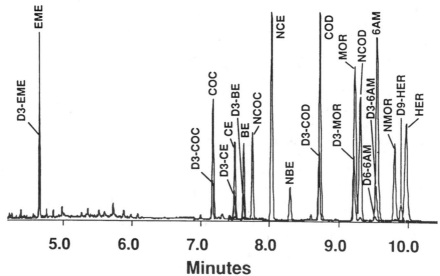

Figure 3
Selective ion chromatogram recordings of hair and standard extracts. Panel A represents an extract of a cocaine and heroin user's hair. Panel B represents an extract of cocaine, heroin, and metabolites added to drug-free control hair. (Abbreviations: COC, cocaine; BE, benzoylecgonine; EME, ecgonine methyl ester; CE, cocaethylene; NBE, norbenzoylecgonine; NCOC, norcocaine; NCE, norcocaethylene; COD, codeine; MOR, morphine; 6AM, 6-acetylmorphine; NCOD, norcodeine; NMOR, normorphine; HER, heroin). (Adapted from Cone, E. J., Darwin, W. D., and Wang, W.-L., *Forensic Sci. Int.*, 63, 55, 1993. With permission.)

Table 1 provides a summary of methods utilized for the analysis of cocaine analytes in hair.

2.   *Opioids*

Opioids in hair are readily detected with a variety of commercial opioid immunoassays including RIA, FPIA, and Emit. Most opioid immunoassays are formulated to detect morphine and morphine glucuronide, but also exhibit significant cross

TABLE 1.

Summary of Methods Utilized for the Analysis of Cocaine Analytes in Hair

Investigator	Wash procedure	Isolation procedure	Extraction technique	Detection technique	Ref.
Cone et al.	Methanol	0.05 M H₂SO₄	SPE (Clean Screen)	GC/MS (TMS deriv)	1
Fritch et al.	Ethanol; Phosphate buffer	0.1 M HCl	LLE (chloroform:isopropanol: isoamyl alcohol; 90:9.5:0.5)	RIA (Coat-a-Count); GC/MS (TMS deriv)	3
Balabanova et al.	Water/ethanol	0.1 M HCl	—	RIA (Abuscreen)	4
Martz et al.	Methanol	0.5 M HCl	LLE (Toxi-Tube A)	MS/MS	5
Springfield et al.	Methanol	0.05 M H2SO4	SPE	GC/MS (TMS deriv)	6
Henderson et al.	1% SDS/water/methanol	Enzymatic digestion	SPE	GC/MS (dimethylsilyl deriv)	7
Harkey et al.	1% SDS/water/methanol	Dithiothreitol, proteinase K	SPE (Bond Elut Certify)	GC/MS (dimethylsilyl deriv)	8
Welch et al.	0.05% SDS/water/ethanol	0.1 M HCl (or) dithiothreitol, proteinase K	SPE (Bond Elut Certify)	GC/MS (TMS deriv); GC/MS/MS (TMS deriv)	9
Smith and Liu	0.05% SDS	Ethanol	—	RIA (manufacturer not specified)	10
Michalodimitrakis	1% SDS/water	1.5 M NaOH	LLE (isopropanol:chloroform;1:9)	RIA (Abuscreen)	11
Nakahara et al.	0.1% SDS/water	Proteinase K	SPE (Bond Elut Certify)	GC/MS (pentafluoropropyl EME-deriv; HFIP BE-deriv)	12
Baumgartner et al.	Shampoo/water	Ethanol	—	RIA (Abuscreen)	13
Martinez et al.	Soap/water	0.1 M HCl	—	RIA (Coat-a-Count)	14
Moeller et al.	Water/acetone	β-Glucuronidase-aryl-sulfatase	SPE (Chromabond C-18 column)	GC/MS (pentafluoropropyl deriv)	15
Kidwell	Pentane	—	—	MS/MS	16
Valente et al.	Not indicated	0.1 M NaOH (or) 0.1 M HCl (or) methanol (or) direct RIA	—	RIA (Abuscreen)	19
Balabanova and Homoki	Not indicated	0.1 M HCl	Chloroform (for GC/MS only)	RIA (manufacturer not specified); GC/MS	20
Nakahara and Kikura	0.1% SDS/water	Proteinase K	SPE (Bond Elut Certify)	GC/MS (pentafluoropropyl EME-deriv; HFIP BE-deriv)	21
Cone et al.	Methanol	Methanol	SPE (Clean Screen)	GC/MS (TMS deriv)	31
Klein et al.	Ethanol	Methanol	—	RIA (Coat-a-Count and Abuscreen)	38
Reuschel and Smith	0.05% SDS	Ethanol	—	RIA (Abuscreen); GC/MS (TMS deriv)	39
Ferko et al.	Water	0.1 M HCl	SPE (C-18 column)	GC/MS (TMS deriv)	78
Koren et al.	Ethanol	Methanol	—	RIA (Abuscreen and Coat-a-Count)	79
DiGregorio et al.	Water	0.1 M HCl	SPE (C-18 column)	GC/MS (TMS)	80

Note: TMS is trimethylsilyl.

reactivity with other opioid compounds such as codeine, hydromorphone, and hydrocodone. Other RIAs are also available from DPC including a highly specific free morphine assay, a 6-AM assay, a methadone assay, and a fentanyl assay.

The first published report of opioids in hair was by Baumgartner et al.[23] in 1979. In a study of morphine-treated mice and heroin users, morphine was tested by Abuscreen morphine RIA. Additional studies with RIA for morphine were performed by Valente et al.,[19] Offidani et al.,[29] and Püschel et al.[43] Sachs and Arnold[25] compared the results of DPC morphine-specific RIA analyses with GC/MS analyses of 50 hair samples and observed acceptable correlation.

The performance of the FPIA with negative and positive hair samples was studied by Franceschin et al.[28] Based upon an evaluation of the data, a screening cutoff concentration of 0.2 ng/mg was proposed. All results compared favorably with results of HPLC analyses.

Emit and FPIA assays were studied concurrently in a study of heroin users by Kintz and Mangin.[26] The performance of the Emit assay was not satisfactory; false negative results were observed in nearly 45% of the subjects studied. The problems were attributed to interference by color and turbidity of the hair extract. The results of the FPIA were acceptable, corresponding well to results obtained with GC/MS (sum of morphine and codeine concentrations). In a subsequent report, Kintz et al.[44] further studied the FPIA with samples obtained from a medical examiner's office. In a total of 40 subjects, 14 were positive for opioids in hair by FPIA and GC/MS. The FPIA results were similar to concentrations previously reported in the literature.

Balabanova and Wolf[45,46] reported the use of RIA for the analysis of methadone in hair. Although cross-reactivity data were not provided, it was stated that the results obtained by RIA reflected the sum concentration of methadone and its metabolites. The results were compared with GC/MS, and acceptable correlation was observed.[47]

Marsh et al.[17] and Marsh and Evans[27] utilized the DPC morphine-specific RIA to study the effectiveness of wash procedures and self-reporting of drug use. The assay was found to be highly specific, accurate, and precise. Evaluation of hair from drug-free controls produced results <0.3 ng/mg. In the study, the results of segmental hair tests for morphine for a period of 15 months were used to challenge a subject's declaration of heroin abstinence. In similar studies, Marsh and Evans[18] evaluated the effectiveness of the DPC methadone RIA. The assay was also found to be highly specific, accurate, and precise.

In a study by Wang et al.,[33] hair samples from surgery patients were assayed for fentanyl by DPC RIA. Fentanyl was detected in hair samples following moderate to high doses of fentanyl. The concentration in hair did not correlate with dose.

HPLC is less commonly used by laboratories performing hair tests. HPLC with fluorescence detection has been evaluated by Marigo et al.[48] Following LLE, extracts were derivatized with a solution of dansyl chloride. The derivatized extract was back-extracted into toluene, evaporated, and reconstituted in chromatographic eluent. The limit of detection was <1 ng/mg. Also, HPLC with coulometric detection was evaluated by Kintz[49] for the analysis of buprenorphine. The detection limits were about 0.01 and 0.02 ng/mg for buprenorphine and norbuprenorphine, respectively.

Analysis of heroin does not require derivatization; however, analysis of other opioids requires derivatization prior to GC/MS analysis. Commonly used derivatization reagents include BSTFA with 1% TMCS, PFPA, MSTFA, acetic anhydride (AA), trifluoroacetic anhydride (TFAA), and N-methyl-bis-trifluoroacetamide (MBTFA). For ease of analysis and stability of the derivatives, the authors prefer BSTFA with 1% TMCS or MBTFA.

In the original heroin assay by Goldberger et al.,[30] heroin, 6-AM, morphine, and codeine were extracted by LLE and assayed simultaneously by GC/MS. The performance of the assay including analyte recovery and stability was not optimized, and the assay was limited by interference. Selective ion chromatograms of hair and standard extracts are illustrated in Figure 4. In an improved procedure by Goldberger et al.,[35] two sets of SPE extracts were assayed independently by GC/MS; one set for heroin, and other set for the trifluoroacetyl derivatives of 6-AM and morphine. The extraction process, in combination with the use of aprotic solvents, mild elution solvents, and an enzyme inhibitor, provided maximum analyte stability for the recovery of heroin from hair with minimal (<5%) hydrolysis. The procedure was modified further by Cone et al.[31] for the analysis of cocaine, heroin, and metabolites in hair.

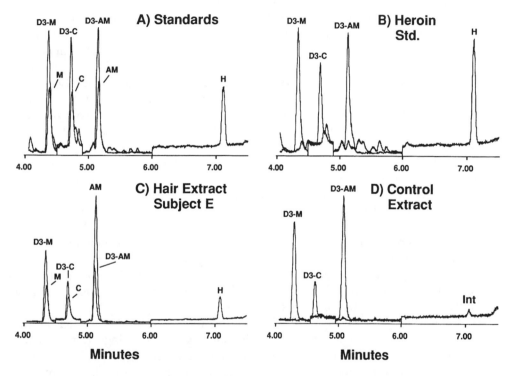

Figure 4

Selective ion chromatogram recordings of standard and hair extracts. Panel A represents an extract of morphine (M), codeine (C), 6-acetylmorphine (AM), and heroin (H) and internal standards (D_3-M, D_3-C, and D_3-AM) from drug-free control hair. Panel B represents an extract of heroin and internal standards from drug-free control hair. Panel C represents an extract of a heroin user's hair. Panel D represents an extract of a drug-free control subject. Note the interfering (Int) peak in the control extract at the retention time for heroin (Panel D). (From Goldberger, B. A., Caplan, Y. H., Maguire, T., and Cone, E. J., *J. Anal. Toxicol.*, 15, 226, 1991. With permission of Preston Publications, a division of Preston Industries, Inc.)

Methods utilizing newer techniques including ITMS, CE, and GC/MS/MS have recently been developed. ITMS analysis by Curcuruto et al.[50] and Traldi et al.[42] readily demonstrated the presence of morphine in hair. Free zone CE with amperometric detection was sensitive and highly selective.[41] A novel procedure by Polettini et al.[51] describes the treatment of methanol-washed samples with a silylating solution, followed by injection into GC. The silylating reagent was prepared by mixing MSTFA with dithioerythritol and ammonium iodide. Analysis of heroin, 6-AM, acetylcodeine, morphine, and codeine was performed by GC/MS/MS using multiple selected reaction monitoring. Under the conditions described, recovery of analytes from hair was comparable to prolonged methanolic incubation, and hydrolysis of heroin to 6-AM was <10%. The limit of detection was approximately 25 pg/mg.

Table 2 provides a summary of methods utilized for the analysis of opioids in hair.

III. COCAINE FINDINGS

Despite the rapid degradation of cocaine during metabolism and the predominance of benzoylecgonine in most biological tissue, cocaine is the primary analyte found in hair. Other cocaine metabolites (illustrated in Figure 1) found in hair include benzoylecgonine, ecgonine methyl ester, ecgonine, cocaethylene, norcocaine, norcocaethylene, and anhydroecgonine methyl ester. This observation is also true when hair is allowed to soak in an aqueous solution containing equal quantities of cocaine and benzoylecgonine and is attributed to the greater affinity of "binding sites" to cocaine rather than to the major metabolite, benzoylecgonine. A wide variety of assay techniques have been utilized to detect cocaine analytes in hair. Most of the methods employ a preliminary wash, isolation, and extraction of the analytes from the keratinous matrix of hair, followed by chromatographic and spectrometric analysis.

Many cocaine users escape detection by conventional urine analysis since cocaine analytes are usually metabolized and excreted within 48 h of ingestion. Application of hair analysis with cocaine users allows for an increased detection window between the time of the last drug administration and drug detection because cocaine analytes are stable once incorporated into hair compared to their stability in biological fluids. Mieczkowski et al.[52] showed that analysis of hair from jail arrestees also identified infrequent users of cocaine and that self-reported cocaine use was generally underreported. Studies conducted by Hindin et al.[53] comparing self-reported drug use by patients in a drug rehabilitation center and RIA analysis of hair indicate that RIA is relatively accurate for identifying frequent or chronic users; however, it is not reliable, especially for low to moderate users of cocaine.

Hair analysis can be useful for collaborating the cause of death, such as that reported by Martz et al.,[5] when the subject dies subsequent to the ingestion of a drug, and by which time blood and urine tests would be negative. In general, hair analysis can be used whenever blood or urine samples are not collected in time to indicate drug use.

Cocaine use among young women is an increasing problem. Abuse of drugs often leads to neonatal complications. For example, children born to women who used cocaine during pregnancy often suffer from prematurity, growth retardation and microcephaly, and other developmental and medical conditions. Self-reported drug use is generally unreliable, and urine analysis for cocaine is often unable to detect periodic drug users. Analysis of maternal or neonatal hair can be used to identify maternal drug use during gestational development. In a report by Graham et al.,[54] an RIA of maternal hair produced positive test results for benzoylecgonine in seven pregnant women who all admitted cocaine use during pregnancy. Hair from all seven infants was also positive for benzoylecgonine, although as the babies grew older their hair tested negative, corresponding to loss of fetal hair. Other studies comparing maternal hair analysis and cocaine use have been conducted by Callahan et al.,[55] Forman et al.,[56] Marques et al.,[57] DiGregorio et al.,[58] and Grant et al.[59] In all studies, hair samples of neonates that test positive for cocaine analytes indicate intrauterine exposure. It has also been observed that the amount of cocaine used during the last trimester is reflected in the amount of cocaine found in the infant's hair.

Cocaine analytes can be easily detected in hair using a number of different sample preparation methods and detection techniques. Results of the analyses must be regarded with caution, however, due to the possibility that an environmentally contaminated sample can still produce detectable amounts of cocaine analytes, despite prewashing the hair, leading to false positive test results. Baumgartner et al.[2] described a "washout kinetics" protocol which uses the extract-to-wash ratio to establish with reasonable certainty whether a drug found in hair is derived from interior or exterior sources. Cone et al.[1] proposed that norcocaine and cocaethylene detected in a hair sample are sufficient evidence of cocaine use. This suggestion is based on the assumption that neither analyte is able to produce environmental contamination, since they are either absent in illicit cocaine or present in only trace amounts and can only arise through metabolism of the parent drug. However, a rare incident that occurred in Canada where illicit cocaine was found to contain 20% cocaethylene has been reported by Janzen.[60]

Differences in hair texture, inherent to race or hair color, sex, or diseased states of the donor can lead to varying amounts of cocaine that is incorporated into hair and subsequently detected. Hair that is damaged will also behave in a unique way and could lead to unexpected results. Harkey et al.[8] note that there is little correlation between the concentration of cocaine in hair and self-reported drug use; the highest concentration of cocaine was found in a moderate user, while no cocaine was detected in the hair of a heavy user of the drug.

It is still unclear how cocaine analytes are incorporated into hair. Throughout the literature, investigators have consistently reported finding higher concentrations of cocaine in hair than of the other analytes. Cone et al.[1] found the mean ratio of cocaine to benzoylecgonine was 10.5 in cocaine users while ecgonine methyl ester was detected in only trace amounts. Henderson et al.[7] found cocaine to be present at concentrations approximately five times higher than benzoylecgonine and twelve times higher than ecgonine methyl ester. Moeller et al.[15] demonstrated the mean ratio of cocaine to benzoylecgonine was three, cocaine to ecgonine methyl ester being six. Nakahara and Kikura[21] report similar values to Cone with ratios of cocaine to benzoylecgonine and ecgonine methyl ester of 10:1 and 20:1, respectively. These results are surprising, since benzoylecgonine and ecgonine methyl ester have much longer plasma half-lives compared to cocaine, and are present in higher concentrations than cocaine in blood. In an attempt to help elucidate drug incorporation into hair, Nakahara and Kikura[21] experimented with trideuterated benzoylecgonine and ecgonine methyl ester and their incorporation into rat hair. It was found that rats dosed with cocaine and with the deuterated analytes simultaneously had more cocaine than benzoylecgonine or ecgonine methyl ester in their fur, and they had higher concentrations of undeuterated benzoylecgonine and ecgonine methyl ester than of the deuterated analogs, particularly the former. This suggested that most of the benzoylecgonine that is detected in hair analyses arises from hydrolysis of cocaine incorporated into hair, and the incorporation rates of benzoylecgonine and ecgonine methyl ester, themselves, into hair are very low. Based upon this information, the theoretical distribution of cocaine and its metabolites in hair is illustrated in Figure 5. It was previously thought that drugs incorporated into hair were chemically changed while still in the blood and that they entered the hair by diffusion from the blood into growing cells at the base of the hair follicle. A report by Henderson[61] proposes a model in which drugs are incorporated into hair at various times during the hair growth cycle via multiple sites and multiple mechanisms.

Table 3 provides a summary of cocaine analyte results in hair.

TABLE 2.
Summary of Methods Utilized for the Analysis of Opioids in Hair

Investigator	Wash procedure	Isolation procedure	Extraction technique	Detection technique	Ref.
Welch et al.	0.05% SDS/water/ethanol	0.1 M HCl (or) dithiothreitol, proteinase K	SPE (Bond Elut Certify)	GC/MS (TMS deriv) GC/MS/MS (TMS deriv)	9
Marsh et al.	None (or) ethanol/water (or) acetone-water (50:50) (or) 2% shampoo/water (or) 1% SDS/water	0.1 M HCl	—	RIA (Coat-a-Count)	17
Marsh and Evans	Methanol (or) acetone/water (or) 1% SDS/water	0.1 M HCl	—	RIA (Coat-a-Count)	18
Valente et al.	Not indicated	0.1 M NaOH (or) 0.1 M HCl (or) methanol (or) water (or) pH 7.4 buffer (or) NaCl (or) direct RIA	—	RIA (Abuscreen)	19
Baumgartner et al.	Detergent/water	Methanol	—	RIA (Abuscreen)	23
Arnold	Acetone/water/acetone	1 M NaOH	—	RIA (Coat-Count)	24
Kintz and Mangin	Dichloromethane	1 M NaOH	None (Emit and FPIA) (or) LLE (chloroform:isopropanol: heptane; 50:17:33)	Emit FPIA GC/MS (TMS deriv)	26
Marsh and Evans	1% SDS/water	0.1 M HCl	—	RIA (Coat-a-Count)	27
Franceschin et al.	Not indicated	0.1 M HCl	LLE (Toxi-Lab A)	FPIA	28
Offidani et al.	Methylene chloride	1 M NaOH (or) pronase	—	RIA (Coat-a-Count)	29
Goldberger et al.	Methanol	Methanol	LLE (toluene:heptane:isoamyl alcohol; 70:20:10)	GC/MS (TFA deriv)	30
Cone et al.	Methanol	Methanol	SPE (Clean Screen)	GC/MS (TMS deriv)	31

Reference	Wash	Hydrolysis	Extraction	Analysis	No.
Nakahara et al.	0.1% SDS/water	Methanol:trifluoroacetic acid (9:1)	SPE (Bond Elut Certify)	GC/MS (TMS deriv)	32
Wang et al.	Methanol	Methanol	—	RIA (Coat-a-Count)	33
Edder et al.	Water/0.1 M HCl/methanol	None	Subcritical fluid extraction	GC/MS (propionyl deriv)	37
Püschel et al.	Not indicated	NaOH	—	RIA (Abuscreen)	43
Kintz et al.	Ethanol	1 M NaOH	LLE (chloroform:isopropanol: heptane; 50:17:33) (for GC/MS only)	FPIA / GC/MS (TFA deriv)	44
Balabanova and Wolf	Water/ethanol	0.1 M HCl	—	RIA (Biermann)	45,46
Marigo et al.	Ethyl ether/0.01 M HCl	0.1 M HCl / 0.6 M HCl / 0.6 M NaOH	LLE (dichloroethane:dichloromethane:heptane; 18:18:64)	HPLC with fluorimetric detection (dansyl chloride deriv)	48
Kintz	Dichloromethane, 2 min	0.1 M HCl	LLE (toluene)	HPLC with coulometric detection	49
Goldberger et al.	Methanol	Methanol	SPE (Clean Screen)	GC/MS (TFA deriv)	64
Nakahara et al.	0.1% SDS/water	10% HCl (morphine) Methanol, 40C, 14 h (6-acetylmorphine)	LLE (chloroform:isopropanol w/ NH4OH (morphine) SPE (Bond Elut Certify) (6-acetylmorphine)	GC/MS (TMS deriv)	65
Moeller et al.	Not indicated	Not indicated	Not indicated	Not indicated	66
Kintz and Mangin	Dichloromethane	1 M NaOH	LLE (chloroform:isopropanol:heptane; 50:17:33)	GC/MS (TMS deriv)	67,68
Kintz et al.	Dichloromethane	1 M NaOH	LLE (chloroform:isopropanol:heptane; 50:17:33)	GC/MS (TMS deriv)	69,70, 74
Offidani et al.	0.1% Tween 80/water/acetone	Dithiothreitol/pronase	—	RIA (Coat-a-Count)	71
Sachs et al.	Acetone	0.75 M NaOH, followed by 25% HCl	LLE (toluene:butanol; 7:3; followed by dichloromethane:propanol; 9:1)	GC/MS (heptafluorobutyryl deriv)	73
Goldberger et al.	Methanol	Methanol	SPE (Clean Screen)	GC/MS	75
Moeller et al.	Water/acetone	β-Glucuronidase-aryl-sulfatase	SPE (Chromabond C-18)	GC/MS (pentafluoropropyl)	76

Note: TMS is trimethylsilyl.

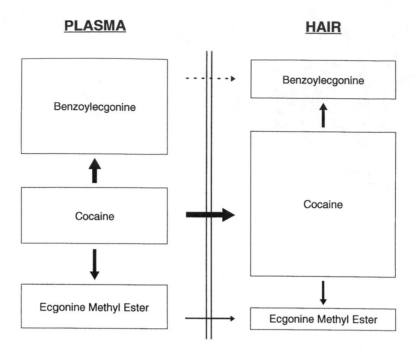

PLASMA   **HAIR**

Figure 5
Theoretical distribution of cocaine in hair. (Adapted from Nakahara, Y., Ochiai, T., and Kikura, R., *Arch. Toxicol.*, 66, 446, 1992.)

IV. OPIOID FINDINGS

In a study of morphine in hair by Baumgartner et al.,[23] mice were administered morphine by i.p. injection. An area on the back of the mice was plucked free of hair prior to morphine treatment, and hair grown during and after treatment was collected from the plucked area. Morphine was detected in hair one week following treatment.

Püschel et al.[43] assayed samples with Abuscreen RIA obtained from drug-related deaths, heroin users, patients with cancer administered morphine on a regular basis, and guinea pigs administered codeine. In 60 drug-related deaths, opioids were detected in the hair of 90% of the subjects; only 66% were positive for opioids in other tissues. Hair samples obtained from 13 heroin users were tested and concentrations compared to dose information gathered through self-report. The opioid concentrations in hair were highly variable and did not correlate well to reported dose. In addition, the concentration of opioids in hair of cancer patients administered morphine and guinea pigs administered codeine did not correlate to dose. Based on this information, the authors concluded that no dose-concentration relationship was apparent. These studies are limited by a lack of clinical information including use of other drugs and use of a relatively nonspecific RIA for quantitative determinations.

In further studies by Püschel et al.,[43] codeine was administered to drug-free volunteers. Beard hair samples were collected and assayed by Abuscreen RIA. Codeine was detected within 24 h of drug administration and was present for a period of 6 to 8 d. Cone[62] also studied the time course of appearance of morphine and codeine in beard hair following single dose administration of morphine and codeine in two human subjects. Data were obtained with the DPC morphine-specific RIA and GC/MS. Morphine and codeine appeared in beard hair about 7 to 8 d after

TABLE 3.

Summary of Cocaine Analyte Results in Hair

Investigator	Source of hair sample	Cocaine	Benzoylecgonine	Other	Ref.
Cone et al.	Cocaine users	6.4–19.2	0.3–2.5	0–1.9 (ecgonine methyl ester) 0–2.6 (cocaethylene) 0–0.7 (norcocaine) 0–trace (norcocaethylene)	1
Fritch et al.	Random human individuals	27.9–1361 (RIA) 0.13–6.70 (GC/MS)	0–3.3 (GC/MS)	0–2.7 (ecgonine methyl ester)	3
Balabanova et al.	Cocaine users	6.3–98.0	0.3–6.4	—	4
Martz et al.	Accidental poisoning	0.11–0.45	—	—	5
Springfield et al.	Mummies	1.0–28.9	0.7–3.9	0–3.5 (ecgonine methyl ester)	6
Henderson et al.	Coca-leaf chewers	0–5.70	0.3–4.4	0–4.4 (ecgonine methyl ester)	7
Harkey et al.	Cocaine users	X = 6.96, 7.31	0–1.1	—	8
Welch et al.	Nist reference material	—	X = 3.99, 1.57	—	9
Smith and Liu	Alleged sexual assault victim	—	6.2	—	10
Michalodimitrakis	Cocaine-treated rats	X = 16.4	0.024–0.047	—	11
Nakahara et al.	Cocaine-treated rats	0.25–2.5	X = 1.7	X = 0.8 (ecgonine methyl ester)	12
Baumgartner et al.	Cocaine users	—	0.007–6.4	—	13
Martinez et al.	Cocaine users	1.7–45.2	—	—	14
Moeller et al.	Coca-leaf chewers	2.0–98.0	0.6–14.3	1.0–10.4 (ecgonine methyl ester)	15
Kidwell	Cocaine users	—	1.9–8.9	1.1–9.2 (ecgonine)	16
Valente et al.	Cocaine users	—	2.0 (NaOH) 11.5 (HCl) 9.4 (methanol) 3.6 (direct RIA)	—	19
Balabanova and Homoki	Cocaine user 24 h after death and cocaine-treated sheep	—	7.3 (human) 2.7 (sheep)	—	20
Nakahara and Kikura	Cocaine-treated rats	X = 15.4	X = 2.2	X = 0.6 (ecgonine methyl ester)	21
Cone et al.	Cocaine/heroin users	0.4–76.0	0–15.8	—	31
Klein et al.	Maternal hair	18.0	1.8	—	38
Reuschel and Smith	Jail detainees	—	0.26–18.0	—	39
Ferko et al.	Cocaine-treated rats	X = 0.72–2.30	X = 0.32–0.58	—	78
Koren et al.	Cocaine users	1.2–28.2	0.2–10.0	—	79
DiGregorio et al.	Maternal hair, Neonatal hair	0–421.0, 0–243.5	0–183.6, 0–55.4	—	80

Note: All results are in ng/mg.

drug administration at a time when drug was not present in plasma, urine, and saliva. Drug concentrations in hair appeared to be dose related.

In preliminary studies by Rollins et al.,[63] single and multiple oral doses of codeine were administered to drug-free volunteers. After a single dose, codeine was present in the hair bulb within 30 min and peak concentrations were observed at 2 h. The concentration of codeine in the hair bulb decreased over 48 h following a single dose and over 4 weeks following multiple doses. The concentration of codeine in distal hair segments reached a relatively steady concentration within 3 weeks. It was concluded that codeine is rapidly distributed into the hair bulb and a portion is distributed back into the plasma; only a fraction is permanently incorporated into hair.

For the purposes of investigating the disposition of heroin and its metabolites in heroin users, Goldberger et al.[30] developed a GC/MS assay for heroin, 6-AM and morphine in biological fluids, including hair. Hair samples were obtained from subjects who had been enrolled in an outpatient maintenance and detoxification study. 6-AM and morphine were present in all samples; heroin was present in 35% of the samples. Codeine was also present in 75% of the samples. The concentrations of 6-AM in hair generally predominated over those of heroin, morphine, and codeine.

In a post-mortem study by Goldberger et al.,[64] tissue and hair samples from the head, axillary, and pubic regions were collected during autopsy from two drug-related deaths. Heroin was present in the urine samples and 6-AM and morphine were present in other tissue, including blood and urine. Negative or trace amounts of opiates were found in hair samples of one subject, suggesting limited heroin exposure. The amount of heroin and metabolites in the hair of the other subject were much higher, suggesting considerable prior heroin exposure.

Nakahara et al. [65] studied the disposition of 6-AM in monkey and human hair. 6-AM and morphine were present in all hair samples, and heroin was not detected. In addition, the concentration of 6-AM was greater than the concentration of morphine. Nakahara et al.[65] also evaluated the total morphine content (heroin and 6-AM were hydrolyzed to morphine) in monkey hair following subcutaneous administration of heroin (2.5 mg) and morphine (10 mg). Although the morphine dose was 4 times greater than the heroin dose, the mean morphine concentration was only about two thirds of that found following heroin administration. The results indicate that heroin and 6-AM are incorporated more readily into hair than morphine. Based upon this information, the theoretical distribution of heroin and its metabolites in hair is illustrated in Figure 6.

Criteria for the differentiation of heroin use from other forms of opioid exposure based upon hair analysis have been proposed by Moeller et al.[66] Data from over 1000 hair analyses were evaluated, and the following criteria must be met in order to establish heroin use: if the morphine concentration <1.0 ng/mg, the morphine-to-codeine ratio must be >5:1; if the morphine concentration >1.0 ng/mg, the morphine-to-codeine ratio must be >2:1. In addition, the presence of 6-AM is definitive evidence of heroin exposure, since it can only be derived from metabolism of heroin.

In a study of maternal drug use by Kintz and Mangin,[67,68] hair samples were collected from neonates with confirmed *in utero* exposure to heroin. Morphine was present in all samples in a concentration range similar to previous reports based on analysis of hair from adults. Codeine was also present in all samples, but in low concentration. Maternal hair was not tested. Urine samples were also collected and tested for opioids. Four (44%) samples were positive for morphine and 6-AM. The data indicate that drug and/or drug metabolite is transferred to the fetus through the placenta and retained in hair.

The variability of opioid hair test results based upon anatomic origin has been studied by Kintz and Mangin.[69,70] Head, pubic, and axillary hair samples were

PLASMA   **HAIR**

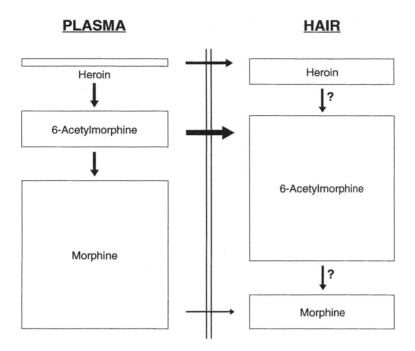

Figure 6

Theoretical distribution of heroin in hair.

collected from drug-related deaths and assayed for morphine and codeine content. The analysis did not include heroin and 6-AM. The highest concentration of morphine was found in hair collected from the pubis; lower concentrations were observed in head and axillary hair. In addition, 6-AM was assayed in several sets of samples, and in all cases, the concentration of 6-AM predominated over morphine and codeine concentrations. The differences in morphine and codeine concentration among sample sources were attributed to growth rate, blood circulation, number of apocrine glands, and exposure to sweat and urine (pubis only). Similar findings were also reported for morphine in head, pubic, and axillary hair by Offidani et al.[71] in drug-related deaths. In the study, the concentration of morphine in other biological tissues including blood, urine, and bile did not correlate to the morphine hair concentration.

In further studies of anatomic origin by Cone et al.[31] with heroin users, 20 paired head and arm hair samples were assayed for heroin and its metabolites. Heroin was detected in two head hair samples and 6-AM was present in 14 head hair and six arm hair samples. If present, the concentration of 6-AM was generally greater in arm hair than head hair. Morphine was present in three head hair samples and was not present in arm hair samples. Selective ion chromatograms of hair and standard extracts adapted from Cone et al.[31] are illustrated in Figure 3.

In a study performed by Carpenter et al.[72] designed to evaluate whether or not poppy seed consumption produces morphine-positive hair test results, five subjects consumed between 25 to 35 g of poppy seed containing approximately 112 µg/g of morphine and 11 µg/g of codeine. Samples were collected as follows: all consecutive urine samples for 4 d; sequential 3-d cumulative beard hair samples beginning prior to consumption of poppy seeds through 15 d; and head hair samples beginning prior to consumption of poppy seeds through 30 d. Urine samples were screened by Emit. Urine samples, beard hair, and head hair samples were extracted by SPE and assayed by GC/MS for 6-AM, morphine, and codeine. Emit results for urine were positive up to 72 h following ingestion. Analysis by GC/MS yielded morphine and codeine results as high as 6123 and 446 ng/mL, respectively. Using a 0.1-ng/mg cutoff concentration, all head and beard hair samples were negative for 6-AM, morphine, and codeine by GC/MS. This indicates that analysis of hair for 6-AM and morphine could be used to differentiate opiate exposure from ingestion of food containing poppy seeds.

Based upon information available in the literature, the following conclusions can be made regarding positive opioid hair test results:

1. heroin, 6-AM, and morphine are usually present in hair following heroin use; the concentration of 6-AM in hair generally predominates over heroin, morphine, and codeine concentrations

2. morphine and codeine in hair indicates prior codeine, or morphine and codeine use

3. morphine only in hair indicates prior morphine use

4. codeine only in hair indicates prior codeine use

Individual opioid test results should be interpreted with caution on a case-by-case basis since multiple-drug combinations are possible.

Other opioids such as buprenorphine, norbuprenorphine, dihydrocodeine, ethylmorphine, fentanyl, methadone, methadone primary metabolite, and pentazocine have been identified in hair of drug users. Data from the literature are summarized below:

1. Hair samples were collected from subjects who reported daily use of buprenorphine, a partial opioid agonist. Buprenorphine and its metabolite, norbuprenorphine, were identified in the samples using HPLC with coulometric detection. The concentrations of norbuprenorphine were lower than buprenorphine.[49]

2. Dihydrocodeine, an opioid commonly abused by heroin users in Germany, is present in hair in concentrations similar to morphine and codeine. Since many opioid immunoassays cross react with dihydrocodeine, a more specific method of analysis such as GC/MS must be utilized.[73]

3. Tissue samples were collected from a drug-related death and assayed by GC/MS. Ethylmorphine was detected in all tissues, including hair from the head, pubis, and axillae.[74]

4. Hair samples were collected from surgical patients who were administered fentanyl during anesthesia. Analysis by RIA revealed the presence of fentanyl following moderate to high doses of fentanyl.[33]

5. Methadone and its primary metabolite were detected by GC/MS in hair samples obtained from methadone users.[75,76]

6. Hair samples were collected from an anesthesiologist suspected of diverting pentazocine. Hair analysis revealed the presence of pentazocine. Hair collected 8 months later were also positive for pentazocine. Results of hair analysis were negative one year later.[66]

Like cocaine users, heroin users also escape detection by conventional urine analysis since heroin is rapidly metabolized and its major metabolite, morphine, is excreted within several days following ingestion. Magura et al.[77] compared urine test results and self-report data with RIA hair tests in patients in a methadone-maintenance treatment program. Positive or negative RIA results were corroborated by urine analysis and/or self-report in 84% of the subjects. However, studies conducted by Hindin et al.[53] comparing self-reported drug use by patients in a drug rehabilitation center and RIA analysis of hair indicate that RIA is not reliable, since a significant number of subjects who self-reported drug use were missed by RIA.

Table 4 provides a summary of opioid results in hair.

TABLE 4.
Summary of Opioid Results in Hair

Investigator	Source of hair sample	Morphine	Codeine	Other	Ref.
Welch et al.	NIST reference material	X = 4.32, 11.9	X = 2.87, 6.69	—	9
Marsh et al.	Heroin users	1.2–16.3	—	—	17
Marsh and Evans	Fortified drug-free hair	—	—	Positive (methadone)	18
Valente et al.	Heroin users	1.3–2.9 (NaOH) / 0.4–2.8 (HCl) / 2.1 (methanol) / 0.7 (water) / 0.8–1.2 (buffer) / 0.4 (NaCl) / 0.4 (direct)	—	—	19
Baumgartner et al.	Heroin users	0–7	—	—	23
Arnold	Heroin user	0–3.2	—	—	24
Kintz and Mangin	Heroin users	0–21.19 (Emit) / 0.48–23.06 (FPIA) / 0.24–21.61 (GC/MS)	0.08–3.07 (GC/MS)	—	26
Marsh and Evans	Heroin users	0.06–9.4	—	—	27
Franceschin et al.	Heroin users / Occasional users / Methadone maintenance	X = 1.34 / X = 0.41 / X = 0	—	—	28
Offidani et al.	Heroin users	0.3–11.8 (NaOH) / 0.1–10.9 (pronase)	—	—	29
Goldberger et al.	Heroin users	Trace to 2.85	0–2.03	Trace to 67.11 (6-acetylmorphine) / 0–15.46 (heroin)	30
Cone et al.	Cocaine/heroin users	0–0.91	—	0–0.78 (6-acetylmorphine) / 0–0.54 (heroin)	31
Nakahara et al.	Heroin users	0–4.1	—	0–22.8 (6-acetylmorphine)	32
Wang et al.	Surgery patients	—	—	0–0.08 (fentanyl)	33
Edder et al.	Drug users / Drug-related deaths / Patients admin. morphine	0.55–3.88 / 0.1–6.0 / 1.5–1.8	0–2.07	—	37
Püschel et al.	Heroin users	0–3.2	—	0.23–3.33 (6-acetylmorphine)	43

TABLE 4. (continued)

Summary of Opioid Results in Hair

Investigator	Source of hair sample	Morphine	Codeine	Other	Ref.
Kintz et al.	Drug-related deaths	> 0.05 (FPIA) 0.41–11.74 (GC/MS)	0–4.21 (GC/MS)	—	44
Balabanova and Wolf	Methadone users	—	—	Head 0.5–2.7 (methadone) Pubis 1.0–4.0 (methadone) Axillae 1.3–8.0 (methadone)	45,46
Marigo et al.	Heroin users	0.08–15.7	—	—	48
Kintz	Buprenorphine users	—	—	0.48–0.59 (buprenorphine) 0.06–0.15 (norbuprenorphine)	49
Goldberger et al.	Drug-related deaths	0–0.09	—	0–0.22 (6-acetylmorphine)	64
Nakahara et al.	Heroin users	0–27.5	—	0.7–7.2 (6-acetylmorphine)	65
Moeller et al.	Heroin users	1.0–21.8	0.1–4.0	2.0–74.2 (6-acetylmorphine)	66
Kintz and Mangin	Neonates with *in utero* exposure to heroin	0.61–3.47	—	—	67,68
Kintz et al.	Drug-related deaths	Head 0.62–27.10 Pubis 0.80–41.34 Axillae 0.40–24.20	Head 0.15–1.87 Pubis 0.22–2.34 Axillae 0.12–1.56	Head 13.54 (6-acetylmorphine) Pubis 26.72 (6-acetylmorphine) Axillae 8.56 (6-acetylmorphine) Head 0.12 (ethylmorphine) Pubis 0.18 (ethylmorphine) Axillae 0.08 (ethylmorphine)	69,70,74
Offidani et al.	Drug-related deaths	Head 0.23–5.3 Pubis 0.18–31.7 Axillae 0.18–1.6	—	—	71
Sachs et al.	Drug users	0–12.3	0–5.3	1.2–31.2 (dihydrocodeine)	73
Goldberger et al.	Methadone maintenance	—	—	0–15.0 (methadone) Positive (methadone primary metabolite)	75
Moeller et al.	Methadone maintenance	—	—	0–42.0 (methadone) 0–2.4 (methadone primary metabolite)	76

Note: All results are in ng/mg.

V. CONCLUSIONS

Results of hair tests for abused drugs such as cocaine and opioids can provide useful information regarding drug use. Compared to urine analysis, hair testing provides an extended window of analyte (parent and metabolites) detection which cannot be affected by short periods of abstinence. In addition, collection of hair is considered by most a noninvasive technique, and retesting through a second sample is possible. Also, in contrast to urine, hair is not readily adulterated.

Further research is required in this relatively new field to establish definitive data regarding differentiation between internal incorporation (ingestion) and external contamination (environmental), effect of hair type on analyte incorporation, time course of analyte appearance, dose vs. analyte concentration relationships, and the mechanism of drug entry into hair. Already this novel technique has proved useful in a wide variety of applications, and will unquestionably become more popular in future years.

REFERENCES

1. Cone, E. J., Yousefnejad, D., Darwin, W. D., and Maguire, T., Testing human hair for drugs of abuse. II. Identification of unique cocaine metabolites in hair of drug abusers and evaluation of decontamination procedures, *J. Anal. Toxicol.*, 15, 250, 1991.
2. Baumgartner, W. A., Hill, V. A., and Blahd, W. H., Hair analysis for drugs of abuse, *J. Forensic Sci.*, 34, 1433, 1989.
3. Fritch, D., Groce, Y., and Rieders, F., Cocaine and some of its products in hair by RIA and GC/MS, *J. Anal. Toxicol.*, 16, 112, 1992.
4. Balabanova, S., Brunner, H., and Nowak, R., Radioimmunological determination of cocaine in human hair, *Z. Rechtsmed.*, 98, 229, 1987.
5. Martz, R., Donnelly, B., Fetterolf, D., Lasswell, L., Hime, G. W., and Hearn, W. L., The use of hair analysis to document a cocaine overdose following a sustained survival period before death, *J. Anal. Toxicol.*, 15, 279, 1991.
6. Springfield, A. C., Cartmell, L. W., Aufderheide, A. C., Buikstra, J., and Ho, J., Cocaine and metabolites in the hair of ancient Peruvian coca leaf chewers, *Forensic Sci. Int.*, 63, 269, 1993.
7. Henderson, G. L., Harkey, M. R., Zhou, C., and Jones, R. T., Cocaine and metabolite concentrations in the hair of South American coca chewers, *J. Anal. Toxicol.*, 16, 199, 1992.
8. Harkey, M. R., Henderson, G. L., and Zhou, C., Simultaneous quantitation of cocaine and its major metabolites in human hair by gas chromatography/chemical ionization mass spectrometry, *J. Anal. Toxicol.*, 15, 260, 1991.
9. Welch, M. J., Sniegoski, L. T., Allgood, C. C., and Habram, M., Hair analysis for drugs of abuse: evaluation of analytical methods, environmental issues, and development of reference materials, *J. Anal. Toxicol.*, 17, 389, 1993.
10. Smith, F. P. and Liu, R. H., Detection of cocaine metabolite in perspiration stain, menstrual bloodstain, and hair, *J. Forensic Sci.*, 31, 1269, 1986.
11. Michalodimitrakis, M., Detection of cocaine in rats from analysis of hair, *Med. Sci. Law*, 27, 13, 1987.
12. Nakahara, Y., Ochiai, T., and Kikura, R., Hair analysis for drugs of abuse. V. The facility in incorporation of cocaine into hair over its major metabolites, benzoylecgonine and ecgonine methyl ester, *Arch. Toxicol.*, 66, 446, 1992.
13. Baumgartner, W. A., Black, C. T., Jones, P. F., and Blahd, W. H., Radioimmunoassay of cocaine in hair: concise communication, *J. Nucl. Med.*, 23, 790, 1982.
14. Martinez, F., Poet, T. S., Pillai, R., Erickson, J., Estrada, A. L., and Watson, R. R., Cocaine metabolite (benzoylecgonine) in hair and urine of drug users, *J. Anal. Toxicol.*, 17, 138, 1993.
15. Moeller, M. R., Fey, P., and Rimbach, S., Identification and quantitation of cocaine and its metabolites, benzoylecgonine and ecgonine methyl ester, in hair of Bolivian coca chewers by gas chromatography/mass spectrometry, *J. Anal. Toxicol.*, 16, 291, 1992.

16. Kidwell, D. A., Analysis of phencyclidine and cocaine in human hair by tandem mass spectrometry, *J. Forensic Sci.*, 38, 272, 1993.

17. Marsh, A., Carruthers, M. E., Desouza, N., and Evans, M. B., An investigation of the effect of washing upon the morphine content of hair measured by a radioimmunoassay technique, *J. Pharm. Biomed. Anal.*, 10, 89, 1992.

18. Marsh, A. and Evans, M. B., Radioimmunoassay of drugs of abuse in hair. Part I. Methadone in human hair, method adaptation and the evaluation of decontamination procedures, *J. Pharm. Biomed. Anal.*, 12, 1123, 1994.

19. Valente, D., Cassini, M., Pigliapochi, M., and Vansetti, G., Hair as the sample in assessing morphine and cocaine addiction, *Clin. Chem.*, 27, 1952, 1981.

20. Balabanova, S. and Homoki, J., Determination of cocaine in human hair by gas chromatography/mass spectrometry, *Z. Rechtsmed.*, 98, 235, 1987.

21. Nakahara, Y. and Kikura, R., Hair analysis for drugs of abuse. VII. The incorporation rates of cocaine, benzoylecgonine and ecgonine methyl ester into rat hair and hydrolysis of cocaine in rat hair, *Arch. Toxicol.*, 68, 54, 1994.

22. Cassani, M. and Spiehler, V., Analytical requirements, perspectives and limits of immunological methods for drugs in hair, *Forensic Sci Int.*, 63, 175, 1993.

23. Baumgartner, A. M., Jones, P. F., Baumgartner, W. A., and Black, C. T., Radioimmunoassay of hair for determining opiate-abuse histories, *J. Nucl. Med.*, 20, 748, 1979.

24. Arnold, W., Radioimmunological hair analysis for narcotics and substitutes, *J. Clin. Chem. Clin. Biochem.*, 25, 753, 1987.

25. Sachs, H. and Arnold, W., Results of comparative determination of morphine in human hair using RIA and GC/MS, *J. Clin. Chem. Clin. Biochem.*, 27, 873, 1989.

26. Kintz, P. and Mangin, P., Analysis of opiates in human hair with FPIA, EMIT, and GC/MS, *Adli Tip Derg.*, 7, 129, 1991.

27. Marsh, A. and Evans, M. B., Challenging declarations of abstinence by the determination of morphine in hair by radioimmunoassay, *J. Pharm. Biomed. Anal.*, 11, 693, 1993.

28. Franceschin, A., Morosini, L., and Dell'Anna, L., Detection of morphine in hair with the Abbott TDx, *Clin. Chem.*, 33, 2125, 1987.

29. Offidani, C., Carnevale, A., and Chiarotti, M., Drugs in hair: a new extraction procedure, *Forensic Sci. Int.*, 41, 35, 1989.

30. Goldberger, B. A., Caplan, Y. H., Maguire, T., and Cone, E. J., Testing human hair for drugs of abuse. III. Identification of heroin and 6-acetylmorphine as indicators of heroin use, *J. Anal. Toxicol.*, 15, 226, 1991.

31. Cone, E. J., Darwin, W. D., and Wang, W-L., The occurrence of cocaine, heroin and metabolites in hair of drug abusers, *Forensic Sci. Int.*, 63, 55, 1993.

32. Nakahara, Y., Kikura, R., and Takahashi, K., Hair analysis for drugs of abuse. VIII. Effective extraction and determination of 6-acetylmorphine and morphine in hair with trifluoroacetic acid-methanol for the confirmation of retrospective heroin use by gas chromatography-mass spectrometry, *J. Chromatogr. B*, 657, 93, 1994.

33. Wang, W.-L., Cone, E. J., and Zacny, J., Immunoassay evidence for fentanyl in hair of surgery patients, *Forensic Sci. Int.*, 61, 65, 1993.

34. Krishnan, T. R. and Ibraham, I., Solid-phase extraction technique for the analysis of biological samples, *J. Pharm. Biomed. Anal.*, 12, 287, 1994.

35. Goldberger, B. A., Darwin, W. D., Grant, T. M., Allen, A. C., Caplan, Y. H., and Cone, E. J., Measurement of heroin and its metabolites by isotope-dilution electron-impact mass spectrometry, *Clin. Chem.*, 39, 670, 1993.

36. Sachs, H. and Raff, I., Comparison of quantitative results of drugs in human hair by GC/MS, *Forensic Sci. Int.*, 63, 207, 1993.

37. Edder, P., Staub, C., Veuthey, J. L., Pierroz, I., and Haerdi, W., Subcritical fluid extraction of opiates in hair of drug addicts, *J. Chromatogr. B*, 658, 75, 1994.

38. Klein, J., Greenwald, M., Becker, L., and Koren, G., Fetal distribution of cocaine: case analysis, *Pediatr. Pathol.*, 12, 463, 1992.

39. Reuschel, S. A. and Smith, F. P., Benzoylecgonine (cocaine metabolite) detection in hair samples of jail detainees using radioimmunoassay (RIA) and gas chromatography/mass spectrometry (GC/MS), *J. Forensic Sci.*, 36, 1179, 1991.

40. Tagliaro, F., Antonioli, C., Moretto, S., Archetti, S., Ghielmi, S., and Marigo, M., High-sensitivity low-cost methods for determination of cocaine in hair: high-performance liquid chromatography and capillary electrophoresis, *Forensic Sci. Int.*, 63, 227, 1993.

41. Tagliaro, F., Poiesi, C., Aiello, R., Dorizzi, R., Ghielmi, S., and Marigo, M., Capillary electrophoresis for the investigation of illicit drugs in hair: determination of cocaine and morphine, *J. Chromatogr.*, 638, 303, 1993.

42. Traldi, P., Favretto, D., and Tagliaro, F., Ion trap mass spectrometry, a new tool in the investigation of drugs of abuse in hair, *Forensic Sci. Int.*, 63, 239, 1993.

43. Püschel, K., Thomasch, P., and Arnold, W., Opiate levels in hair, *Forensic Sci. Int.*, 21, 181, 1983.

44. Kintz, P., Ludes, B., and Mangin, P., Detection of drugs in human hair using Abbott ADx with confirmation by gas chromatography/mass spectrometry (GC/MS), *J. Forensic Sci.*, 37, 328, 1992.

45. Balabanova, S. and Wolf, H. U., Determination of methadone in human hair by radioimmunoassay, *Z. Rechtsmed.*, 102, 1, 1989.

46. Balabanova, S. and Wolf, H. U., Methadone concentrations in human hair of the head, axillary and pubic hair, *Z. Rechtsmed.*, 102, 293, 1989.

47. Balabanova, S., Arnold, P. J., Brunner, H., Luckow, V., and Wolf, H. U., Detection of methadone in human hair by gas chromatography/mass spectrometry, *Z. Rechtsmed.*, 102, 495, 1989.

48. Marigo, M., Tiagliaro, F., Poiesi, C., Lafisca, S., and Neri C., Determination of morphine in the hair of heroin addicts by high performance liquid chromatography with fluorimetric detection, *J. Anal. Toxicol.*, 10, 158, 1986.

49. Kintz, P., Determination of buprenorphine and its dealkylated metabolite in human hair, *J. Anal. Toxicol.*, 17, 443, 1993.

50. Curcuruto, O., Guidugli, F., Traidi, P., Sturaro, A., Tagliaro, F., and Marigo, M., Ion-trap mass spectrometry applications in forensic sciences. I. Identification of morphine and cocaine in hair extracts of drug addicts, *Rapid Commun. Mass Spectrom.*, 6, 434, 1992.

51. Polettini, A., Groppi, A., and Montagna, M., Rapid and highly selective GC/MS/MS detection of heroin and its metabolites in hair, *Forensic Sci. Int.*, 63, 217, 1993.

52. Mieczkowski, T., Barzelay, D., Gropper, B., and Wish, E., Concordance of three measures of cocaine use in an arrestee population: hair, urine, and self-report, *J. Psychoactive Drugs*, 23, 241,1991.

53. Hindin, R., McCusker, J., Vickers-Lahti, M., Bigelow, C., Garfield, F., and Lewis, B., Radioimmunoassay of hair for determination of cocaine, heroin, and marijuana exposure: comparison with self-report, *Int. J. Addict.*, 29, 771, 1994.

54. Graham, K., Koren, G., Klein, J., Schneiderman, J., and Greenwald, M., Determination of gestational cocaine exposure by hair analysis, *JAMA*, 262, 3328, 1989.

55. Callahan, C. M., Grant, T. M., Phipps, P., Clark, G., Novack, A. H., Streissguth, A. P., and Raisys, V. A., Measurement of gestational cocaine exposure: sensitivity of infants' hair, meconium, and urine, *J. Pediatr.*, 120, 763, 1992.

56. Forman, R., Schneiderman, J., Klein, J., Graham, K., Greenwald, M., and Koren, G., Accumulation of cocaine in maternal and fetal hair; the dose response curve, *Life Sci.*, 50, 1333, 1992.

57. Marques, P. R., Tippetts, A. S., and Branch, D. G., Cocaine in the hair of mother-infant pairs: quantitative analysis and correlations with urine measures and self-report, *Am. J. Drug Alcohol Abuse*, 19, 159, 1993.

58. DiGregorio, G. J., Barbieri, E. J., Ferko, A. P., and Ruch, E. K., Prevalence of cocaethylene in the hair of pregnant women, *J. Anal. Toxicol.*, 17, 445, 1993.

59. Grant, T., Brown, Z., Callahan, C., Barr, H., and Streissguth, A. P., Cocaine exposure during pregnancy: improving assessment with radioimmunoassay of maternal hair, *Obstet. Gynecol.*, 83, 524, 1994.

60. Janzen, K. E., Ethylbenzoylecgonine: a novel component in illicit cocaine, *J. Forensic Sci.*, 36, 1224, 1991.

61. Henderson, G. L., Mechanisms of drug incorporation into hair, *Forensic Sci. Int.*, 63, 19, 1993.

62. Cone, E. J., Testing human hair for drugs of abuse. I. Individual dose and time profiles of morphine and codeine in plasma, saliva, urine, and beard compared to drug-induced effects on pupils and behavior, *J. Anal. Toxicol.*, 14, 1, 1990.

63. Rollins, D. E., Wilkins, D. G., Foltz, R. L., Haughey, H. M., Gygi, S. P., and Krueger, G. G., Disposition of codeine in hair, Workbook — SOFT Conference on Drug Testing in Hair, Tampa, FL, 1994.

64. Goldberger, B. A., Cone, E. J., Grant, T. M., Caplan, Y. H., Levine, B. S., and Smialek, J. E., Disposition of heroin and its metabolites in heroin-related deaths, *J. Anal. Toxicol.*, 18, 22, 1994.

65. Nakahara, Y., Takahashi, K., Shimamine, M., and Saitoh, A., Hair analysis for drugs of abuse. IV. Determination of total morphine and confirmation of 6-acetylmorphine in monkey and human hair by GC/MS, *Arch. Toxicol.*, 66, 669, 1992.

66. Moeller, M. R., Fey, P., and Sachs, H., Hair analysis as evidence in forensic cases, *Forensic Sci. Int.*, 63, 43, 1993.

67. Kintz, P. and Mangin, P., Evidence of gestational heroin or nicotine exposure by analysis of fetal hair, *Forensic Sci. Int.*, 63, 99, 1993.

68. Kintz, P. and Mangin, P., Determination of gestational opiate, nicotine, benzodiazepine, cocaine and amphetamine exposure by hair analysis, *J. Forensic Sci. Soc.*, 33,139, 1993.

69. Kintz, P. and Mangin, P., Opiate concentrations in human head, axillary, and pubic hair, *J. Forensic Sci.*, 38, 657, 1993.

70. Mangin, P. and Kintz, P., Variability of opiates concentrations in human hair according to their anatomical origin: head, axillary and pubic regions, *Forensic Sci. Int.*, 63, 77, 1993.

71. Offidani, C., Rossi, S. S., and Chiarotti, M., Drug distribution in the head, axillary and pubic hair of chronic addicts, *Forensic Sci. Int.*, 63, 105, 1993.

72. Carpenter, P. C., Goldberger, B. A., and Cone, E. J., Preliminary results.

73. Sachs, H., Denk, R., and Raff, I. Determination of dihydrocodeine in hair of opiate addicts by GC/MS, *Int. J. Leg. Med.*, 105, 247, 1993.

74. Kintz, P., Jamey, C., and Mangin, P., Ethylmorphine concentrations in human sample in an overdose case, *Arch. Toxicol.*, 68, 210, 1994.

75. Goldberger, B. A., Darraj, A. G., Cone, E. J., and Caplan, Y. H., Detection of methadone, methadone metabolites, and other illicit drugs of abuse in hair of methadone treatment subjects, 2nd International Meeting on Clinical and Forensic Aspects of Hair Analysis, Genova, Italy, 1994.

76. Moeller, M. R., Fey, P., and Wennig, R., Simultaneous determination of drugs of abuse (opiates, cocaine, and amphetamine) in human hair by GC/MS and its application to a methadone treatment program, *Forensic Sci. Int.*, 63, 185, 1993.

77. Magura, S., Freeman, R. C., Siddiqi, Q., and Lipton, D. S., The validity of hair analysis for detecting cocaine and heroin use among addicts, *Int. J. Addict.*, 27, 51, 1992.

78. Ferko, A. P., Barbien, E. J., DiGregorio, G. J., and Ruch, E. K., The accumulation and disappearance of cocaine and benzoylecgonine in rat hair following prolonged administration of cocaine, *Life Sci.*, 51, 1823, 1992.

79. Koren, G., Klein, J., Forman, R., and Graham, K., Hair analysis of cocaine: differentiation between systemic exposure and external contamination, *J. Clin. Pharmacol.*, 32, 671, 1992.

80. DiGregorio, G. J., Ferko, A. P., Barbieri, E. J., Ruch, E. K., Chawla, H., Keohane, D., Rosenstock, R., and Aldano, A., Determination of cocaine usage in pregnant women by a urinary EMIT drug screen and GC/MS analyses, *J. Anal. Toxicol.*, 18, 247, 1994.

Chapter **7**

CANNABIS AND AMPHETAMINE DETERMINATION IN HUMAN HAIR

Vincent Cirimele

CONTENTS

I. CANNABIS DETERMINATION IN HUMAN HAIR

A. Introduction

According to statistical information, cannabis is today the most widely abused illicit drug in the world.[1] Cannabis has been used for its euphoric effects for over

4000 years. Δ9-tetrahydrocannabinol (THC), the primary psychoactive analyte, is found in the flowering or fruity tops, leaves, and resin of the plant. THC, like cannabidiol (CBD) and cannabinol (CBN), are three constituents among the sixteen compounds presently isolated from the *Cannabis sativa* plant.

Cannabis is extensively metabolized in the human to 11 hydroxy- and 8β hydroxy-Δ9-tetrahydrocannabinol (11 OH and 8β OH-THC) and finally to 11 nor-Δ9-tetrahydrocannabinol-9-carboxylic acid (THC-COOH), which is conjugated with glucuronic acid to a variable extent.

THC and its metabolites can be detected in plasma or urine samples only within a few hours to days after cannabis intake, which was documented by numerous extraction procedures.[2,3] At the opposite, hair appears to be an interesting substrate for the investigation of chronic exposure.[4-6]

The aim of this review is to present the different methodologies reported for cannabis detection in human hair (sample, decontamination, hydrolysis, internal standards [IS], extraction, derivatization, system of detection) and the analytical performances observed.

B. Specimen Collection

Generally, the samples tested were head hair obtained from expertises or from subjects deceased from fatal heroin or cocaine overdose. Collection procedures have not been standardized, but hair is often collected from the area at the back of the head (vertex posterior), cut as close as possible to the scalp and stored in dry tubes. In this area, there is less variability in hair growth rate, the number of hairs in the growing phase is more constant, and the hair is less subject to age and sex-related influences.[7] Occasionally, axillary[8] or pubic hair[8,9] have also been tested for cannabinoids.

The sample size considerably varies among laboratories and ranged from 5 to 200 mg (see Table 1).

TABLE 1.

Summary of Pretreatment of Hair for Hair Analysis

Compound	Sample size (mg)	Decontamination	Ref
Marijuana	5 to 10	Methanol and phosphate buffer	4
Cannabinoids	50	Not published (N.P.)	8
THC	150 to 200	Water, acetone, and petroleumbenzin	10
THC-COOH	50	Water and acetone	11
Cannabis	50	Ethanol	12
THC/THC-COOH	10 to 30	Water and acetone	13
THC-COOH	10 to 25	Water	14
THC/THC-COOH	N.P.	Methylene chloride	15
THC/THC-COOH	N.P.	N.P.	16
THC/THC-COOH	100	Methylene chloride	9
CBD/CBN/THC	50	Methylene chloride	17
THC-COOH	100	Methylene chloride	18

C. Decontamination Procedure

In most cases, laboratories included a washing step before extraction; however, there is no consensus or uniformity in the procedure (Table 1). The most crucial issue facing hair analysis is the avoidance of evidentiary false positives caused by passive exposure to cannabis smoke, which contains THC, CBD, and CBN. More decontamination procedures exist, but none of them was presented as of absolute safety to prevent false positives.[19]

Only the detection in hair of metabolites would provide convincing evidence of cannabis abuse. The presence in hair of THC-COOH, which is not detected in cannabis smoke, can be considered as a potential marker of chronic cannabis exposure and evidence that THC excretion occurs in hair after active use.

D. Cannabinoid Extraction

1. Hydrolysis

Different methods have been reported for hydrolysis and are listed in Table 2.

TABLE 2.

Summary of Hydrolysis Methods

Compound	Hydrolysis	Ref.
Marijuana	Dissolution	4
Cannabinoids	0.1 M HCl	8
THC	—	10
THC-COOH	2 M NaOH, 2 M HCl	11
Cannabis	1 N NaOH	12
THC/THC-COOH	2 M NaOH	13
THC-COOH	Methanol + water + 10 N NaOH	14
THC/THC-COOH	11.8 N KOH	15
THC/THC-COOH	Enzymatic	16
THC/THC-COOH	1 N NaOH	9
CBD/CBN/THC	1 N NaOH	17
THC-COOH	1 N NaOH	18

2. Extraction

Like the hydrolysis methods, many extraction procedures have been described. Table 3 presents the various procedures and the compounds used as IS. With the development of the technology and the introduction of gas chromatography coupled with mass spectrometry (GC/MS), deuterated IS have been used for the quantification of cannabis in hair. Presently, this technique represents the state of the art. Most authors use liquid-liquid extraction; only Moeller[11] and Moeller et al.[11,13] have reported solid-phase extraction (SPE) procedures.

E. Drug Analysis

1. Derivatization and Methods of Detection

In Table 4 are listed each compound detected with their specific conditions of derivatization and instrument requirements. Electronic impact (EI) detection is widely documented. To enhance the limit of detection (LOD), negative chemical ionization (NCI) and tandem MS were recently proposed.[14,16,18]

2. Limit of Detection

According to the different procedures and technologies used, various LOD were published. In Table 5 are listed the LOD reported for the analysis of cannabinoids in

TABLE 3.

Summary of the Different Extraction Procedures and Internal Standards Used

Compound	IS	Extraction	Ref.
Marijuana	N.P.	N.P.	4
Cannabinoids	—	—	8
THC	—	*n*-Hexane	8
THC	Methaqualone	Methanol	10
THC-COOH	Levallorphan	SPE	11
Cannabis	—	—	12
THC-COOH	SKF 525A	Chloro./isopro./*n*-hept.	12
THC/THC-COOH	THC-d$_3$ and THC-COOH-d$_3$	SPE	13
THC-COOH	THC-COOH-d$_3$	*n*-Hexane/ethylacetate (9:1 v/v)	14
THC/THC-COOH	THC-d$_3$ and THC-COOH-d$_3$	*n*-Hexane/ethylacetate (9:1 v/v)	15
Cannabinoids	—	—	16
THC/THC-COOH	THC-d$_3$ and THC-COOH-d$_3$	*n*-Hexane/ethylacetate	16
THC/THC-COOH	THC-d$_3$ and THC-COOH-d$_3$	*n*-Hexane/ethylacetate (9:1 v/v)	9
CBD/CBN/THC	THC-d$_3$	*n*-Hexane/ethylacetate (9:1 v/v)	17
THC-COOH	THC-COOH-d$_3$	*n*-Hexane/ethylacetate (9:1 v/v)	18

TABLE 4.

Summary of Derivatization and Detection Mode
for the Analysis of Cannabinoids

Compound	Derivatization	Detection	Ref.
Marijuana	N.P.	RIA and GC/MS	4
Cannabinoids	—	RIA	8
THC	—	GC/MS-EI	8
THC	PFP	GC/MS-EI	10
THC-COOH	PFP	GC/MS-EI	11
Cannabis	—	FPIA	12
THC-COOH	TFA	GC/MS-EI	12
THC/THC-COOH	Iodomethane	GC/MS-EI	13
THC-COOH	HFBA-HFIP	GC/MS/MS-NCI	14
THC/THC-COOH	HFBA-HFPOH	GC/MS-EI	15
Cannabinoids	—	RIA	16
THC/THC-COOH	HFBA-HFIP	GC/MS/MS-NCI	16
THC/THC-COOH	PFP	GC/MS-EI	9
CBD/CBN/THC	—	GC/MS-EI	17
THC-COOH	PFP	GC/MS-NCI	18

hair by GC/MS. This table clearly shows that GC/MS-NCI and GC/MS/MS-NCI are the most sensitive detectors. The LOD were more than 20 times lower when compared to those obtained by GC/MS-EI. For the detection of cannabinoids in hair, and particularly for the detection of THC-COOH, these instruments appear to be more suitable, although the general routine method for the detection of drugs is the GC/MS operating in EI mode.

3. *Reported Concentrations in Hair*

A compendium of the published concentrations obtained by GC/MS is presented in Table 6. Except for Moeller et al.,[13] the THC concentrations were higher than the THC-COOH concentrations. This was also observed for opiates and cocaine, where parent drug concentrations always exceeded metabolite concentrations.[19]

When compared with opiates and cocaine,[20] the concentrations of THC and THC-COOH were particularly low. The weak incorporation rate of THC-COOH in hair is

TABLE 5.

Summary of the LOD Reported
for the Analysis of Cannabinoids
in Hair by GC/MS

Compound	LOD (ng/mg)	Ref.
Cannabinoids	—	8
THC	N.P.	8
THC	0.01	10
THC	N.P.	11
THC-COOH	N.P.	11
Cannabis	—	12
THC-COOH	N.P.	12
THC/THC-COOH	N.P.	13
THC-COOH	0.02 pg/mg	14
THC	0.014	15
THC-COOH	0.01	15
Cannabinoids	—	16
THC/THC-COOH	0.02 pg/mg	16
THC	0.1	9
THC-COOH	0.1	9
CBD	0.02	17
CBN	0.01	17
THC	0.1	17
THC-COOH	0.005	18

TABLE 6.

Summary of the Results Concerning Cannabinoids in Hair

Compound	Number of cases	Number of positive	Range (ng/mg)	Mean (ng/mg)	Ref.
THC	31	N.P.	0.009–9.9	0.939	10
THC-COOH	N.P.	N.P.	N.P.	N.P.	11
THC-COOH	40	27	0.4–2.7	N.P.	12
THC	57	10	0.4–6.2	2.0	13
THC-COOH	57	2	1.7–5.0	3.3	13
THC-COOH	N.P.	N.P.	N.P.	0.71 pg/mg	14
THC	70	49	0.06–7.63	0.97	15
THC-COOH	70	45	0.06–3.87	0.50	15
THC	93	15	0.003–0.038	0.0431	16
THC-COOH	93	3	0.03–1.53 pg/mg	0.322 pg/mg	16
THC	43	15	0.1–2.17	0.74	9
THC-COOH	43	6	0.07–0.33	0.16	9
CBD	30	23	0.03–3.00	0.44	17
CBN	30	22	0.01–1.07	0.13	17
THC	30	5	0.1–0.29	0.15	17
THC-COOH	30	17	0.05–0.39	0.10	18

not surprising taking into consideration the three main factors which influence drug incorporation in hair (lipophilicity, melanine affinity, and membrane permeability).[21] In spite of its high melanin affinity, THC-COOH is less incorporated into hair than basic drugs, as membrane permeability is based on the pH gradient between blood (pH 7.4) and hair matrix (pH < 5).

Pubic hair was shown to concentrate opiates more than head hair.[22,23] Balabanova et al.[8] published some results for cannabis in pubic and axillary hair using radioimmunoassay (RIA). The higher concentrations were found generally in pubic hair, and axillary hair showed the lower concentrations (Table 7). Cirimele et al.[9] found that in most cases, THC and THC-COOH concentrations in pubic hair were higher when compared with head hair.

TABLE 7.

Summary of Cannabinoid Concentrations in Head, Pubic, and Axillary Hair

Type of hair	Compoud	Ref. 8 Cannabinoids	Ref. 9 THC	Ref. 9 THC-COOH
Head	Range (ng/mg)	0.0–3.1	0.28–2.17	0.10–0.18
		(n = 4)	(n = 7)	(n = 3)
	Mean (ng/mg)	1.68	0.84	0.06
Pubic	Range (ng/mg)	0.5–3.8	0.34–3.91	0.07–0.83
		(n = 6)	(n = 7)	(n = 7)
	Mean (ng/mg)	1.75	1.35	0.28
Axillary	Range (ng/mg)	0.4–1.9	—	—
		(n = 6)		
	Mean (ng/mg)	1.05	—	—

4. Comparison with Urine Investigations

Baumgartner et al.[4] and Mieczkowski and Newel[24] compared cannabinoid detection in urine and hair samples. They concluded that hair analysis and urinalysis were complementary rather than competing tests. Urinalysis provides short-term information of an individual's drug use, whereas long-term histories are accessible through hair analysis. Moreover, in contrast with qualitative information from urinalysis, hair analysis provides quantitative information on the severity and pattern of an individual's drug use.

In contrast with opiates and cocaine, an important number of cases were observed where urine tested positive for cannabinoids, whereas hair remained negative.[9,24] This could be explained by an inadequate sensitivity of the hair assay or by the slow clearance rate of the cannabinoids in people taking cannabis for the first time.

5. Interlaboratory Comparison

Interlaboratory comparisons have been organized for opiates and cocaine determination in hair.[25] Recently, the first interlaboratory comparison was organized between a Spanish laboratory and a French laboratory.[26] Ten identical samples were analyzed in both laboratories using their respective procedures.[9,26] For an identical sample tested, the concentrations determined varied from 2 to 43% (mean 25.7%) and from 0 to 73.7% (mean 36.8%) for THC and THC-COOH, respectively, showing good accuracy between the two different procedures used.

More comparison studies between laboratories must be organized to determine the limitations and advantages of each method, as American authors reported concentrations in the low pg range and European in the ng range.

6. Cannabis Screening Procedure

To validate data on cannabis use, two complementary methods were investigated. The first procedure was used as screening for simultaneous identification of THC, CBD, and CBN by GC/MS-EI.[17] To confirm cannabis exposure, THC-COOH was detected by GC/MS-NCI.[18]

II. AMPHETAMINE DETERMINATION IN HUMAN HAIR

Until March 1994, 23 papers dealing with the detection of amphetamine in hair have been published. A characteristic point is that 18 out of the 23 papers about amphetamines in hair have been reported by Japanese researchers, which obviously marks the prominence of such drug consumption in the Far East. Recently, Nakahara[27] published an excellent review on the detection of amphetamine in hair. Table 8 lists the workup and the main findings of each paper.

TABLE 8.

Summary of Pretreatment of Hair Samples, Extraction, Derivatization, and Ranges of Concentration for Amphetamines in Hair

Compound	Wash	Extraction	Derivatization	Range (ng/mg)	Ref.
MA	5% Extran + 0.1% SDS + 0.01 M HCl	NaOH 1.5 N + HCl 1.5 N + org. phase (pH 10)	TFA	4–120	28
MA/AP	50% aq. MeOH ultrasonication	NaOH 2.5 N 80°C 10 min n-heptane	TFA	0.8–12	29
MA/AP	50% aq. MeOH ultrasonication	NaOH 2.5 N 80°C 30 min n-heptane	TFA	0.03–10.2	30
MA/AP	0.1% SDS + H_2O ultrasonication	HCl 5 M/MeOH (1:20)	TFA	0.9–56	31–32
AP	EtOH 37°C 15 min	Chloro./isoprop./n-hept.	TFA	0.96–12.71	12
MA	0.1% SDS +H_2O ultrasonication	N.P.	N.P.	N.P.	34
MA/AP	50% aq. MeOH ultrasonication	0.6 M HCl	TFA	0.5–15.8	35
MOP	0.1% SDS + H_2O ultrasonication	HCl 5 M/MeOH (1:20)	TFA	1.1-10.3	36
MA	H_2O + methanol	NaOH 2.5 N 80°C 20 min CH_2Cl_2	TFA	0.7–10.6	37
MDMA	H_2O + acetone	Gluculase 2 h 40°C SPE	PFP	0.6	38
MOP/ DMMOP/ MA/AP	0.1% SDS + H_2O ultrasonication	HCl 5 M/MeOH (1:20)	TFA	Approx. 45	39
AP	H_2O + acetone	Gluculase 2 h 40°C SPE	PFP	N.P.	40
DPN	0.1% SDS + H_2O ultrasonication	HCl 5 M/MeOH (1:20)	N.P.	1	41
MA/AP		HCl 5 M/MeOH (1:20)	TFA	13–14	41
MDMA	H_2O + acetone	Glucuronidase-sulfatase 2 h 40°C SPE	PFP	0.6	42

ABBREVIATIONS

AP	Amphetamine
Chloro	Chloroform
DMMOP	O-Desmethylmethoxyphenamine
DPN	Deprenyl
FPIA	Fluorescence polarization immunoassay
HFBA	Heptafluorobutyric anhydride
HFIP	1,1,1,3,3,3-hexafluoro-2-propanol
HFPOH	Hexafluoropropanol

Isopro	Isopropanol
MA	Methamphetamine
MDMA	Methylenedioxymethamphetamine
MOP	Methoxyphenamine
n-Hept	*n*-Heptane
PFP	Pentafluoropropanol
TFA	Trifluoroacetic anhydride

REFERENCES

1. Huestis, M. A., Cannabis monograph SOFT committees on driving under influence of drug, 1, 1994.
2. Burney, L. J., Bobbie, B. A., and Sepp, L. A., GC/MS and EMIT analyses for Δ^9-tetrahydrocannabinol metabolites in plasma and urine of human subjects, *J. Anal. Toxicol.*, 10, 56, 1986.
3. Huestis, M. A., Henningfield, J. E., and Cone, E. J., Blood cannabinoids. I. Absorption of THC and formation of 11-OH-THC and THC-COOH during and after smoking marijuana, *J. Anal. Toxicol.*, 16, 276, 1992.
4. Baumgartner, W. A., Hill, V. A., and Blahd, W. H., Hair analysis for drugs of abuse, *J. Forensic Sci.*, 34, 1433, 1989.
5. Kintz, P., Tracqui, A., and Mangin, P., Detection of drugs in human hair for clinical and forensic applications, *Int. J. Leg. Med.*, 105, 1, 1992.
6. Moeller, M. R. and Fey, P., Identification and quantification of cocaine and its metabolites, benzoylecgonine and ecgonine methyl ester, in hair of Bolivian coca chewers by gas chromatography/mass spectrometry, *J. Anal. Toxicol.*, 16, 291, 1992.
7. Saitoh, M., Uzaka, M., Sakamoto, M., and Kobori, T., Rate of hair growth, in *Advances in Biology of Skin: Hair Growth*, 9, Montana and Dobson, Oxford, 1969, 183.
8. Balabanova, S., Arnold, P. J., Luckow, V., Brunner, H., and Wolf, H. U., Tetrahydrocannabinole im Haar von Haschischrauchern, *Z. Rechtsmed.*, 102, 503, 1989.
9. Cirimele, V., Kintz, P., and Mangin, P., Testing human hair for cannabis, *Forensic Sci. Int.*, 70, 175, 1995.
10. Kauert, G. F., Drug analysis in hair samples: applications and experiences with a new rapid analytical procedure, *Z. Rechtsmed.*, 40, 229, 1993.
11. Moeller, M. R., Drug determination in hair by chromatographic procedures, *J. Chromatogr.*, 580, 125, 1992.
12. Kintz, P., Ludes, B., and Mangin, P., Detection of drugs in human hair using Abbott ADx, with confirmation by gas chromatography/mass spectrometry (GC/MS), *J. Forensic Sci.*, 37, 328, 1992.
13. Moeller, M. R., Fey, P., and Sachs, H., Hair analysis as evidence in forensic cases, *Forensic Sci. Int.*, 63, 43, 1993.
14. Hanes, G., Scholtz, H., Matusi, P., Baumgartner, W., and Kippenberger, D. J., Determination of carboxy-THC in hair by tandem mass spectrometry, presented at the second International Meeting on Clinical and Forensic Aspect of Hair Analysis, Genoa, Italy, June 6–8, 1994.
15. Jurado, C., Giménez, M. P., Menéndez, M., and Repetto, M., Simultaneous quantification of opiates, cocaine and cannabinoids in hair, *Forensic Sci. Int.*, 70, 165, 1995.
16. Mieczkowski, T., A research note: the outcome of GC/MS/MS confirmation of hair assays on 93 cannabinoid (+) cases, *Forensic Sci. Int.*, 70, 83, 1995.
17. Cirimele, V., Sachs, H., Kintz, P., and Mangin, P., Testing human hair for cannabis. III. Rapid screening procedure for simultaneous identification of Δ^9-tetrahydrocannabinol, cannabinol and cannabidiol, *J. Anal. Toxicol.*, in press.
18. Kintz, P., Cirimele, V., and Mangin, P., Testing human hair for cannabis. II. Identification of THC-COOH by GC/MS-NCI as an unique proof, *J. Forensic Sci.*, 40, 619, 1995.
19. Blank, D. L. and Kidwell, D. A., Decontamination procedures for drugs of abuse in hair: are they sufficient?, *Forensic Sci. Int.*, 70, 13, 1995.
20. Kintz, P. and Mangin, P., What constitutes a positive result in hair analysis: proposal for the establishment of cut-off values, *Forensic Sci. Int.*, 70, 1, 1995.
21. Nakahara, Y., Effect of the physicochemical properties on the incorporation rates of drugs into hair, presented at the second International Meeting on Clinical and Forensic Aspect of Hair Analysis, Genoa, Italy, June 6–8, 1994.

22. Kintz, P., Tracqui, A., and Mangin, P., Opiate concentrations in human head, axillary and pubic hair, *J. Forensic Sci.*, 88, 657, 1993.

23. Offidani, C., Strano Rossi, S., and Chiarotti, M., Drug distribution in the head, axillary and pubic hair of chronic addicts, *Forensic Sci. Int.*, 63, 105, 1993.

24. Mieczkowski, T. and Newel, N., Comparing hair and urine assays for cocaine and marijuana, *Fed. Prob.*, 67, 59, 1993.

25. Kintz, P., Interlaboratory comparison of quantitative determinations of drugs in hair samples, *Forensic Sci. Int.*, 70, 105, 1995.

26. Jurado, C., Menéndez, M., Repetto, M., Kintz, P., Cirimele, V., and Mangin, P., Hair testing for cannabis in Spain and France: is there a difference in consumption?, presented at the TIAFT-SOFT Joint Meeting, Tampa, FL, October 31 to November 4, 1994.

27. Nakahara, Y., Detection and diagnostic interpretation of amphetamines in hair, *Forensic Sci. Int.*, 70, 135, 1995.

28. Ishiyama, I., Nagai, T., and Toshida, S., Detection of basic drugs (methamphetamine, antidepressants and nicotine) from human hair, *J. Forensic Sci.*, 28, 380, 1983.

29. Suzuki, O., Hattori, H., and Asano, M., Detection of methamphetamine and amphetamine in a single human hair by gas chromatography/chemical ionization mass spectrometry, *J. Forensic Sci.*, 29, 611, 1984.

30. Suzuki, O., Hattori, H., and Asano, M., Nails as useful materials for detection of methamphetamine or amphetamine abuse, *Forensic Sci. Int.*, 24, 9, 1984.

31. Nagai, T., Sato, M., Nagai, T., Kamiyama, S., and Miura, Y., New analytical method for stereoisomers of methamphetamine and amphetamine and its application to forensic toxicology, *Clin. Biochem.*, 22, 439, 1989.

32. Nakahara, Y., Takahashi, K., Shimamine, M., and Takeda, Y., Hair analysis for drug abuse. I. Determination of methamphetamine and amphetamine in hair by stable isotope dilution gas chromatography/mass spectrometry method, *J. Forensic Sci.*, 36, 70, 1991.

33. Nakahara, Y., Takahashi, K., Takeda, Y., Kunoma, K., Fukui, S., and Tokui, T., Hair analysis for drug abuse. Part II. Hair analysis for monitoring of methamphetamine abuse by stable isotope dilution gas chromatography/mass spectrometry, *Forensic Sci. Int.*, 46, 243, 1990.

34. Nakahara, Y., Ishigami, A., Takeda, Y., Usagawa, T., and Uda, T., Enzyme Linked Immunosorbent Assay (ELISA) using monoclonal antibody to detect methamphetamine in urine and hair, *Jpn. J. Toxicol. Environ. Health*, 35, 333, 1989.

35. Suzuki, S., Inoue, T., Hori, H., and Inayama, S., Analysis of methamphetamine in hair, nail, sweat, and saliva by mass fragmentography, *J. Anal. Toxicol.*, 13, 176, 1989.

36. Moriya, F., Miyaishi, S., and Ishizu, H., Presumption of an history of methamphetamine abuse by postmortem analyses of hair and nails: a case report, *Jpn. J. Alcohol Drug Depend.*, 27, 152, 1992.

37. Nakahara, Y., Kikura, R., Takahashi, K., and Kunoma, K., GC/MS analysis of drugs and metabolites in hair for diagnosis of chronic methamphetamine abuse, *Adv. Chem. Diagnos. Treat. Metab. Disord.*, Matsumoto, Ed., London, 1994, 187.

38. Nakahara, Y., Kikura, R., Takahashi, K., Wakutani, K., and Numano, H., Hair analysis of an infant born from maternal methamphetamine abuser during pregnancy, *Jpn. J. Forensic Toxicol.*, 12, 160, 1994.

39. Moeller, M. R., Fey, P., and Wennig, R., Simultaneous determination of drugs of abuse (opiates, cocaine and amphetamine) in human hair by GC/MS and its application to a methadone treatment program, *Forensic Sci. Int.*, 63, 185, 1993.

40. Kintz, P. and Mangin, P., Determination of gestational opiate, nicotine, benzodiazepine, cocaine and amphetamine exposure by hair analysis, *J. Forensic Sci. Soc.*, 33, 139, 1993.

41. Nakahara, Y., Takahashi, K., and Shimamine, M., Hair analysis for drug abuse. III. Movement and stability of methoxyphenamine (as a model compound of methamphetamine) along hair shaft with hair growth, *J. Anal. Toxicol.*, 16, 253, 1992.

42. Moeller, M. R., Maurer, H. H., and Roesler, M., MDMA in blood, urine and hair: a forensic case, in *Proc. 30th TIAFT Meeting*, Takeaki Nagata, Ed., Yoyodo Printing Kaisha Ltd., Fukuoka, Japan, 1992, 347.

Chapter **8**

UNUSUAL DRUGS IN HAIR

Antoine Tracqui

CONTENTS

I. INTRODUCTION

Since the early 1980s, the development of highly sensitive and specific analytical methods such as radioimmunoassay (RIA) or gas chromatography/mass spectrometry (GC/MS) has allowed the determination of organic, exogenous compounds trapped in hair. This theoretically offered the possibility of revealing an individual's recent history of drug exposure beginning at sampling day and dating back over a period of weeks to months. Hair has been initially regarded as a tool of interest in the framework of forensic toxicology, especially to broaden the diagnostic time window for the retrospective assessment of illicit drug consumption. As a consequence, most works carried out during the 1980–90 decade have been focused on the detection in hair of *three groups of drugs of abuse:* opiates (morphine, codeine, 6-monoacetylmorphine) and cocaine (plus metabolites) in the U.S. and European countries, and amphetamine/methamphetamine in Japan. In this regard, the year 1990 stands for a kind of frontier between this early, somewhat restrictive range of applications, and a much wider meaning of hair analysis, as the number of organic compounds susceptible to be assayed has increased in an explosive way from this timepoint (Figure 1).

To date more than 50 pharmaceuticals or drugs of abuse have been reported to be detectable in hair after oral or parenteral administration, in addition to the "classical" drugs mentioned above. The present chapter aims to review the papers devoted to these "unusual" compounds, that have been categorized in 11 classes: opioids (i.e., semisynthetic or synthetic morphine derivatives), hallucinogens, psychostimulants (including nicotine), barbiturates, benzodiazepines, other sedatives/hypnotics,

0-8493-8112-6/96/$0.00+$.50
© 1996 by CRC Press, Inc.

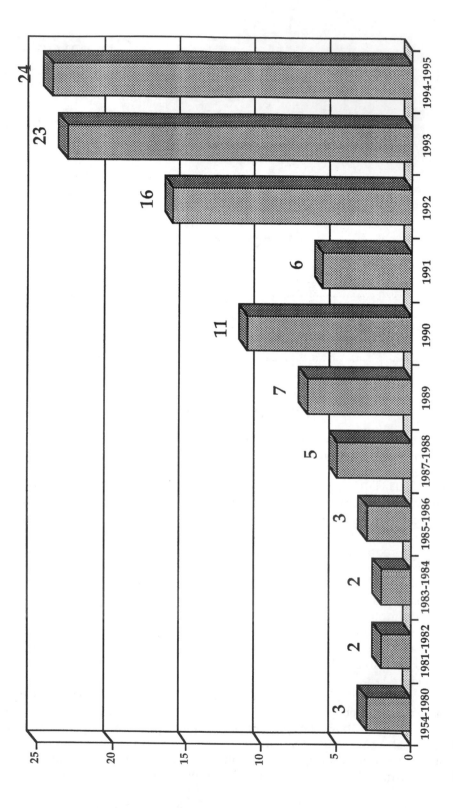

Figure 1
Time-related evolution of the number of communications devoted to hair analysis for "unusual" compounds.

antidepressants, neuroleptics (antipsychotics), cardiovascular drugs, anti-infectious drugs, and miscellaneous compounds. The articles are listed in Table 1. For each one, the main data on hair decontamination, preparation, and analysis are systematically indicated, since this review is primarily intended to serve as a "cookbook" for new investigators wishing to work on similar substances.

TABLE 1.

Unusual Drugs in Hair

Compound	Procedure	Drug levels	Comments	Ref.
I. Opioids				
Buprenorphine	D: CH_2Cl_2 (5 mL, 2 min) × 2 P: MP, then HCl 0.1 N (12 h, 56°C), EX: toluene A: LC-ECD	n = 3 abusers r = 0.48-0.59 m = 0.53	norbuprenorphine 0.06-0.14	1
Buprenorphine	Same as Ref. 1, except A: RIA, GC/MS (EI mode), LC-ECD	n = 14 abusers r = 0.02-0.59 m = 0.25 SD = 0.20	norbuprenorphine n.d.-0.15	2
Dextromoramide	D: CH_2Cl_2 (5 mL, 2 min) × 2 P: MP, then HCl 0.1 N (12 h, 56°C), EX: CPH A: GC/MS (EI mode)	n = 1 abuser r = 1.09-1.93 m = 1.50	CS analysis (3 × 2.5-cm segments)	3
Dihydrocodeine	D: acetone P: NaOH (30 g/l), then 25-% HCl (30 min, 100°C), EX: SPE, DR: HFBA A: GC/MS (EI mode)	n = 28 hair segments from 13 abusers r = 1.2-31.2 m = 8.21 SD = 7.97	CS analysis on 3-cm segments	4
Dihydrocodeine	D, P: ? A: GC/MS	n = 19 abusers r = 0.3-273.0 m = 27.5		5
Ethylmorphine	D: CH_2Cl_2 (5 mL, 2 min) P: HCl 0.1 N (12 h, 56°C), EX: CPH, DR: BSTFA + TMCS 1% A: GC/MS (EI mode)	n = 1 fatality r = head 0.12, pubis 0.18, axilla 0.08	Blood ethylmor- phine 488 ng/mL	6
Fentanyl	D: MeOH (3 × 1 mL) P: MeOH (12 h, 40°C) A: RIA	n = 13 patients r = n.d-0.048		7
Hydromorphone	D: none P: microtoming in paraffin A: FT-IR microscopy	n = 1 overdose r = 12.7	GC/MS used for quantification	8
Methadone	D, P: ? A: RIA	?		9
Methadone	D: H_2O (10 mL, 3 x), EtOH (10 mL, 3 x) P: HCl 0.1 N (12 h, 45°C) A: RIA	n = 2 patients r = 4.9-5.2		10
Methadone	Same as Ref. 10	Same as Ref. 10		11
Methadone	Same as Ref. 10	n = 10 patients r = head 0.5-2.7, pubis 1.0-4.0, axilla 1.3-8.0		12
Methadone	D: ? P: HCl 0.1 N (12 h, 45°C), EX: CHCl3, DR: none A: GC/MS (EI mode)	n = 7 patients r = head 0.5-2.9, pubis 1.0-3.9, axilla 2.5-6.9	RIA used for quantification; see controversy[14]	13
Methadone	Same as Ref. 10, except P (24 h at 50°C)	?	Compar. external contamination vs. internal transfer	15

TABLE 1. (continued)

Unusual Drugs in Hair

Compound	Procedure	Drug levels	Comments	Ref.
Methadone	D: warm H_2O (5 min), acetone (1 min) P: MP, EX: SPE, DR: PFPA-PFPOH A: GC/MS (EI mode)	n = 94 hair segments from 5 patients r = 0.8-42.0	EDDP: 0.0-2.4	16
Pentazocine	D, P: ? A: GC/MS (EI mode)	n = 1 abuser r = 200		5
Pholcodine	D: acetone P: MP, then 1) Sörensen buffer (48 h), A: RIA, or 2) EX: DPE, DR: AAA/pyridine (3: 2), A: GC/MS (EI mode)	n = 3 subjects r = head 0.5-0.6, beard 0.6-1.7	RIA used for quantification	17
Zipeprol	D: CH_2Cl_2 (5 mL, 2 min) × 2 P: HCl 0.1 N (12 h, 56°C), EX: CPH, DR: BSTFA + TMCS 1% A: GC/MS (EI mode)	n = 1 fatality r = 24.3-38.6 m = 33.0	CS analysis on 2-cm segments Blood zipeprol 6.69 mg/l	18

II. Hallucinogens

Compound	Procedure	Drug levels	Comments	Ref.
Cannabis (THC)	?	?	Animal study with tritiated drugs	19
Cannabis	D,P: ? A: RIA	n = 28 abusers r = 0.008->0.1	Unpublished data from Baumgartner; no GC/MS confirmation	20
Cannabis	D,P: ? A: RIA (?)	n = ? r = up to 0.5	Results given in ng marijuana/10 mg hair	21
Cannabis (Δ9-THC, Δ6-THC, Δ9-THC-COOH + metabolites)	D: H_2O (10 mL, 3 ×), EtOH (10 mL, 3 ×) P: HCl 0.1 N (12 h, 45°C) A: RIA, or GC/MS after EX by n-hexane	n = 11 abusers r = head 0.0–3.1 (m = 1.7, SD = 1.1), pubis 0.5–3.8, axilla 0.4–1.9	n = 6 for pubis and axilla RIA used for quantification See remarks[14,23]	22
Cannabis (Δ9-THC-COOH)	D: warm H_2O and acetone P: MP, 2-% NaOH (30 min, 60°C), then 2-% HCl (12 h, room temp.), EX: SPE, DR: PFPA-PFPOH A: GC/MS (EI mode)	n = 1 fatality (from opiates) r = 2.0		24
Cannabis (Δ9-THC-COOH)	D: EtOH (15 min, 37°C) P: NaOH 1 N (60 min, 100°C) EX: CPH, DR: TFA A: GC/MS (EI mode)	n = 27 abusers r = 0.4–2.7		25
Cannabis (Δ9-THC-COOH)	D: 5 mL EtOH (15 min, 37°C) P: NaOH 1 N (10 min, 100°C) EX: CPH, DR: BSTFA + TMCS 1% A: GC/MS (EI mode)	n = 32 abusers r = 0.27–2.91		26
Cannabis (Δ9-THC-COOH)	Same as Ref. 25	Same as Ref. 25	Comparison FPIA[27] (Abbott ADx) vs. GC/MS	
Cannabis	D, P: ? A: RIA, GC/MS (EI mode)	n = 4 r = 0.8–4.1	THC claimed to be found in 4 ancient Egyptian mummies	28

TABLE 1. (continued)

Unusual Drugs in Hair

Compound	Procedure	Drug levels	Comments	Ref.
Cannabis (Δ9-THC, Δ9-THC-COOH)	D, P: ? A: GC/MS (EI mode)	n = 10 r = THC 0.4–6.2 (m = 2.0), THC-acid 1.7–5.0 (n = 2)		5
Cannabis (Δ9-THC-COOH)	D: H_2O, acetone, n-hexane P: acetone (US, 2 h, then 40°C, 12 h) A: RIA, KIMS-immunoassay		*In vitro* study on spiked hair, RIA vs. Abuscreen On-line®	29
Cannabis (Δ9-THC, Δ9-THC-COOH)	1) *Sevilla procedure:* D: CH_2Cl_2 (15 min, 37°C) P: KOH 11.8 N (10 min, room temp.), EX: n-hexane: ethyl acetate (9:1), DR: HFBA + HFPOH A: GC/MS (EI mode) 2) *Strasbourg procedure:* D: CH_2Cl_2 (5 mL, 2 min, room temp.) × 2 P: MP, NaOH 1 N (30 min, 95°C), EX: n-hexane: ethyl acetate (9:1), DR: PFPA-PFPOH A: GC/MS (NCI mode)	n = 10 abusers r = 1) *Sevilla* THC 0.34–1.60 (m = 0.71, SD = 0.35) THC-COOH 0.02–0.29 (m = 0.12, SD = 0.08) 2) *Strasbourg* THC 0.29–1.32 (m = 0.60, SD = 0.30) THC-COOH 0.04–0.26 (m = 0.15, SD = 0.08)	Comparison of procedures for cannabinoids in Sevilla vs. Strasbourg	30
Cannabis (Δ9-THC-COOH)	D: H_2O (1 mL, 30 min) P: NaOH 10 N (0.2 mL, 1 h, 70°C), EX: n-hexane: ethyl acetate (9:1), DR: HFBA + HFPOH A: GC/MS/MS (NCI mode)	n = 1 abuser r = 0.18 pg/mg	Detection limit for THC-COOH: 20 fg/mg	31
Cannabis (Δ9-THC-COOH)	Same as Ref. 31	Same as Ref. 31	Same as Ref. 31	32
Cannabis (Δ9-THC, Δ9-THC-COOH , cannabinol, cannabidiol)	*Method I:* D: CH_2Cl_2 (5 mL, 2 min, room temp.) × 2 P: MP, NaOH 1 N (10 min, 95°C), EX: n-hexane: ethyl acetate (9:1), DR: none A: GC/MS (EI mode) *Method II:* Same, except DR: PFPA-PFPOH A: GC/MS (NCI mode)	n = 43 abusers r = THC 0.10–2.17 (m = 0.63), THC-COOH 0.005–0.39 (m = 0.10), cannabinol 0.02–0.25 (m = 0.08), cannabidiol 0.04–0.35 (m = 0.26)	Detection limits: THC, cannabinol, cannabidiol (Method I): 0.1, 0.01, 0.02 ng/mg, THC-COOH; (Method II): 5 pg/mg	33
Cannabis (Δ9-THC, Δ9-THC-COOH)	Same as Method I in Ref. 33, except DR: PFPA, PFPOH	n = 43 abusers r = 1) *head:* THC 0.26–2.17 (m = 0.74), THC-COOH 0.07–0.33 (m = 0.16) 2) *pubis:* THC 0.34–3.91 (m = 1.35), THC-COOH 0.07–0.83 (m = 0.28)		34
Cannabis (Δ9-THC, Δ9-THC-COOH)	Same as Ref. 30 (Sevilla procedure)	n = 298 hair segments from 70 abusers r = THC 0.06–7.63 (m = 0.97), THC-COOH 0.06–3.87 (m = 0.50)		35

TABLE 1. (continued)

Unusual Drugs in Hair

Compound	Procedure	Drug levels	Comments	Ref.
Cannabis (Δ9-THC-COOH)	Same as Method II in Ref. 33	n = 30 abusers r = 0.02–0.39 m = 0.12	Detection limit: 5 pg/mg	36
Cannabis (Δ9-THC, Δ9-THC-COOH)	Same as Ref. 31, except A: GC/MS/MS (NCI mode) vs. RIA	n = 93 abusers r = THC 0.003–0.44 (m = 0.04), THC-COOH 30–1530 fg/mg	Comparison GC/MS/MS vs. RIA	37
Phencyclidine	D: surgical detergent P: MP, then MeOH (5 mL, 3 h) A: RIA	n = 8 abusers r = 0.3–2.8 m = 2.0 SD = 0.8		38
Phencyclidine	D,P: ? A: RIA	n = 11 abusers r = 0.14–5.1	CS analysis on 5 subjects	39
Phencyclidine	D,P: ? A: RIA	n = ? r = up to 8		21
Phencyclidine	D: pentane (some min) P: none (hair put directly into probe cup of the GC) A: GC/MS/MS	n = 1 abuser r = 16		40
Phencyclidine + metabolites (PCHP, PPC)	D: ? P: MeOH (acid), EX: SPE, DR: BSA A: GC/MS (EI mode)	r > 0.5 for PCP, PCHP, PPC	Experim. study on rats given 0.1–2.0 mg/kg PCP intraperitoneally	41
LSD-25 + metabolite (norLSD)	D: ? P: MeOH (acid), EX: CH_2Cl_2/toluene (3:7), DR: BSTFA-TMCS A: GC/MS (EI mode)	Samples positive for LSD-25 and norLSD	Experim. study on rats given 0.1–2.0 mg/kg LSD intraperitoneally	42

III. Psychostimulants

Compound	Procedure	Drug levels	Comments	Ref.
Caffeine	D: H_2O (10 mL, 3 ×), EtOH (10 mL, 3 ×) P: MP (at - 180°C), then HCl 0.1 N (12 h, 45°C) A: EMIT®	n = 18 subjects r = head 1.5–66.2 (m = 19.5, SD = 16), pubis 5.0–71.8 (m = 25.1, SD = 19.6), axilla 3.4–124.4 (m = 40.1, SD = 36.0)		43
Caffeine	Same as Ref. 43, except A: EMIT®, and GC/MS (EI mode) after EX by ethyl acetate	Same as Ref. 43		44
Caffeine, theophylline, theobromine	D: CH_2Cl_2 (15 min, 37°C) P: NaOH 1 N (1 mL, 10 min, 100°C), EX: $CHCl_3$ A: HPLC-DAD	n = 35 subjects r = caffeine 0.1–2.3, theophylline 0.1–0.6, theobromine 0.3–10.0		45
Fenfluramine	D: EtOH (15 min, 37°C) P: NaOH 1 N (60 min, 100°C) EX: CPH, DR: TFA A: GC/MS (EI mode)	n = 1 r = 14.1		25
Fenfluramine	Same as Ref. 25	Same as Ref. 25		26,46, 47

TABLE 1. (continued)

Unusual Drugs in Hair

Compound	Procedure	Drug levels	Comments	Ref.
Nicotine	D: detergent, 0.1-% SDS, HCl 0.01 N, acetone P: NaOH 1.5 N (12 h, room temp.), EX: ?, DR: TFA A: GC, GC/MS (EI mode)	n = ? r = 18.0–177.2		48
Nicotine + cotinine	D: shampoo (before sampling), n-hexane (after) P: acetone (10 mL, 30 min, US, room temp.) A: RIA	n = 10 smokers, 10 nonsmokers r = 1) *smokers*: nicotine 3.0–38.7 (m = 15.8, SD = 13.2), cotinine n.d.–1.4 (m = 0.5, SD = 0.53) 2) *nonsmokers*: nicotine n.d.–11.3 (m = 2.42, SD = 3.41), cotinine n.d.–1.0 (m = 0.12, SD = 0.31)	Nicotine and cotinine significantly differ ($p < 0.005$ and $p < 0.05$) between smokers and nonsmokers	49
Nicotine	D: none vs. shampoos (before sampling) P: NaOH 5 M, EX: diethyl ether, DR: none A: GC/NPD	n = 24 smokers, 24 nonsmokers r = smokers 1.3–2.8, nonsmokers 36–60	CS analysis on 2-cm segments Effect of washing studied	50
Nicotine	D: H_2O (10 mL, 3 ×), EtOH (10 mL, 3 ×) P: MP, then HCl 0.1 N (12 h, 45°C) A: RIA, GC/MS (EI mode)	n = 19 smokers (A), 3 passive smokers (B), 5 nonsmokers (C) r = (A) head 0.7–11.1 (m = 4.3, SD = 3.5), pubis 0.4–23.0 (m = 5.0, SD = 5.2), axilla 0.4–19.2 (m = 5.3, SD = 5.4) (B,C): n.d. in all samples	Correlations head/axilla, head/pubis, and pubis/axilla highly signif. ($p < 0.001$, r = 0.827, 0.753, 0.890)	51
Cotinine	Same as Ref. 51	n = 10 smokers (A), 5 nonsmokers (B) r = (A) head 0.9–11.1, pubis 0.8–53.8, axilla 1.9–22.4 (B): n.d. in all samples		52
Nicotine	Same as Ref. 51	Same as Ref. 51	Same as Ref. 51	53
Nicotine + cotinine	D: EtOH (15 min, 37°C) P: NaOH 1 N (60 min, 100°C) EX: CPH, DR: none A: GC/MS (EI mode)	n = 42 smokers, 22 nonsmokers r = 1) *smokers*: nicotine 0.91–33.89, cotinine 0.09–4.99 2) *nonsmokers*: nicotine 0.06–1.82, cotinine 0.01–0.13		25
Nicotine	Same as Ref. 25	n = 42 smokers (A), 12 nonsmokers exposed (B), 10 nonsmokers nonexposed (C) r = (A) 0.91–33.89 (B) 0.54–1.82 (C) 0.06–0.33	Sets cutoff at 0.5 ng/mg between exposed and nonexposed nonsmokers	26

TABLE 1. (continued)

Unusual Drugs in Hair

Compound	Procedure	Drug levels	Comments	Ref.
Nicotine + cotinine	D: none P: NaOH 1 N (60 min, 100°C) EX: diethyl ether, DR: none A: GC/MS (EI mode)	n = 42 smokers (A), 12 nonsmokers exposed (B), 10 nonsmokers nonexposed (C) r = (A) nicotine 0.91–33.89, cotinine 0.09–4.99, (B) nicotine 0.54–1.82, cotinine 0.01–0.13, (C) nicotine 0.06–0.33, cotinine 0.01–0.13	Sets cutoffs for nicotine at 2 ng/mg between smokers and nonsmokers, and 0.5 ng/mg between exposed and nonexposed nonsmokers	54
Nicotine + cotinine	Same as Ref. 54	n = 56 smokers (A), 31 nonsmokers (B), 9 neonates from smoking mothers (C) r = (A) nicotine 0.91–38.27, cotinine 0.09–4.99, (B) nicotine 0.06–1.82, cotinine 0.01–0.13, (C) nicotine 0.27–1.37 (m = 0.70, SD = 0.42)	1st work showing nicotine in hair from neonates exposed *in utero*	55
Nicotine	Same as Ref. 54, except: D: CH$_2$Cl$_2$ (5 mL, 15 min, 37°C)	n = 40 smoking mothers (A) and 40 neonates (B) r = (A) 0.37–63.50 (B) 0.15–11.80	1st work showing mother/neonate correlation for nicotine in hair ($p < 0.001$, r = 0.83)	56,57
Nicotine + cotinine	D: detergent P: NaOH 0.6 N (12 h, 50°C) A: RIA	n = 10 smoking mothers (A), their neonates (B), 11 nonsmoking mothers (C), their neonates (D) 1) *Nicotine*: m = (A) 21 (B) 6 (C) 0.9 (D) 0.7 2) *Cotinine*: m = (A) 3.7 (B) 2.1 (C) 0.3 (D) 0.3	Correlations (A)/(B): $p < 0.01$, r = 0.78 (nicotine), $p < 0.05$, r = 0.64 (cotinine); for nicotine and cotinine, (A) > (C) ($p < 0.001$) and (B) > (D) ($p < 0.001$)	58
Nicotine	Same as Ref. 56	n = 34 neonates from smoking mothers r = 0.15–11.80		59
Nicotine	Same as Ref. 56,57	Same as Ref. 56,57		60,61
Nicotine + cotinine	Same as Ref. 58	Same as Ref. 58	Same as Ref. 58	62
Nicotine	D: 1-% SDS (2 ×), H$_2$O (2 ×) P: NaOH 2.5 N (30 min, 80°C), EX: diethyl ether A: GC-NPD	n = 36 smokers, 14 nonsmokers + animal study (rats) r = ?	Correlation hair nicotine/no. of cigarettes daily smoken: $p < 0.01$, r = 0.685	63
Nicotine	Same as Ref. 63, except D: ?	n = 22 smokers r = about 20 to about 60	CS analysis on 1-cm segments	64
Nicotine + cotinine	Same as Ref. 58	Same as Ref. 58	Same as Ref. 58	65

TABLE 1. (continued)

Unusual Drugs in Hair

Compound	Procedure	Drug levels	Comments	Ref.
Nicotine	Same as Ref. 54	n = ? r = ?		66
Nicotine + cotinine	D, P: ? A: RIA (?)	n = smoking mothers (A), their neonates (B), nonsmoking mothers (C), their neonates (D) (total: 94 pairs) 1) *Nicotine*: m = (A) 19.2 (B) 2.4 (C) 1.2 (D) 0.4 2) *Cotinine*: m = (A) 6.3 (B) 2.8 (C) 0.3 (D) 0.26	Correlations mother/infant for nicotine ($p <$ 0.001, r = 0.49) and cotinine ($p <$ 0.0001, r = 0.85) For nicotine and cotinine, (A) > (C) ($p <$ 0.0001) and (B) > (D) ($p <$ 0.01)	67

IV. Barbiturates

Compound	Procedure	Drug levels	Comments	Ref.
Barbiturate (pentobarbital ?)	D, P: ? A: RIA (?)	n = 1 abuser r = about 1 to about 4		68
Amobarbital	D: EtOH (15 min, 37°C) P: NaOH 1 *N* (60 min, 100°C) EX: CPH, DR: none A: GC/MS (EI mode)	n = 1 r = 41.6		25
Amobarbital	Same as Ref. 25	Same as Ref. 25		26,27, 46
Phenobarbital	D, P: ? A: UV spectrophotometry	n = 5 guinea pigs (A), 6 guinea pigs (B) r = (A) 159–4760 (B) 235–1545	(A) given 16.2–32.4 mg/d phenobarb., (B) given 32.4 mg/d 1st paper on hair analysis for organic compounds	69
Phenobarbital	D: H_2O (10 ×) P: 0.1-% SDS (0.2 mL, 24 h) A: RIA	n = 1 patient r = 22.7	Analysis performed on 1 single hair (0.39 mg)	70
Phenobarbital	D: H_2O (10 mL, 3 ×), EtOH (10 mL, 3 ×) P: HCl 0.1 *N* (12 h, 45°C) A: RIA	n = ? (sheep) r = 7–50	Animal study on sheep given 0.1 g/d for 1 month	10
Phenobarbital	D: EtOH (15 min, 37°C) P: NaOH 1 *N* (60 min, 100°C) EX: CPH, DR: none A: GC/MS (EI mode)	n = 2 r = 51.4–137.3		25
Phenobarbital	Same as Ref. 25	Same as Ref. 25		27
Phenobarbital	Same as Ref. 25	n = 4 r = 21.7–137.3		26,46
Phenobarbital	D: CH_2Cl_2 (10 min) P: MP, EX: SPE, DR: none A: GC/MS (EI mode)	n = 40 patients r = 1.5–194.0 m = 36.4 SD = 37.9	Hair/serum correlation: $p < 0.001$, r = 0.779 (n = 23)	71
Secobarbital	D: EtOH (15 min, 37°C) P: NaOH 1 *N* (60 min, 100°C) EX: CPH, DR: none A: GC/MS (EI mode)	n = 2 r = 21.6–58.9		25
Secobarbital	Same as Ref. 25	Same as Ref. 25		27

TABLE 1. (continued)

Unusual Drugs in Hair

Compound	Procedure	Drug levels	Comments	Ref.
Secobarbital	Same as Ref. 25	n = 3 r = 21.6–58.9		26,46
V. Benzodiazepines				
Diazepam + nordiazepam	D: EtOH (15 min, 37°C) P: NaOH 1 N (60 min, 100°C) EX: CPH, DR: none A: GC/MS (EI mode)	*Diazepam*: n = 1, r = 1.37 *Nordiazepam*: n = 2, r = 1.04–1.47		25
Diazepam + nordiazepam	Same as Ref. 25	Same as Ref. 25		27,46
Diazepam + nordiazepam	Same as Ref. 25	*Diazepam*: n = 1, r = 1.37 *Nordiazepam*: n = 3, r = 1.04–2.41		26
Diazepam	D: CH_2Cl_2 (5 mL, 15 min, 37°C) P: NaOH 1 N (10 min, 100°C) EX: CPH, DR: none A: GC/MS (EI mode)	n = ? r = 3.36–17.55	Hair from neonates with mother under benzodiaz. during pregnancy	57
Diazepam	D: EtOH, phosphate buffers (37°C, 3 ×) P: proprietary method (!) A: RIA	n = 7 patients r = 0.20–1.64	CS analysis in 2 subjects	72
Diazepam	Same as Ref. 57, except P: HCl 1 N (12 h, 50°C), EX : CPH, DR: none	n = 8 r = 3.36–17.55	Same as Ref. 57	59
Nordiazepam	D: H_2O, acetone, *n*-hexane P: acetone (US, 2 h, then 40°C, 12 h) A: RIA, KIMS-immunoassay		*In vitro* study on spiked hair, RIA vs. Abuscreen On-line®	29
Flunitrazepam	D: EtOH (15 min, 37°C) P: NaOH 1 N (60 min, 100°C) EX: CPH, DR: none A: GC/MS (EI mode)	n = 1 r = 0.41		25
Flunitrazepam	Same as Ref. 25	Same as Ref. 25		26,27
Nitrazepam	D: EtOH (15 min, 37°C) P: NaOH 1 N (60 min, 100°C) EX: CPH, DR: none A: GC/MS (EI mode)	n = 1 r = 0.37		25
Nitrazepam	Same as Ref. 25	Same as Ref. 25		26,27
Oxazepam	D: CH_2Cl_2 (5 mL, 15 min, 37°C) P: NaOH 1 N (10 min, 100°C) EX: CPH, DR: none A: GC/MS (EI mode)	n = ? r = 0.78–31.83	Hair from neonates with mother under benzodiaz. during pregnancy	57
Oxazepam	Same as Ref. 57, except P: HCl 1 N (12 h, 50°C), EX : CPH, DR: none	n = 3 r = 0.78–31.83	Same as Ref. 57	59
VI. Other Sedatives/Hypnotics				
Meprobamate	D: CH_2Cl_2 (5 mL, 15 min, 37°C) P: HCl 1 N (12 h, 60°C) EX: CHCl3, DR: TFMPTMA A: GC/MS (EI mode)	n = 2 patients r = 3.32–4.21	Subjects given 200 mg/d meprobamate for 6 months	73

TABLE 1. (continued)

Unusual Drugs in Hair

Compound	Procedure	Drug levels	Comments	Ref.
Meprobamate	Same as Ref. 73	n = 16 subjects in 3 groups (A, B, C) r = (A) 4.27–6.08 (B) 8.54–13.01 (C)11.89–17.64	Subjects given a single dose of (A) 400, (B) 800, (C) 1200 mg Meprobamate levels measured daily in beard hair	74
Methaqualone	D, P: ? A: RIA (?)	n = 1 abuser r = about 5		68
Methaqualone	D: Acetone, H_2O, acetone P: NaOH 1 N (100°C) A: RIA	n = 1 abuser r = ?		75
Methaqualone	D,P: ? A: RIA (?)	n = ? r = about 2 to about 12	Animal study (mice)	21
VII. Antidepressants				
Amineptine	D: CH_2Cl_2 (5 mL, 2 min) × 2 P: MP, then HCl 0.1 N (12 h, 56°C), EX: ethyl acetate, DR: TFMPTMA A: GC/MS (EI mode)	n = 1 abuser r = 8.06–12.23	CS analysis on 2 segments (root to 3 cm, 3 cm to end)	76
Amitriptyline + nortriptyline	D: detergent, 0.1-% SDS, HCl 0.01 N, acetone P: NaOH 1.5 N (12 h, room temp.), EX: ?, DR: none (?) A: GC	n = 3 patients r = ami. 12.8–23.9, nor. 11.2–118.5		48
Amitriptyline + nortriptyline	D: EtOH (15 min, 37°C) P: NaOH 1 N (60 min, 100°C) EX: CPH, DR: none A: GC/MS (EI mode)	n = 1 r = ami. 0.42, nor. 0.91		25
Amitriptyline + nortriptyline	Same as Ref. 25	Same as Ref. 25		27
Amitriptyline	Same as Ref. 25	n = 6 r = 0.04–1.89		46
Amitriptyline	Same as Ref. 25	n = 14 r = 0.04–1.89		26
Amitriptyline	D: EtOH (10 min) P: NaOH 1 N (30 min, 100°C), EX: n-heptane/ isoamylalcohol (98.5: 1.5), DR: none A: GC/MS (EI mode)	n = 30 patients r = 0.00–17.21 m = 4.06 SD = 4.70	Correlation dose given/hair level: $p < 0.002$, r = 0.563	77,78
Clomipramine	D: EtOH (15 min, 37°C) P: NaOH 1 N (60 min, 100°C) EX: CPH, DR: none A: GC/MS (EI mode)	n = 2 r = 0.37–0.79		25
Clomipramine	Same as Ref. 25	Same as Ref. 25		26,27
Imipramine + desipramine	D: detergent, 0.1-% SDS, HCl 0.01 N, acetone P: NaOH 1.5 N (12 h, room temp.), EX: ?, DR: none (?) A: GC	n = 3 patients r = imi. 16.5–69.2, desi. 16.5–22.6		48

TABLE 1. (continued)

Unusual Drugs in Hair

Compound	Procedure	Drug levels	Comments	Ref.
Amitriptyline + nortriptyline + imipramine + desipramine + dothiepin + nordothiepin	D: H_2O (5 min) then MeOH (3×) P: NaOH (30 min, 70°C) vs. HCl 0.1 N (18 h, 55°C) vs. MeOH (18 h, 55°C) vs. 10 g/l Subtilisin (18 h, 55°C), EX: n-hexane/butanol (95: 5) A: HPLC-UV	n = 9 fatalities r = ?	Comparison between 4 digestion procedures: best recovery with NaOH for antidepressants	79

VIII. Neuroleptics (Antipsychotics)

Compound	Procedure	Drug levels	Comments	Ref.
Chlorpromazine	?	?	Animal study with tritiated drugs	19
Chlorpromazine	D: 0.1-% SDS, then H_2O P: NaOH 2 N (1 mL, 30 min, 80°C), EX: n-hexane/isoamylalcohol (98.5: 1.5) A: HPLC-ECD	n = 23 patients r = 1.6–27.5	Correlation hair/ plasma levels: $p <$ 0.001, r = 0.902 Correlation hair level/dose given: $p < 0.001$, r = 0.788	80
Clozapine	D: H_2O, acetone, petrol ether (10 mL each) P: MeOH A: GC/MS	n = 1 fatality r = 3.6	Hair taken from exhumed child	81
Haloperidol	D: 0.1-% SDS, then H_2O P: NaOH 2.5 N (30 min, 80°C), EX: n-hexane/ isoamylalcohol (98.5: 1.5) A: RIA	n = 40 patients r = 2.33–245.00 (in hair), 0.67–16.89 (in nails)	Correlation hair/ plasma levels: $p <$ 0.001, r = 0.772 Correlation hair level/daily dose: $p < 0.001$, r = 0.555 Correlation nail level/daily dose: $p < 0.05$, r = 0.525	82
Haloperidol	Same as Ref. 82	n = 8 patients (test 1), 36 rats (test 2) r = ?	CS analyses	83
Haloperidol + metabolite	Same as Ref. 82, except A: HPLC-ECD, RIA	n = 59 patients r = 3.44–208.11 (haloperidol), 4.11–106.52 (metabolite)	Correlation hair/ plasma levels: $p < 0.001$, r = 0.558 (haloperidol) $p < 0.001$, r = 0.563 (metabolite) Correlation hair level/daily dose: $p < 0.001$, r = 0.682 (haloperidol) $p < 0.001$, r = 0.813 (metabolite)	84
Haloperidol	Same as Ref. 82	n = 7 patients + rat study r = 0–190	Study of hair levels albino vs. pigmented rats (animal test), and black vs. white hair (human test)	85
Haloperidol + metabolite	Same as Ref. 82, except A: HPLC-ECD	n = 3 neonates + mothers r = ?	CS analysis neonate vs. mother	86

TABLE 1. (continued)

Unusual Drugs in Hair

Compound	Procedure	Drug levels	Comments	Ref.
Haloperidol + metabolite	Same as Ref. 82, except A: HPLC-ECD	n = 10 patients r = about 10 to about 70 (haloperidol)	Study of correlation haloperidol (or metab.) vs. daily dose, C_{max}, AUC, plasma levels	87
Haloperidol	D: H_2O (5 min) then MeOH (3 ×) P: NaOH (30 min, 70°C) vs. HCl 0.1 N (18 h, 55°C) vs. MeOH (18 h, 55°C) vs. 10 g/l Subtilisin (18 h, 55°C), EX: n-hexane/butanol (95: 5) A: HPLC-UV	n = 9 fatalities r = ?	Comparison between 4 digestion procedures: best recovery with NaOH for haloperidol	79
Haloperidol	D: ? P: Case 1: NaOH 1.5 N (12 h), Case 2: HCl 0.1 N (12 h), EX: liquid-liquid A: GC-NPD, GC/MS (EI mode)	n = 2 r = 0–0.033 (case 1), 0.12–0.81 (case 2)	CS analyses on 3 segments (case 1) and 45 3-mm segments (case 2)	88

IX. Cardiovascular Drugs

Compound	Procedure	Drug levels	Comments	Ref.
Atenolol	D: EtOH (15 min, 37°C) P: NaOH 1 N (10 min, 100°C) EX: ether/CH_2Cl_2 (80: 20) A: HPLC-UV	n = 1 patient r = 0.9		89
Betaxolol	Same as Ref. 89	n = 3 r = 0.6–2.8		25
Betaxolol	Same as Ref. 89	n = 5 r = 0.6–2.8		26
Betaxolol	D: EtOH (15 min, 37°C) P: NaOH 1 N (10 min, 100°C) EX: ether/CH_2Cl_2 (80:20) A: HPLC-UV	n = 3 patients r = 1.2, 2.4, 2.7	CS analysis on 1-cm segments for 2 subjects	89
Digoxin	D, P: ? A: RIA	n = ? r = 0.002–0.006		9
Digoxin	Same as Ref. 9	Same as Ref. 9		21
Propranolol	D: EtOH (15 min, 37°C) P: NaOH 1 N (10 min, 100°C) EX: ether/CH_2Cl_2 (80:20) A: HPLC-UV	n = 2 patients r = 1.6–2.4		89
Sotalol	D: EtOH (15 min, 37°C) P: NaOH 1 N (10 min, 100°C) EX: ether/CH_2Cl_2 (80:20) A: HPLC-UV	n = 2 patients r = 4.4–5.3		89

X. Anti-infectious Drugs

Compound	Procedure	Drug levels	Comments	Ref.
Chloroquine + metabolite	D: detergent, H_2O P: 60-% KOH (2 mL, 5 to 10 min, 100°C), EX: diethylether A: TLC, GC/MS (EI mode)	n = 20 patients r = chloroquine 145–310 (n = 2), metabolite 11–23 (n = 2)	Daily dose 100 mg 6 d/7 for several months	90
Cypermethrin	?	?		91

TABLE 1. (continued)

Unusual Drugs in Hair

Compound	Procedure	Drug levels	Comments	Ref.
Ofloxacin	D: 0.1-% SDS, then H_2O P: NaOH 1 N (30 min, 80°C), EX: CHCl3 A: HPLC-fluorimetry	n = 12 subjects (human test), 14 rats (animal test) r = 0 to about 45 (humans), 0 to about 400 (rats)	Correlation hair level/dose given: $p < 0.001$, r = 0.868 (humans) $p < 0.001$, r = 0.853 (rats)	92
Ofloxacin	Same as Ref. 92	n = 14 patients r = 0 to about 200	CS analyses on 1-cm segments	93
Ofloxacin	Same as Ref. 92	n = 4 subjects (human test), 27 rats (animal test) r = 1.9–71.1 (humans), up to ≈ 1000 (rats)	Study of hair levels albino vs. pigmented rats (animal test), and black vs. white hair (human test)	94
Ofloxacin	Same as Ref. 92	n = 3 subjects r = ?	CS analysis on 1-cm segments	64
Temafloxacin	Same as Ref. 92	n = 11 subjects r = ?	CS analysis on 1-cm segments	95
Temafloxacin	Same as Ref. 92	n = 12 subjects in 2 groups (A, B) of 6 each m = (A) 31.7 (B) 226.3 SD = (A) 15.0 (B) 99.4	Subjects given (A) a single dose of 600 mg, or (B) 900 mg/d for 6.5 d CS analysis on 1-cm segments	96
Q-35 (fluoroquinolone)	Same as Ref. 92	n = 18 subjects in 3 groups (A, B, C) of 6 each r = (A) 0–36 (B) 0 to about 175 (C) 0 to about 400	Subjects given (A) single dose of 400 mg (B) 400 mg/d for 6.5 d (C) 600 mg/d for 6.5 d CS analysis on 1-cm segments	97
XI. Miscellaneous				
ɪ-dopa (antiparkinson.) α-methyldopa (antihypertens.) Isoproterenol (bronchodilat.)	?	?		98
Methoxyphenamine (bronchodilat.) + metabolites	D: none vs. 0.1-% SDS then H_2O (3 × each) vs. 0.1-% SDS then H_2O (3 × each) then EtOH (30 s, US) P: MeOH/HCl 5 N (20: 1) (1 h, US, then 12 h, room temp), DR: TFA A: GC/MS (EI mode)	n = 6 patients r = 0 to about 13 (unwashed hair)	Subjects given 250 mg/d methoxyphenami ne for 8 d Drug levels measured daily in beard hair	99
Carbamazepine (antiepileptic) + metabolites: carbamazepine- 10,11-epoxide (CEO), acridine (ACR)	D: acetone (US, 5 min) P: MeOH (US, 8 h, room temp to 45–48°C) vs. NaOH 1 N (30 min, 100°C) then EX: SPE A: GC/MS (EI mode)	n = 30 patients r = carbamazepine 7.6–205.0 (m = 70.0), CEO 1.1–23.0 (m = 6.8), ACR 0.2–1.8 (m = 0.8)		100

TABLE 1. (continued)

Unusual Drugs in Hair

Compound	Procedure	Drug levels	Comments	Ref.
Ftorafur *(antineoplastic)*	D: 0.1-% SDS, then H_2O P: NaOH 1 N (30 min, 80°C), DR: BMDMC, EX: liquid/liquid (3 steps) A: HPLC-fluorimetry	n = 13 patients in 3 groups (A, B, C) m = (A) 0.24 (B) 1.35 (C) 2.85 SD = (A) 0.07 (B) 0.39 (C) 0.74	Subjects (A), (B), (C) given 5, 15, 50 mg/kg/d for 4 weeks Correlation hair level/dose given: $p < 0.001$, r = 0.914	101
Ethyl glucuronide *(ethanol metabolite)*	D, P: ? A: GC/MS	n = ? r = 10–70	Possible marker of chronic alcoholism	102

Note: If units are not specified, all drug levels are in ng/mg hair.

List of Abbreviations:

D	Decontamination	FT-IR	Fourier transformed infrared
P	Preparation	GC	Gas chromatography
A	Analysis	HFBA	Heptafluorobutyric anhydride
n	Number of cases	HFPOH	Hexafluoropropanol
r	Result (if 1 case) or range of results (if > 1 case)	HPLC	High performance liquid chromatography
m	Mean of results	KIMS	Kinetic interaction of microparticles in solution
SD	Standard deviation of results		
AAA	Acetic acid anhydride	LC	Liquid chromotography
AUC	Area under the curve	MeOH	Methanol
BMDMC	4-bromomethyl-6,7-dimethoxycoumarin	MP	Mechanical pulverization (e.g., ball mill)
BSA	Bis(trimethylsilyl)acetamide	MS	Mass spectrometry
BSTFA	Bis(trimethylsilyl)trifluoroacetamide	NCI	Negative chemical ionization
C_{max}	Peak plasma concentration	NPD	Nitrogen-phosphorous detection
CPH	Chloroform/2-propanol/n-heptane (50:17:33, v/v/v)	PFPA	Pentafluoropropionic anhydride
		PFPOH	Pentafluoropropanol
CS	Cross sectional	RIA	Radioimmunoassay
DAD	Diode-array detection	SDS	Sodium dodecyl sulphate
DPE	Dichloromethane/2-propanol/ethyl acetate (1:1:3, v/v/v)	SPE	Solid-phase extraction
		TFA	Trifluoroacetic acid anhydride
DR	Derivatization	TFMPTMA	(m-Trifluoromethylphenyl)trimethyl-ammonium hydroxide
ECD	Electrochemical detection		
EDDP	Methadone metabolite	THC	Tetrahydrocannabinol
EI	Electron impact	TLC	Thin-layer chromatography
EtOH	Ethanol	TMCS	Chlorotrimethylsilane
EX	Extraction	US	Ultrasonication
FPIA	Fluorescence polarization immunoassay	UV	Ultraviolet

II. DISCUSSION

It is not surprising that most of these papers still concern psychotropic compounds: such drugs are well known to account for the great majority of acute poisonings in developed countries (75 to 90%, depending on sources), and most of them constitute, at least from a medical point of view, authentic or potent substances of abuse. A more unexpected — and interesting — observation is the emergence of studies involving *nonpsychotropic* drugs, such as cardiovascular or anti-infectious compounds: in fact, when considering the abundance and diversity of the investigations that have been performed on hair analysis over the past 5 years, it appears more and more obvious that the usefulness of this technology can no longer be limited to the criminal prosecution in individual drug abuse cases. On the contrary, a number of exciting, alternative applications have been suggested, evaluated, or implemented, including:

- Appraisal of criminal liability (e.g., defendants under psychotropic medications);
- Licit or illicit drug fatalities, especially when "classical" biological samples are lacking (fragmented and/or highly decayed corpses);
- Systematic/oriented drug screening at school/military service/jail/workplace;
- Toxicological screening of hair of neonates, when clinical features suggest an intake of definite medications or drugs of abuse by the mother during pregnancy;
- Withdrawal supervision and/or control of compliance to methadone (or other substitutes) for addicts in rehabilitation centers;
- Assessment of occupational exposure to toxicants in workers asking for compensations, or other civil investigations (e.g., divorce, life insurance, etc.);
- Retrospective and long-term control of doping in athletes, racehorses, etc.;
- Control of medical compliance in patients under long-term therapies (antihypertensive drugs, antipsychotics, etc.).

In the same way, the time where hair analysis was restricted to purely descriptive cases ("By Jove! Heroin addicts have morphine in hair!") is over, and now most authors present experimental works with serious methodologies in order to:

- Understand the mechanisms of drug incorporation into hair, and the reasons for its inter-/intraindividual variability;
- Optimize the procedures of decontamination;
- Optimize the procedures of drug release.

The evolution observed in the techniques chosen for analysis also provides a clear indication of these changes: in place of the RIA that was widely used in early times of hair analysis, but has become obsolete since it cannot be practically applied to unequivocal confirmations or study of drug metabolisms, GC/MS is presently employed by almost all investigators as the unique method of reference — for routine as well as for experimental purposes.

Thanks to these efforts, hair analysis has definitely proven to be particularly reliable for all applications requiring only qualitative or semiquantitative determinations (e.g., workplace testing, doping control) where the situation is quite similar to drug abuse testing in arrestees (a "yes/no" answer being sufficient), and in some clinical applications based upon quantitative measurements, such as the determination of tobacco fumes exposure by assaying patients' hair — where unequivocal cutoffs could be demonstrated between active smokers, passive smokers (environmentally exposed), and authentic nonsmokers. In spite of some prematurely optimistic papers, the applicability of hair analysis to other investigations presently remains more dubious, notably in the field of therapeutic compliance monitoring;[103] however, it is believable that the continuous progress in the knowledge of hair toxicology will progressively overcome this last obstacle.

REFERENCES

1. Kintz, P., Determination of buprenorphine and its dealkylated metabolite in human hair, *J. Anal. Toxicol.*, 17, 443, 1993.
2. Kintz, P., Cirimele, V., Edel, Y., Jamey, C., and Mangin, P., Hair analysis for buprenorphine and its dealkylated metabolite by RIA and confirmation by LC/ECD, *J. Forensic Sci.*, 39, 1497, 1994.
3. Kintz, P., Cirimele, V., Edel, Y., Tracqui, A., and Mangin, P., Characterization of dextromoramide (Palfium) abuse by hair analysis in a denied case, *Int. J. Leg. Med.*, 107, 269, 1995.
4. Sachs, H., Denk, R., and Raff, I., Determination of dihydrocodeine in hair of opiate addicts by GC/MS, *Int. J. Leg. Med.*, 105, 247, 1993.
5. Möller, M. R., Fey, P., and Sachs, H., Hair analysis as evidence in forensic cases, *Forensic Sci. Int.*, 63, 43, 1993.
6. Kintz, P., Jamey, C., and Mangin, P., Ethylmorphine concentrations in human samples in an overdose case, *Arch. Toxicol.*, 68, 210, 1994.
7. Wang, W. L., Cone, E. J., and Zacny, J., Immunoassay evidence for fentanyl in hair of surgery patients, *Forensic Sci. Int.*, 61, 65, 1993.
8. Kalasinsky, K. S., Magluilo, J., Jr., and Schaefer, T., Hair analysis by infrared microscopy for drugs of abuse, *Forensic Sci. Int.*, 63, 253, 1993.
9. Baumgartner, W. A., Hill, V. A., Baer, J. D., Lyon, I. W., Charuvastra, V. C., Sramek, J. J., and Blahd, W. H., Detection of drug use by analysis of hair, *J. Nucl. Med.*, 29, 980, 1988.
10. Balabanova, S. and Wolf, H. U., Bestimmung von Drogen und Medikamenten im menschlichen Haar mittels Radioimmunoassay, *Laboratoriumsmedizin*, 12, 332, 1988.
11. Balabanova, S. and Wolf, H. U., Determination of methadone in human hair by radioimmunoassay, *Z. Rechtsmed.*, 102, 1, 1989.
12. Balabanova, S. and Wolf, H. U., Methadone concentrations in human hair of the head, axillary and pubic hair, *Z. Rechtsmed.*, 102, 293, 1989.
13. Balabanova, S., Arnold, P. J., Brunner, H., Luckow, V., and Wolf, H. U., Detection of methadone in human hair by gas chromatography/mass spectrometry, *Z. Rechtsmed.*, 102, 495, 1989.
14. Käferstein, H. and Sticht, G., Anmerkungen zu: "Detection of methadone in human hair by gas chromatography/mass spectrometry" (Balabanova, S., Arnold, P. J., Brunner, H., Luckow, V., and Wolf, H. U.), und "Tetrahydrocannabinole im Haar von Haschischrauchern" (S. Balabanova, P. J. Arnold, V. Luckow, H. Brunner, und H. U. Wolf) (Letter to the Editor), *Z. Rechtsmed.*, 103, 393, 1990.
15. Balabanova, S., Investigations of the cocaine and methadone transfer in and out the hair *in vitro*, in *Proc. 29th Int. Meeting TIAFT*, Kaempe, B., Ed., Mackeenzie, Copenhagen, 1991, 27.
16. Möller, M. R., Fey, P., and Wennig, R., Simultaneous determination of drugs of abuse (opiates, cocaine and amphetamine) in human hair by GC/MS and its application to a methadone treatment program, *Forensic Sci. Int.*, 63, 185, 1993.
17. Maurer, H. H. and Fritz, C. F., Toxicological detection of pholcodine and its metabolites in urine and hair using radio immunoassay, fluorescence polarisation immunoassay, enzyme immunoassay and gas chromatography-mass spectrometry, *Int. J. Leg. Med.*, 104, 43, 1990.
18. Kintz, P., Tracqui, A., Potard, D., Petit, G., and Mangin, P., An unusual death by zipeprol overdose, *Forensic Sci. Int.*, 64, 159, 1994.
19. Forrest, I. S., Otis, L. S., and Serra, M. T., Passage of ^{3}H-chlorpromazine and ^{3}H-Δ9-tetrahydrocannabinol into the hair (fur) of various mammals, *Proc. West. Pharmacaol. Soc.*, 15, 83, 1972.
20. Harkey, M. R. and Henderson, G. L., Hair analysis for drugs of abuse, in *Advances in Analytical Toxicology*, Vol. 2, Baselt, R. C., Ed., Year Book Medical Publishers, Chicago, 1987, 298.
21. Baumgartner, W. A., Hill, V. A., and Blahd, W. H., Hair analysis for drugs of abuse, *J. Forensic Sci.*, 34, 1433, 1989.
22. Balabanova, S., Arnold, P. J., Luckow, V., Brunner, H., and Wolf, H. U., Tetrahydrocannabinole im Haar von Haschischrauchern, *Z. Rechtsmed.*, 102, 503, 1989.
23. Bogusz, M., Anmerkungen zu: "Tetrahydrocannabinole im Haar von Haschischrauchern" (S. Balabanova et al.) und zu der Antwort Dr. Balabanova auf die Bemerkungen von H. Käferstein und G. Sticht (Letter to the Editor), *Z. Rechtsmed.*, 103, 621, 1990.
24. Möller, M. R. and Fey, P., Detection of drugs in hair by GC/MS, in *Proc. Int. Congr. Clinical Toxicology, Poison Control and Analytical Toxicology (LUX TOX '90)*, in *Bull. Soc. Sci. Med. Grand-Duche Luxemb.*, 127, 1990, 460.
25. Kintz, P. and Mangin, P., L'Analyse des médicaments et des stupéfiants dans les cheveux — Intérêt et limites pour le diagnostic clinique et la toxicologie médico-légale, *J. Méd. Strasbourg*, 22, 518, 1991.

26. Kintz, P., Tracqui, A., and Mangin, P., Detection of drugs in human hair for clinical and forensic applications, *Int. J. Leg. Med.*, 105, 1, 1992.

27. Kintz, P., Ludes, B., and Mangin, P., Detection of drugs in human hair using Abbott ADx with confirmation by gas chromatography/mass spectrometry (GC/MS), *J. Forensic Sci.*, 37, 328, 1992.

28. Balabanova, S., Parsche, F., and Pirsig, W., First identification of drugs in Egyptian mummies, *Naturwissenschaften*, 79, 358, 1992.

29. Skopp, G. and Aderjan, R., Comparison of RIA and KIMS-immunoassay as tools for the detection of drug abuse patterns in human hair, in *Proc. 31st Int. Meeting TIAFT*, Müller, R. K., Ed., Molinapress, Leipzig, 1993, 446.

30. Jurado, C., Menéndez, M., Repetto, M., Kintz, P., Cirimele, V., and Mangin, P., Hair testing for cannabis in Spain and France: is there a difference in consumption?, presented at TIAFT/SOFT Joint Congress, Tampa, October 31 to November 4, 1994, paper 16.

31. Hayes, G., Scholtz, H., Matusi, P., Baumgartner, W., and Kippenberger, D. J., Determination of carboxy-THC in hair by tandem mass spectrometry, presented at the 2nd Int. Meeting on Clinical and Forensic Aspects of Hair Analysis, Genoa, June 6 to 8, 1994.

32. Kippenberger, D., Hayes, E., Schultz, H., Gordon, A. M., and Baumgartner, W., The detection and quantitation of 11-nor-Δ9-tetrahydrocannabinol-9-carboxylic acid in hair using tandem mass spectrometry, presented at TIAFT/SOFT Joint Congress, Tampa, October 31 to November 4, 1994, paper 52.

33. Cirimele, V., Kintz, P., and Mangin, P., Identification des cannabinoïdes dans les cheveux, *Toxicorama*, in press.

34. Cirimele, V., Kintz, P., and Mangin, P., Testing human hair for cannabis, *Forensic Sci. Int.*, 70, 175, 1995.

35. Jurado, C., Giménez, M. P., Menéndez, M., and Repetto, M., Simultaneous quantification of opiates, cocaine and cannabinoids in hair, *Forensic Sci. Int.*, 70, 165, 1995.

36. Kintz, P., Cirimele, V., and Mangin, P., Testing human hair for cannabis. II. Identification of THC-COOH by GC/MS-NCI as an unique proof, *J. Forensic Sci.*, 40, 619, 1995.

37. Mieczkowski, T., A research note: the outcome of GC/MS/MS confirmation of hair assays on 93 cannabinoid (+) cases, *Forensic Sci. Int.*, 70, 83, 1995.

38. Baumgartner, A. M., Jones, P. F., and Black, C. T., Detection of phencyclidine in hair, *J. Forensic Sci.*, 26, 576, 1981.

39. Sramek, J. J., Baumgartner, W. A., Tallos, J. A., Ahrens, T. N., Heiser, J. F., and Blahd, W. H., Hair analysis for detection of phencyclidine in newly admitted psychiatric patients, *Am. J. Psychiatry*, 142, 950, 1985.

40. Kidwell, D. A., Analysis of phencyclidine and cocaine in human hair by tandem mass spectrometry, *J. Forensic Sci.*, 38, 272, 1993.

41. Sakamoto, T., Tanaka, A., and Nakahara, Y., Determination of PCP and its major metabolites, PCHP and PPC, in rat hair after administration of PCP, presented at TIAFT/SOFT Joint Congress, Tampa, October 31 to November 4, 1994, paper 51.

42. Nakahara, Y., Kikura, R., Takahashi, K., Foltz, R. L., and Mieczkowski, T., Detection of LSD and NorLSD in rat hair and human hair, presented at TIAFT/SOFT Joint Congress, Tampa, October 31 to November 4, 1994, paper 53.

43. Balabanova, S. and Schneider, E., Nachweis von Coffein in menschlichen Haaren mittels Emit, *Toxichem + Krimtech*, 57(4,5), 130, 1990.

44. Balabanova, S. and Schneider, E., Nachweis von Coffein in menschlichen Haare mittels Enzymimmunoassay und Gaschromatographie/Massenspektrometrie, in *Proc. Wissenschaftl. Symp. Rechstmedizin und Forensische Toxikologie — Neue Analytische Methoden*, Helm, D., Ed., Universität Hamburg, Hamburg, 1990, 118.

45. Kintz, P., Tracqui, A., and Mangin, P., Determination of methylxanthine stimulants in human hair, presented at the 45th Ann. Meeting of the American Academy of Forensic Sciences, Boston, Abstract K25, 1993.

46. Kintz, P., Tracqui, A., and Mangin, P., Toxicological investigations on unusual materials (hair and vitreous humor): interest and limitations, *Arch. Toxicol., Suppl. 15* (Medical Toxicology — Proc. 1991 EUROTOX Congress), 282, 1992.

47. Kintz, P. and Mangin, P., Toxicological findings after fatal fenfluramine self-poisoning, *Hum. Exp. Toxicol.*, 11, 51, 1992.

48. Ishiyama, I., Nagai, T., and Toshida, S., Detection of basic drugs (methamphetamine, antidepressants, and nicotine) from human hair, *J. Forensic Sci.*, 28, 380, 1983.

49. Haley, N. J. and Hoffmann, D., Analysis for nicotine and cotinine in hair to determine cigarette smoker status, *Clin. Chem.*, 31, 1598, 1985.

50. Zahlsen, K. and Nilsen, O. G., Gas chromatographic analysis of nicotine in hair, *Environ. Technol.*, 11, 353, 1990.

51. Balabanova, S., Schneider, E., and Bühler, G., Nachweis von Nicotin in Haaren, *Dtsche. Apoth. Ztg.*, 130, 2200, 1990.
52. Balabanova, S. and Schneider, E., Detection of cotinine in human hair, in *Proc. Int. Congr. Clinical Toxicology, Poison Control and Analytical Toxicology (LUX TOX '90)*, in *Bull. Soc. Sci. Med. Grand-Duche Luxembourg*, 127, 531, 1990.
53. Balabanova, S. and Schneider, E., Nachweis von Nikotin im Kopf-, Achsel- und Schamhaar, in *Medizinrecht — Psychopathologie — Rechtsmedizin*, Schütz, H., Kaatsch, H. J., and Thomsen, H., Eds., Springer-Verlag, Berlin, 1991, 334.
54. Kintz, P., Ludes, B., and Mangin, P., Evaluation of nicotine and cotinine in human hair, *J. Forensic Sci.*, 37, 72, 1992.
55. Kintz, P., Gas chromatographic analysis of nicotine and cotinine in hair, *J. Chromatogr.*, 580, 347, 1992.
56. Kintz, P. and Mangin, P., Detection of prenatal nicotine exposure in newborn infant by hair analysis, in *Proc. 30th Int. Meeting TIAFT*, Nagata, T., Ed., Yoyodo Printing Kaisha, Fukuoka, 1992, 97.
57. Kintz, P., Tracqui, A., and Mangin, P., Tabac, médicaments et stupéfiants pendant la grossesse — evaluation de l'exposition *in utero* par analyse des cheveux du nouveau-né, *Presse Med.*, 21, 2139, 1992.
58. Koren, G., Klein, J., Forman, R., Graham, K., and Phan, M. K., Biological markers of intrauterine exposure to cocaine and cigarette smoking, *Dev. Pharmacol. Ther.*, 18, 228, 1992.
59. Kintz, P. and Mangin, P., Determination of gestational opiate, nicotine, benzodiazepine, cocaine and amphetamine exposure by hair analysis, *J. Forensic Sci. Soc.*, 33, 139, 1993.
60. Kintz, P. and Mangin, P., Evidence of gestational heroin or nicotine exposure by analysis of fetal hair, *Forensic Sci. Int.*, 63, 99, 1993.
61. Kintz, P., Kieffer, I., Messer, J., and Mangin, P., Nicotine analysis in neonates' hair for measuring gestational exposure to tobacco, *J. Forensic Sci.*, 38, 119, 1993.
62. Klein, J., Chitayat, D., and Koren, G., Hair analysis as a marker for fetal exposure to maternal smoking, *N. Engl. J. Med.*, 328, 66, 1993.
63. Mizuno, A., Uematsu, T., Oshima, A., Nakamura, M., and Nakashima, M., Analysis of nicotine content of hair for assessing individual cigarette-smoking behavior, *Ther. Drug Monit.*, 15, 99, 1993.
64. Uematsu, T., Utilization of hair analysis for therapeutic drug monitoring with a special reference to ofloxacin and to nicotine, *Forensic Sci. Int.*, 63, 261, 1993.
65. Klein, J., Forman, R., Eliopoulos, C., and Koren, G., A method for simultaneous measurement of cocaine and nicotine in neonatal hair, *Ther. Drug Monit.*, 16, 67, 1994.
66. Goulle, J. P., Noyon, J., and Leroux, P., Nicotine in hair and passive exposure in pediatric pathology, presented at the 16th Congress of the Int. Academy of Legal Medicine and Social Medicine, Strasbourg, May 31 to June 2, 1994.
67. Koren, G., Measurement of drugs in neonatal hair; a window to fetal exposure, *Forensic Sci. Int.*, 70, 77, 1995.
68. Arnold, W. and Bohn, G., Hautanhangsgebilde, ein besonderes Asservat für die forensische Diagnostik und Begutachtung, *Beitr. Gerichtl. Med.*, 45, 261, 1987.
69. Goldblum, R. W., Goldbaum, L. R., and Piper, W. N., Barbiturate concentrations in the skin and hair of guinea pigs, *J. Invest. Dermatol.*, 22, 121, 1954.
70. Smith, F. P. and Pomposini, D. A., Detection of phenobarbital in bloodstains, semen, seminal stains, saliva, saliva stains, perspiration stains, and hair, *J. Forensic Sci.*, 26, 582, 1981.
71. Goulle, J. P., Noyon, J., Layet, A., Rapoport, N. F., Vaschalde, Y., Pignier, Y., Bouige, D., and Jouen, F., Phenobarbital in hair and drug monitoring, *Forensic Sci. Int.*, 70, 191, 1995.
72. Sramek, J. J., Baumgartner, W. A., Ahrens, T. N., Hill, V. A., and Cutler, N. R., Detection of benzodiazepines in human hair by radioimmunoassay, *Ann. Pharmacother.*, 26, 469, 1992.
73. Kintz, P. and Mangin, P., Determination of meprobamate in human plasma, urine, and hair by gas chromatography and electron impact mass spectrometry, *J. Anal. Toxicol.*, 17, 408, 1993.
74. Kintz, P., Tracqui, A., and Mangin, P., Pharmacological studies on meprobamate incorporation in human beard hair, *Int. J. Leg. Med.*, 105, 283, 1993.
75. Arnold, W., Radioimmunological hair analysis for narcotics and substitutes, *J. Clin. Chem. Clin. Biochem.*, 25, 753, 1987.
76. Kintz, P., Amineptine abuse detected by hair analysis, *Toxichem + Krimtech*, 61(3), 70, 1994.
77. Tracqui, A., Kreissig, P., Kintz, P., Pouliquen, A., and Mangin, P., Determination of amitriptyline in the hair of psychiatric patients, *Hum. Exp. Toxicol.*, 11, 363, 1992.
78. Tracqui, A., Kintz, P., Ludes, B., and Mangin, P., Amitriptyline concentrations in hair from psychiatric subjects: a tool for estimating long-term patient compliance to therapy?, presented at the 44th Ann. Meeting of the American Academy of Forensic Sciences, New Orleans, Abstract K57, 1992.
79. Couper, F. J., Mc Intyre, I. M., and Drummer, O. H., Method for quantifying antidepressant and antipsychotic drug levels in post-mortem human scalp hair, in *Proc. 31st Int. Meeting TIAFT*, Müller, R. K., Ed., Molinapress, Leipzig, 1993, 160.

80. Sato, H., Uematsu, T., Yamada, K., and Nakashima, M., Chlorpromazine in human scalp hair as an index of dosage history: comparison with simultaneously measured haloperidol, *Eur. J. Clin. Pharmacol.*, 44, 439, 1993.

81. Sachs, H. and Raff, I., Comparison of quantitative results of drugs in human hair by GC/MS, *Forensic Sci. Int.*, 63, 207, 1993.

82. Uematsu, T., Sato, R., Suzuki, K., Yamaguchi, S., and Nakashima, M., Human scalp hair as evidence of individual dosage history of haloperidol: method and retrospective study, *Eur. J. Clin. Pharmacol.*, 37, 239, 1989.

83. Sato, R., Uematsu, T., Sato, R., Yamaguchi, S., and Nakashima, M., Human scalp hair as evidence of individual dosage history of haloperidol: prospective study, *Ther. Drug Monit.*, 11, 686, 1989.

84. Matsuno, H., Uematsu, T., and Nakashima, M., The measurement of haloperidol and reduced haloperidol in hair as an index of dosage history, *Br. J. Clin. Pharmacol.*, 29, 187, 1990.

85. Uematsu, T., Sato, R., Fujimori, O., and Nakashima, M., Human scalp hair as evidence of individual dosage history of haloperidol: a possible linkage of haloperidol excretion into hair with hair pigment, *Arch. Dermatol. Res.*, 282, 120, 1990.

86. Uematsu, T., Yamada, K., Matsuno, H., and Nakashima, M., The measurement of haloperidol and reduced haloperidol in neonatal hair as an index of placental transfer of maternal haloperidol, *Ther. Drug Monit.*, 13, 183, 1991.

87. Uematsu, T., Matsuno, H., Sato, H., Hirayama, H., Hasegawa, K., and Nakashima, M., Steady-state pharmacokinetics of haloperidol and reduced haloperidol in schizophrenic patients: analysis of factors determining their concentrations in hair, *J. Pharm. Sci.*, 81, 1008, 1992.

88. Mc Mullin, M., Selavka, C. M., Wheeler M. A., Wier, K., Werrell, P., Karbiwnyk, C., and Lipman, J. J., Forensic hair testing for haloperidol — a tale of two cases, presented at the 46th Ann. Meeting of the American Academy of Forensic Sciences, San Antonio, Abstract K33, 1994.

89. Kintz, P. and Mangin, P., Hair analysis for detection of beta-blockers in hypertensive patients, *Eur. J. Clin. Pharmacol.*, 42, 351, 1992.

90. Viala, A., Deturmeny, E., Aubert, C., Estadieu, M., Durand, A., Cano, J. P., and Delmont, J., Determination of chloroquine and monodesethylchloroquine in hair, *J. Forensic Sci.*, 28, 922, 1983.

91. Taylor, S. M., Elliott, C. T., and Blanchflower, J., Cypermetrin concentrations in hair of cattle after application of impregnated ear tags, *Vet. Rec.*, 116, 620, 1985.

92. Miyazawa, N., Uematsu, T., Mizuno, A., Nagashima, S., and Gologolo, M., Ofloxacin in human hair determined by high performance liquid chromatography, *Forensic Sci. Int.*, 51, 65, 1991.

93. Uematsu, T., Miyazawa, N., and Nakashima, M., The measurement of ofloxacin in hair as an index of exposure, *Eur. J. Clin. Pharmacol.*, 40, 581, 1991.

94. Uematsu, T., Miyazawa, N., Okazaki, O., and Nakashima, M., Possible effect of pigment on the pharmacokinetics of ofloxacin and its excretion in hair, *J. Pharm. Sci.*, 81, 45, 1992.

95. Uematsu, T., Nakano, M., Akiyama, H., and Nakashima, M., The measurement of a new antimicrobial quinolone in hair as an index of drug exposure, *Br. J. Clin. Pharmacol.*, 35, 199, 1993.

96. Uematsu, T., Kondo, K., Yano, S., Yamaguchi, T., Umemura, K., and Nakashima, M., Measurement of temafloxacin in human scalp hair as an index of drug exposure, *J. Pharm. Sci.*, 83, 42, 1994.

97. Uematsu, T., Ohsawa, Y., Mizuno, A., and Nakashima, M., Analysis of a new fluoroquinolone derivative (Q-35) in human scalp hair as an index of drug exposure and as a time marker in hair, *Int. J. Leg. Med.*, 106, 237, 1994.

98. Harrison, W. H., Gray, R. H., and Solomon, L. M., Incorporation of L-dopa, L-alpha-methyl-dopa and DL-isoproterenol into guinea-pig hair, *Acta Derm. Venereol.*, 54, 249, 1974.

99. Nakahara, Y., Takahashi, K., and Konuma, K., Hair analysis for drugs of abuse. VI. The excretion of methoxyphenamine and methamphetamine into beards of human subjects, *Forensic Sci. Int.*, 63, 109, 1993.

100. Rothe, M., Pragst, F., Hunger, J., and Thor, S., Determination of carbamazepine and its metabolites in hair of epileptics, presented at TIAFT/SOFT Joint Congress, Tampa, October 31 to November 4, 1994, paper 49.

101. Uematsu, T., Miyazawa, N., and Wada, K., 1-(Tetrahydro-2-furanyl)-5-fluorouracil (Ftorafur) determined in rat hair as an index of drug exposure, *J. Pharm. Sci.*, 82, 1272, 1993.

102. Aderjan, R. E., Besserer, K., Sachs, H., Schmitt, G. G., and Skopp, G. A., Ethyl glucuronide — a nonvolatile ethanol metabolite in human hair, presented at TIAFT/SOFT Joint Congress, Tampa, October 31 to November 4, 1994, paper 12.

103. Tracqui, A., Kintz, P., and Mangin, P., Hair analysis: a worthless tool for therapeutic compliance monitoring, *Forensic Sci. Int.*, 70, 183, 1995.

Chapter 9

FORENSIC APPLICATIONS OF HAIR ANALYSIS

Hans Sachs

CONTENTS

I. INTRODUCTION

The use of hair analysis in forensic cases depends on legal conditions, which differ from country to country and even vary from state to state within the European Union. Whereas legal conditions may be the same for the use of hair analysis in murder cases, the acceptance of hair analytical results varies from region to region, and from court to court concerning drug consumption, drug dealing, general driving ability, certification of drug-dependent physicians, and reliability of witnesses and defendants.

Extraction methods and analytical procedures, mostly gas chromatography/mass spectrometry (GC/MS), are described in other chapters of this book, and have been extensively compared in recent years.[1] It has been shown that immunological methods

0-8493-8112-6/96/$0.00+$.50
© 1996 by CRC Press, Inc.

can only be used as a pretest since nearly all of them show cross reactions with derivatives, metabolites, or the parent drug, respectively. A consensus extists at international conferences that every immunochemical result has to be confirmed by a GC/MS examination if used in forensic cases. The following cases will give an overview of forensic use and acceptance in courts, based on experience in German courts and administrations.

II. FORENSIC CASES

A. Murder Case

Acute poisoning fatalities are normally confirmed by examining body fluids and tissues. There are some rare cases, however, where the examination of blood and stomach contents is useless because the victim did not die immediately following poisoning or suspicion was raised months later, and only hair could be gained from the exhumed body.

In one such case it was necessary to detect the strong antidepressant clozapine (Figure 1). A child had been brought to a clinic with symptoms of intoxication. In the urine, clozapine could be detected. As clozapine is not freely available, the mother of the child was suspected of having intoxicated the child with the drug prescribed to her depressive sister. A year previously a 1-year-old boy in the same family had died. In the absence of an obvious cause of death, it was assumed the child died of sudden infant death syndrome (SIDS). The child was exhumed and hair was the only tissue where toxicological examination promised any success. Since it was known at that time that tricyclic antidepressive drug could be detected,[2] it seemed sensible to extract the hair and look for the drug by GC/MS. From experiments with other drugs it was known that the extraction with buffer at pH 7.4 and the following clean-up of the aqueous phase by solid phase extraction is universally applicable. This is a modification of the method introduced by Moeller et al.[3] and is described in Chapter 2.5 in Sachs and Raff.[1] The selected ion monitoring (SIM) chromatograms are shown in Figure 2.

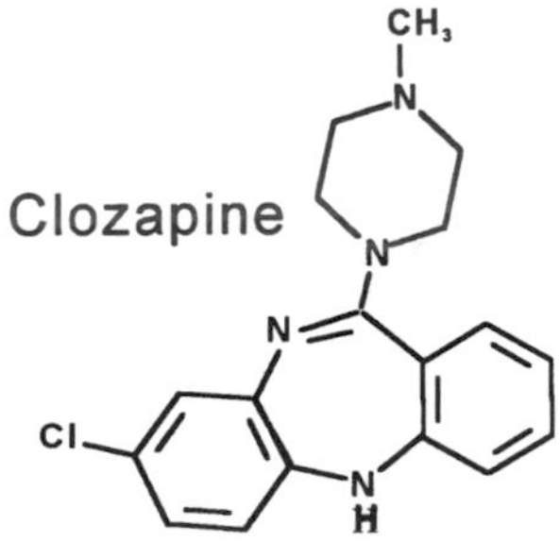

Figure 1
Clozapine molecule.

To the judge, only the qualitative result was of importance, as the child could not have been under antidepressive treatment. It was not important whether the drug application caused the death of the child, but only whether or not it was administered more than once. The mother was sentenced to 4 years only because she eventually admitted giving the drug to the child.

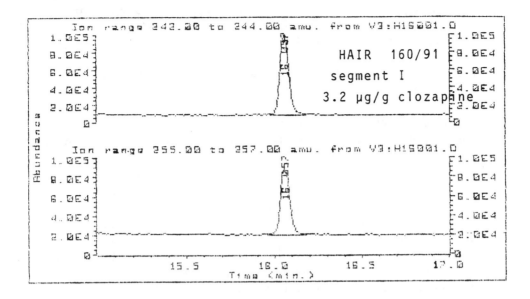

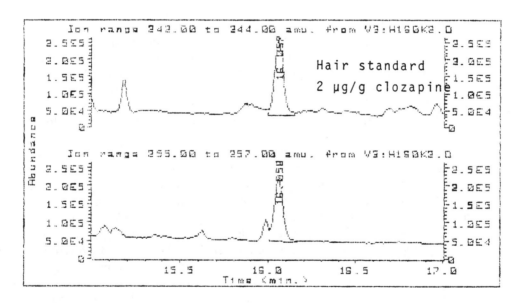

Figure 2
Selected ion monitoring chromatograms of clozapine standard and hair sample extract using the masses
243 and 256.

B. Illegal Drug Abuse

In normal cases, a routine analysis of the following drugs can be expected:

- Cannabinoids (Δ9-tetrahydrocannabinol [THC], possible confirmation by 11-nor-9-carboxy-Δ9-tetrahydrocannabinol [THC-COOH])
- Opiates (heroin, acetylmorphine, morphine, dihydrocodeine, codeine)
- Cocaine (additionally benzoylecgonine and cocaethylene)

- Amphetamine group — amphetamine, methamphetamine, methylenedioxyam-phetamine (MDA), methylenedioxymethylamphetamine (MDMA), methylenedioxy-ethylamphetamine (MDME or MDE)

Three types of questions have to be answered by hair analysis in the forensic cases of illegal drug abuse:

- Can drug consumption be proved by a positive result?
- Can the amount of drug taken be determined or, at least, estimated, i.e., does a dose/concentration relationship exist?
- Does a negative result prove drug abstinence; can drug consumption be excluded?

Which of the questions have to be answered depends in most cases on the country and the kind of court. In Germany, for example, drug possession of heroin or cocaine is illegal, but not its consumption. The question of possession can only be answered indirectly. It is assumed that someone who consumes the drug regularly must also possess it. But since the problems of possible contamination are not yet completely solved, and since most of the illegal drugs can also be smoked, and the previous possession of drug remains unclear, positive results of hair analysis do not lead to sentencing in German courts unless witness statements or urine analysis confirm the results of hair analysis.

Dose/concentration relation could only be found for some legal drugs, for example, methadone by Moeller et al.[3] and meprobamate by Kintz et al.[4] For illegal drugs it will probably not be possible to perform studies to answer this question, because even in countries where such studies are allowed, it is not permitted to administer levels used by drug addicts. An additional difficulty is that most drugs, heroin, cocaine, THC metabolite having a short half-life, and a number of metabolites are detected in the hair.

Therefore the only help that experts can give to the judges is their own statistical findings. These findings have to be compared with the individual case. In one of these cases a woman was accused of having dealt with more than 100 g of heroin. She stated that she needed most of the heroin herself and dealt with the rest to earn enough money to meet her own heroin requirements (Figure 3). Compared with the normal findings of heroin (Table 1) in hair of heroin addicts a very high concentration of acetylmorphine was found in the segment labeled 9–12 in Figure 3, grown during the time period when she was dealing. This confirmed her statement and that of some witnesses that she herself consumed a large amount of heroin. The positive results in the two segments closer to the hair root, a time period when she had been in jail, do not prove a continuing heroin abuse during the time she was in jail. It can be explained by irregular hair growth, which will be discussed later in this chapter.

Generally in criminal courts, such as in the last case, the hair analysis is not used to prove the possession of the drug, but to exclude or to confirm regular drug consumption of defendants accused of something else like robbery, rape, or murder. The results do not always confirm the statements of the defendants. In a case of robbery, the accused claimed to be a heroin addict consuming several grams of heroin every day. In his hair segments (Figure 4), only traces of morphine could be found in some segments. In the segment grown during the robbery, labeled 4–6, not even that trace was detected. It must be emphasized that the possibility of a single dose of heroin could not be excluded, nor whether or not the accused was under the influence of heroin (at the time of the crime). However, it may be assumed that he did not consume the amounts of heroin that a regular addict does. Due to a lack of

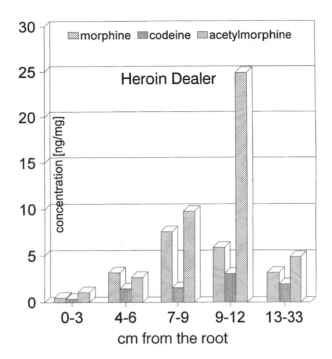

Figure 3

Sectional analysis of a hair sample from a heroin dealer who confessed to personal heroin abuse before she was imprisoned. The hair was taken about 10 months after she was imprisoned.

TABLE 1.

Concentrations of Acetylmorphine, Morphine, and Codeine in Cases of Heroin Abuse

	MAM (ng/mg)	Morphine (ng/mg)	Codeine (ng/mg)
Range	2.0–74.2	1.0–21.8	0.1–4.0
Median	10.8	3.3	1.7
Mean	15.3	4.8	1.2

Note: Number of cases = 116; MAM = monoacetylmorphine.

experiments with drug addicts, however, it will always be difficult to exclude drug consumption after a negative result.

In another case, a cocaine dealer claimed to have consumed 30 to 50 g of cocaine within 6 d around the time when he was dealing with cocaine and was arrested. The dealer's hair was cut in segments of 1 cm lengths just around points corresponding to the time period when he was arrested. In none of these segments could a trace of cocaine be found. In this case the statement could be disproved by comparing the detection limit with the statistics of cocaine results in routine laboratory tests (Table 2). But what could have been said if he had stated that he had consumed 2 or 3 doses of cocaine just before the deal?

For several reasons these difficulties do not exist when examining hair of drug addicts for driving licence authorities. In these cases, the drug consumer has to prove that he is able to drive a car safely. Most of the hair analysis expert reports are prepared for driving licence authorities, who want to be sure that the driver has been abstinent for at least 1 year according to the demands of the Ministry of Transportation (Figure 5). This task previously involved making a couple of urine analyses,

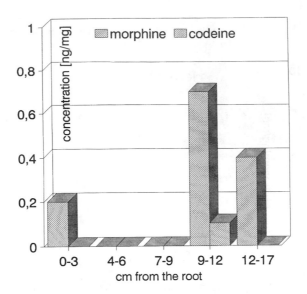

Figure 4
Sectional analysis of a hair sample from a subject accused of robbery, who was imprisoned several months later.

TABLE 2.

Concentrations of Drugs of Abuse other than Opiates found in Hair Samples in Forensic Cases (1989–1992)

Substance	No. cases analyzed	No. positive	Percent positive	Average concentration (ng/mg)	Median (ng/mg)	Concentration range (ng/mg)
Amphetamine	185	9	5	15.3	13.0	0.9–42
Benzoylecgonine	185	34	18	9.3	1.4	0.1–107
Cocaine	185	34	18	20.6	4.7	0.3–127
THC	57	10	17	2.0	1.3	0.4–6.2
THC-COOH	57	2	3	3.3	3.4	1.7–5.0
Dihydrocodeine	60	19	32	27.5	3.8	0.3–273

but it is obvious that hair samples range over a longer time period. The number of these examinations in Germany is estimated to be between 1000 and 1500, while the number of hair analyses for criminal courts count under 300.

Analytical methods with high specificity and sensitivity are required when physicians are suspected of being drug dependent, as they have access to unusual drugs of abuse which are not detected by routine urine controls using mostly immunochemical methods. In one case an anesthetist prescribed more than 4000 ampules of pentazocine in 5 months to various patients. He admitted to have taken some pentazocine himself because of a knee injury. When the health authorities threatened to withdraw his licence, he submitted to a hair analysis, where concentrations of 200 ng/mg of the drug were found. Although dose/concentration studies of pentazocine are not known, it is clear that this amount cannot be explained by a treatment over a few days. His licence was withdrawn temporarily until the evidence showed that he was completely drug free. After 8 months, a second sample was taken. Meanwhile he had cut his hair to 2–3 cm in length. This sample still contained a concentration of ng/mg. This concentration could be explained by the irregular hair

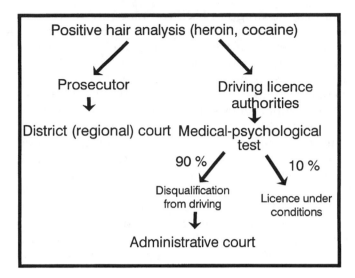

Figure 5
Flowchart showing how the expert reports to criminal and administrative courts.

growth described later. After 1 year, the hair analysis for pentazocine was completely negative.

As fentanyl cannot be easily detected, this drug is sometimes abused by clinical staff. In one such case, the doctor was suspected of abusing fentanyl when the list of drugs did not match the inventory list. Urine analyses of the staff did not lead to positive results. As fentanyl is administered in doses in the order of several micrograms, normal GC/MS procedures failed. But using GC/MS/MS, fentanyl could be detected in the hair of one of the doctors; the concentration was about 100 µg/mg hair. Since the narcotic drug is only used during anesthesia, a qualitative result was sufficient as evidence of drug abuse. Of course, these drugs can only be detected if suspicion is raised concerning a certain drug.

III. GENERAL ASPECTS

A. Hair Sampling

To receive comparable specimens, laboratories should send instructions for collecting hair samples to the police departments. The hair should be taken from the nape of the neck or the cortex posterior. Pencil-thick hair strands should be tied together with a small piece of string before being cut close to the skin. If it is not possible to cut the hair at the skin, the length of the remaining hair should be protocolled.

B. Hair Sample Preparation for Different Purposes

Contrary to examinations for workplace testing or for routine clinical purposes, in forensic cases it might be necessary to determine the drug concentrations in varying lengths of segments. To prove drug abstinence for driving licence authorities, a time period of 1 year is required. Hair grows, on average, at a rate of 1.1 cm/month (0.8–1.4 cm). Finding traces of drugs in a 12-cm hair section, however, does not necessarily mean that the drug has been consumed in the past 12 months.[5] The telogen share of the hair may contain the drug from a consumption which has been totally stopped more than 1 year before the hair sampling. To avoid such complications in interpreting the results, only the last-grown 6 cm of the hair are examined. It can be assumed that this segment will contain at least traces of the drug if a regular intake of heroin or cocaine has taken place 6 to 12 months before the hair sampling. Concerning THC and amphetamines, it must be taken into account that some intakes 6 to 12 months before the hair sampling will not be detected.

The questions in criminal courts can be totally different. One question might be whether or not a drug intake can be excluded before or after a certain date. In such cases it is no use to count 1.1 cm per month and to cut the hair exactly at this length. It has to be taken into account again that the growth rate of the hair can vary between 0.8 and 1.4 cm/month and that the telogen partition of the hair can increase to 20%. The longer the peak consumption period has passed, the more the drug is spread over the hair segment. In many cases this period lies within the past 6 months. The following question may arise: "Did the person take the drug of abuse in the last 4 months?"

In 4 months, for example, the anagen hair can grow 3.2 cm or, perhaps, 5.6 cm. To answer the question, 8 cm of the hair strand should be examined. In Figure 6a–c. three examples are given. In Figure 6a the low concentrations in the last-grown 2 cm do not prove the consumption in the last 4 months, because there has been a regular drug abuse in more than 4 months before. In Figure 6b the consumption in the last 4 months is proven because the concentrations of the last-grown 2 cm are in the same order as the concentration of the segments 4–6 and 6–8. In Figure 6c the consumption in the last 4 months is proven because the last-grown hair segments are positive and the segments 4–6 and 6–8 are not.

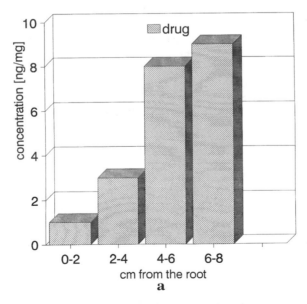

Figure 6
Examples of possible positive findings: (a) the drug consumption in the last 4 months is not proved, although the drug could be detected in the last-grown 4 cm of hair; (b) drug consumption in the last 4 months is proved because the concentrations of the last-grown segment (0–2) lies in the same order as the previously grown hair; (c) drug consumption in the last 4 months is proved because the last-grown segment (0–2) was positive and the previously grown segments (4–8) were negative.

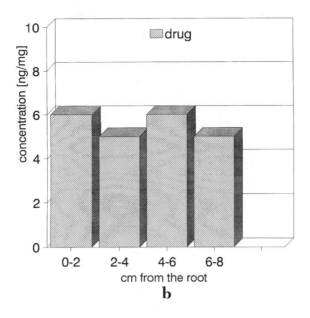

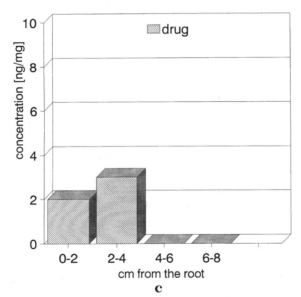

Figure 6 (continued)

In cases where the consumption lies more than 1 year in the past, the evaluation of the results will be more and more complicated. In 1 year the anagen hair might have grown only 9.6 to 16.8 cm. The telogen partition can only be estimated, because it depends on the frequency and length of hair cut in that past year. Additionally, hair cosmetics and hair treatment, bleaching, coloring, and curling might have influenced the drug concentration. Figure 7 shows the hair of a young woman after a codeine abuse over several months. Her hair had been bleached a couple of weeks before her death. The brown, natural hair contained codeine concentrations of over 10 ng/mg, while the bleached strand contained only one fifth of this concentration in nearly every bleached segment. This is the reason why, before cutting the hair in relevant segments, the individual hairs are to be subjected to optic examination of the cuticula under a microscope. Cuticula defects should be described and taken into account when evaluating the quantitative results.

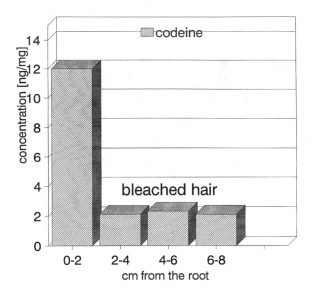

Figure 7
Sectional analysis of a hair sample of a codeine addict. The proximal hair section was natural hair, the distal 3 sections were bleached.

To avoid mistakes caused by contamination of the hair, the total strand is washed with water, petrolether, and methanol. Acute contamination should be excluded by examining the wash solutions. Blank and Kidwell[6] described experiments concerning the contamination of hair with cocaine in the laboratory and found relevant drug concentration even after several washing procedures. Our experience with the hair of technical assistants who are handling heroin, cocaine, and hashish every day shows that these samples contain only traces of drugs. This is confirmed by the studies of Mieczkowski,[7] who examined the hair of undercover narcotic officers.

C. Quality Assurance

For use of hair analysis in forensic cases, internal and external quality control is necessary. In the past 5 years a couple of methods have been developed and evaluated by different laboratories. Interlaboratory comparisons, organized by the National Institute of Science and Technology (NIST)[8] and others performed by the institutes of legal medicine in Strasbourg, France[9] and Munich, Germany,[10] showed that laboratories performing hair analysis on a routine basis gain similar results when using GC/MS, despite the fact that different extraction methods are used. The differences in the results could never justify different evaluations.

D. General Rules for the Evaluation of Hair Analysis in Forensic Cases

Before evaluating the results contamination, interindividual growth rate and cosmetic hair treatment have to be taken into account. Contamination can be detected by washing procedures and examination of the solutions. If necessary, the individual rate of hair growth can be measured after hair sampling. Aggressive hair treatment could be detected by examinating the cuticula of single hairs under the microscope. If heavy contamination or defects in the cuticula are detected, the results should not be evaluated or with criticism only.

Heroin abuse can only be proved if acetylmorphine is found. The concentrations lie in a well-known range up to 100 ng/mg. Higher concentrations are found only in very few cases. Generally, morphine and codeine are detected at the same time, mostly in an acetylmorhine/morphine ratio of about 3 to 1 and an acetylmorphine/codeine ratio of more than 5 to 1. If they are not, the washing solutions should be examined. In criminal courts the *qualitative* results of hair analyses play an important role; the *quantitative* results are only used to support other evidence or to confirm statements. In all studies with regularly given low doses of heroin or cocaine the findings were positive. Therefore, it is certain that the consumptions of over 1 g heroin a day, described in the courts by some defendants, can be disproved if not even a trace of acetylmorphine is found. In administrative courts the question arises whether or not a regular consumption can be proved. Defining the regular consumption with more than one application a week over at least 6 months or 6 cm respectively, it must be proved that the findings lie in the order of the median of total findings of the laboratory.

Cocaine is found in higher concentrations in hair samples than its metabolite benzoylecgonine, generally in a ratio of about 3 to 1. Contamination should always be considered; the consumption can additionally be proved by the detection of cocaethylene.

Amphetamine and cannabinoids can also be detected in routine analyses together with cocaine and opiates. As the frequency of positive findings is relatively low compared to the frequency of positive urine results, it may be assumed that amphetamine and cannabinoids will only be detected after regular consumption. Positive findings of THC can be confirmed by the detection of THC-COOH and OH-THC to exclude the possibility that the hair had only been contaminated by smoke. MDMA, MDE, and MDA can be detected in hair, but there is little experience with quantitative results. In routine analysis on illegal drugs, all listed substances in Figure 8 should be taken into consideration.

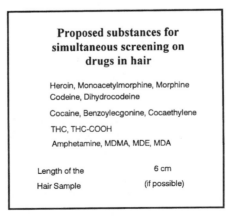

Figure 8

Proposal for routine hair analyis on illegal drugs for criminal courts and driving authorities.

Low-dose narcotic drugs like fentanyl and sufentanil can be detected using radioimmunological methods, and they can be identified using GC/MS/MS.[11] Immunochemical methods, in general, are not able to prove drugs in hair. They may be used as screening procedures, but any positive result is to be confirmed by identification of the single substance using a chromatographical method, mostly GC/MS.

REFERENCES

1. Sachs, H., Raff, I., Comparison of quantitative results of drugs in human hair by GC/MS, *Forensic Sci. Int.* 63, 207, 1992.
2. Ishuiyama, I., Nagai, T., Toshida, S., Detection of basic drugs (methamphetamine, antidepressants, and nicotine) from human hair, *J. Forensic Sci.* 28, 380, 1993.
3. Moeller, M.R., Fey, P., Wennig, R., Simultaneous determination of drugs of abuse (opiates, cocaine, and amphetamine) in human hair by GC/MS and its application to a methadone treatment program, *Forensic Sci. Int.* 63, 185, 1993.
4. Kintz, P., Tracqui, A., Mangin, P., Pharmacological studies on meprobamate incorporation in human beard hair, *Int. J. Leg. Med.* 105, 283–287, 1993.
5. Sachs, H., Theoretical limits of the evaluation of drug concentrations in hair to irregular hair growth, *Forensic Sci. Int.* 70, 53, 1995.
6. Blank, D.L., Kidwell, D.A., Decontamination procedures for drugs of abuse in hair: are they sufficient?, *Forensic Sci. Int.* 70, 13, 1995.
7. Mieczkowski, T., Passive contamination of Undercover Narcotics Officers by cocaine: an assessment to their exposure using hair analysis, *Microgram* 28, 193, 1995.
8. Welch, M.J., Sniegoski, L.T., Allgood, C.C., Interlaboratory comparison studies on the analysis of hair for drugs of abuse, *Forensic Sci. Int.* 63, 295, 1993.
9. Kintz, P., Interlaboratory comparison of quantitative determination of drug in hair samples, *Forensic Sci. Int.* 70, 105, 1995.
10. Sachs, H., Quality assurance during hair testing, 33rd International Congress on Forensic Toxicology (TIAFT), 1995, Tessaloniki, Greece.
11. Sachs, H., Uhl, M., Detection of fentanyl and sufentanil in human hair by GC/MS/MS, *Int. J. Leg. Med.* in press.

Chapter 10

HAIR ANALYSIS FOR ORGANIC ANALYTES: METHODOLOGY, RELIABILITY ISSUES, AND FIELD STUDIES

Werner A. Baumgartner and Virginia A. Hill

CONTENTS

0-8493-8112-6/96/$0.00+$.50
© 1996 by CRC Press, Inc.

I. INTRODUCTION

Hair analysis for drugs of abuse provides long-term information on an individual's drug use; its surveillance window is limited only by the type and length of hair and typically ranges from a week to several months. In this respect, hair analysis is complementary to urinalysis which primarily identifies recent drug use, i.e., that which has occurred during the past few days.

Although the surveillance windows of the two tests do not overlap, hair analysis can be used to validate a positive urinalysis result. This is achieved by allowing sufficient time to pass for hair to grow out from the scalp to permit sampling of the appropriate hair segment corresponding to the approximate time frame of the positive urinalysis result. Hair grows at approximately 1.3 cm/month.[1] Hair analysis can therefore be used for the corroboration of urinalysis results. However, because of the time frame differences, urinalysis cannot provide this kind of support for hair analysis.

Particularly troublesome to urinalysis are challenges that a positive result was caused by subversive actions (e.g., spiked food or drink), by passive drug exposure (e.g., poppy seed ingestion), or by sample contamination. We will show in the present paper that hair analysis, by being able to detect such interpretive false positives, is in a much stronger probative position than urinalysis and that this advantage can also be beneficially applied to the interpretation of challenged urinalysis results. The safety of, and therefore the confidence in, drug testing programs is thereby greatly enhanced, a situation which is appreciated by all involved parties.

We will show that the probative advantages of hair testing depend critically on the methods used for analysis: wash procedures, kinetic analysis of the wash data, the digestion of the hair for the complete release of analytes, the measurement of metabolites, and adoption of appropriate cutoff levels. These analytical procedures will be described in the present paper along with certain unique chemical and physical properties of hair which contribute greatly to the reliability of the analytical result. The advantages of a recently published proprietary method will also be discussed.

Another unique property of hair is the efficient trapping of the short-lived heroin metabolite 6-monoacetylmorphine (6-MAM).[2-4] In contrast to hair, this metabolite is rarely found in opiate-positive urines, thereby causing 90% of such urine results to be overturned by medical review officers.[5] The special procedures required for hair analysis to help resolve the opiate-positive problems of urinalysis will be described.

An additional feature unique to hair analysis, but one which is also critically dependent on appropriate methodologies, is the ability to establish the pattern, severity, and approximate time frame of drug use. The pattern of drug use, i.e., to what degree drug use is increasing, decreasing, or constant, can be ascertained by the analysis of hair segments corresponding to the time frame of interest. These measurements can be performed with a high degree of accuracy, since the patient acts as his own control, thereby overcoming the vexing problem of biochemical individuality.[6] The influence of biochemical individuality, however, precludes that the severity of drug use can be established with the same degree of accuracy as the determination of relative changes in drug use. Nevertheless, it is still possible to distinguish clearly by hair analysis between heavy, intermediate, and light drug use.[7] Urinalysis cannot provide this type of information.

Information on the pattern and severity of drug use is of considerable clinical significance, not only for illicit drugs, but for monitoring licit ones as well, particularly those uses where effective dosing and compliance are major issues, e.g., methadone, antidepressives,[8] and antipsychotic agents.[9] In view of the effects of biochemical individuality on dose correlations, compliance monitoring has to be performed with "calibrated" patients, i.e., with patients whose optimum medication dose and corresponding drug level in hair was first established in a controlled situation (e.g., while an inpatient in a psychiatric hospital). Upon discharge, compliance of medication intake in an outpatient setting can then be measured by deviations from the drug control level that has been set for each patient. Information on severity and pattern of drug use is also useful for appropriate custody referral in the criminal justice system and in many other settings.

Our laboratory has been involved in over 50 field studies evaluating the usefulness and effectiveness of hair analysis in the nine fields listed in Table 1. Most of these studies were performed on a blind basis. The role of our laboratory in these studies has been to develop and provide the appropriate analytical technology and to establish whether a positive result was caused by drug use or external contamination. The objectives, design, and publication of these studies were the responsibilities of the independent investigators. In many instances, the results of hair analysis were validated by comparison with urinalysis and self-reports. The majority of these studies were funded by the National Institute of Justice (NIJ) or the National Institute on Drug Abuse (NIDA). Although preliminary reports are available in most cases, we will focus our selective review on those studies which have been published or where preliminary data are available at least in abstract form.

TABLE 1.

Field Studies Using Hair Analysis

Type of project	Number of studies
1. Drug treatment programs	16
2. Prenatal drug exposure	12
3. Therapeutic drug monitoring	3
4. Criminal justice	7
5. Epidemiology of drug use	6
6. Workplace testing: hair vs. urine	3
7. Forensic cases (over 300)	NA
8. Environmental toxin exposure	1
9. Historical/anthropological studies of drug use	3

However, by far the most extensive work by our laboratory has been in the area of workplace testing, where we have analyzed over half a million hair specimens for cocaine, marijuana, opiates, methamphetamine, and phencyclidine. The reason for this extensive experience with workplace testing is the recognition by the public of the distinct advantages of hair analysis. Because of the narrow window of detection of urinalysis and the scheduled nature of pre-employment testing, a positive urinalysis result can be avoided by temporary abstention from drug use. Because of the 3-month-wide detection window, such evasion is not possible with hair analysis. This window was chosen after extensive consultation with employers and drug treatment professionals. Through these consultations, consensus was reached that a 3-month-wide window is appropriate for demonstrating cessation of drug use by a former user.

Hair analysis has several additional advantages that are attractive for workplace testing. As a scheduled test, it does not suffer from the legal problems of unannounced urine testing. Evasive tactics such as switching or contamination of the sample are precluded by closely supervised, nonembarrassing specimen collection. Multiple sample sources are available, e.g., head hair, body hair, or other keratinized tissue such as fingernail clippings or shavings. The specimen can be matched to an individual by its physical appearance, or by microscopic examination. The sample can be conveniently shipped and stored without refrigeration. The confidence of the tester and testee in a hair testing program is greatly enhanced by the ability to collect a second specimen. Collection of a second sample is also useful in case of a broken chain of custody or any other challenge to the validity of the first result.

We have developed special procedures for the avoidance of false positive results and for the accurate interpretation of positive results in terms of drug use, passive exposure to drugs or specimen contamination. This experience with workplace testing will be described along with an assessment of the efficacy and reliability of hair analysis.

II. METHODOLOGICAL ISSUES

For the accurate interpretation of a positive hair analysis result, it is necessary to subject hair to specially developed wash and extraction procedures. During the development of these procedures, we discovered several important properties of hair, properties which were essential for defining three unique wash kinetic criteria and their cutoff levels which, in conjunction with washing, enhance the accuracy of the interpretation of a positive hair analysis result. Additional certainty can be achieved by the measurement of metabolites or metabolite/drug ratios, by the development of ultrasensitive gas chromatography/tandem mass spectrometry (GC/MS/MS) or ion trap GC/MS procedures,[10] by staining methods to evaluate hair porosity, and by defining appropriate wash kinetic and endogenous cutoff levels.

A. The Extended Wash Procedure

The wash procedures used in our laboratory for differentiating between external drug contamination (exogenous drugs) and drugs deposited as a result of drug use (endogenous drugs) are based on the observation that the hair of drug users exhibit characteristically different wash kinetic/digest profiles than hair that has been contaminated only externally by drugs. This difference, which is illustrated in Figure 1, will be discussed in considerable detail. These wash kinetic/digest profiles are best illustrated by the extended wash kinetic procedure — a procedure which is also the basis for the development of the truncated wash kinetic procedure used for mass production testing (see below).

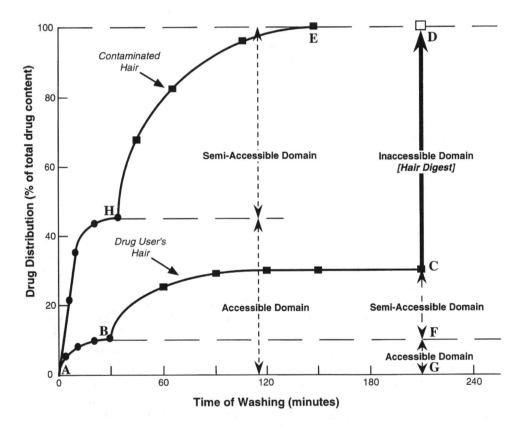

Figure 1
Wash kinetics of contaminated hair and hair from a drug user. Hair treatments: ●, dry isopropanol washes; ■, phosphate buffer; □, enzymatic digestion.

Let us first consider the wash kinetic/digest profile of contaminated hair in Figure 1. The procedure consists of washing approximately 25 mg of hair in 2 mL of dry isopropanol with shaking (100 cycles/min) at 37°C for at least 10–15 min. The wash solution is decanted into another tube and saved for analysis of its drug content. This process is repeated until a plateau is reached, i.e., until no more drug is removed by the isopropanol wash. This occurs usually within 15–30 min in the case of undamaged, i.e., nonporous, hair, and somewhat longer in the case of porous hair. The cumulative plot of this wash data is represented by curve AH in Figure 1. Since isopropanol is a nonhair-swelling solvent, its cleansing action is largely confined to the hair surface. Ethanol can also be used but isopropanol is preferred since it precludes all possible arguments that the presence of cocaethylene was caused by the wash procedure. Cocaethylene is one of the most definitive metabolites indicative of concomitant cocaine and alcohol use.

The alcohol washes are then followed by a series of phosphate buffer washes (pH 5.5). The duration of each of these washes is at least 30 min with shaking (100 cycles/min) at 37°C. Once again, washing is continued until a plateau is attained in the wash kinetics (Figure 1, curve HE). The hair specimen is then dissolved by enzymatic digestion. Enzymatic digestion methods compatible with GC/MS and radioimmunoassay (RIA) have been published.[11,12] In the case of contaminated hair from a nondrug-using individual, no significant quantities of drugs (i.e., drug above the endogenous cutoff level [see below]) are found in the hair digest.

The initial wash with dry isopropanol is performed for two reasons: (1) to remove oily residues from hair, since these can interfere with the effectiveness of the aqueous wash procedure; and (2) to remove possible high quantities of drug contamination from the surface of the hair. If such contaminants are not removed with nonhair-swelling solvents, but, instead, with water, then a fraction of the drugs contained in the first aqueous wash solution will be carried into interior regions of the hair as a result of the "closed system" conditions of the test tube, i.e., the wash solution will then act in both a decontaminating and contaminating capacity. This problem does not arise under conditions of regular hygienic practices, since the constant stream of uncontaminated water from a shower acts only in a decontaminating capacity.

When the above described wash and digestion procedures are performed with the hair from a drug user, then an entirely different wash kinetic/digest profile from that of contaminated hair is obtained. Although the wash kinetics of the alcohol and water washes (Figure 1, curves AB and BC, respectively) resemble those of contaminated hair with respect to the speed with which plateau conditions are attained, they differ dramatically when the kinetics are expressed in terms relative to the drug content of the hair digest (Figure 1, CD). Thus with contaminated hair the ratio of the drug content in the hair digest to the total drugs removed by all water washes approaches zero. In contrast to this, the ratio with the hair of drug users generally has ratio values between 1 and 10 (see below). In Figure 1, the ratio (DC/CF) is approximately 3.5.

These three characteristic features in the wash kinetic/digest profile obtained with the hair from drug users, i.e., curves AB, BC, and CD in Figure 1, provide operational definitions of three characteristic domains for endogenous drug deposition in hair: the accessible, semiaccessible, and inaccessible domains. Contamination experiments with drug vapors suggest that the accessible domain (AB) is confined to surface structures of hair which are readily accessed even by nonhair-swelling solvents such as dry isopropanol. Of course, this domain is also readily accessed by external contamination. However, such contamination (as well as the endogenous drugs present in this domain) can be readily removed by dry isopropanol.

The semiaccessible domain (BC) corresponds to the region which is involved in water-induced swelling of hair. Thus endogenous drugs which are present in this region can be washed out by aqueous solutions. Of course, water-borne drug contamination can enter this deeper-lying structure of hair as well; however, these can also be removed by water — but not by nonhair-swelling solvents such as isopropanol.

With respect to the inaccessible domain (CD), containing exclusively endogenous drugs, we have shown[11] that the inability of water to remove drugs from this region in hair is not caused by stronger binding interactions between the drugs and certain structural components of hair, but instead by the more effective sequestration of drugs into structures which are inaccessible to aqueous wash solutions. Our experience has shown that these structures are also not accessible to water-borne drug contamination, at least under realistic contamination conditions. A water-inaccessible domain is a well-recognized region in hair (the microfibril) that has been identified by various means, including X-ray crystallography.[13]

The critical element for differentiating between endogenous and exogenous drugs is the distribution of drugs between the semiaccessible and inaccessible domains. This distribution is quantitatively defined by the extended safety zone ratio (R_{ESZ}).[11]

$$R_{ESZ} = \frac{\text{amount of drug per 10 mg of hair in digest}}{\text{amount of drug per 10 mg of hair in all phosphate washes}}$$

It should be noted that in this definition all contribution from contamination present in the accessible domain (isopropanol wash) has been excluded. This is done to enhance the precision of the definition, i.e., to shield R_{ESZ} from the possible influence of the widely varying drug concentrations in the accessible domain. These concentrations can range from zero in freshly shampooed hair to a very large number in freshly contaminated hair. The effective differentiation between drugs present in the accessible and semiaccessible domains is made possible by the dramatic difference in the cleansing power of dry isopropanol (as well as ethanol) and water. This distinction, however, cannot be achieved with methanol, since this resembles water in its wash properties.

With thick, nonporous, and uncontaminated hair, the values of R_{ESZ} can be as high as 10. This ratio decreases with increasing porosity and contamination of the hair. On the basis of empirical evidence a cutoff value of 0.25 has been chosen for R_{ESZ}. By choosing this cutoff value, we make allowance for the relatively unlikely possibility of a small amount of penetration into the inaccessible domain by exogenous drugs or for imperfections in the wash kinetic, e.g., imperfect attainment of plateau conditions. Hair with an R_{ESZ} value below the cutoff is judged to be contaminated. Such a ruling can be overturned, however, by the presence of definitive metabolites such as cocaethylene (see below). With contaminated hair from a nondrug user, R_{ESZ} values are close to zero.

B. The Truncated Wash Procedure

There is considerable variation in how fast different hair specimens approach plateau conditions in their wash-out kinetics. Most nonporous uncontaminated hair attains plateau conditions within 1.5 h; heavily contaminated and porous hair specimens may require 3 h or longer. This variability in the wash kinetics of different hair specimens has to be addressed under mass production conditions where individualized monitoring of the attainment of plateau conditions cannot be performed for each hair specimen.

Under mass production conditions, all hair samples are washed in an identical manner by the truncated wash procedure. This consists of one 15-min wash with dry isopropanol and three 1/2-h washes with phosphate buffer. Not all hair samples will have reached plateau conditions by this procedure. To identify those samples which require further washing, we have defined two additional wash kinetic criteria:[11] the curvature ratio, R_C, and the extended wash ratio, R_{EW}. These are defined as follows:

$$R_C = \frac{\text{amount of drug per 10 mg of hair in 3 phosphate washes}}{3 \times \text{amount of drug per 10 mg of hair in last phosphate wash}}$$

$$R_{EW} = \frac{\text{amount of drug per 10 mg of hair in digest}}{\text{amount of drug per 10 mg of hair in last phosphate wash}}$$

R_C measures the extent to which a plateau has been attained. To make $R_C = 1$ under conditions of strictly linear, i.e., noncurving kinetics, the denominator is multiplied by 3, i.e., by the number of phosphate buffer washes. As a plateau is approached, R_C increases in value and in the limit approaches infinity when perfect plateau conditions have been attained.

R_{EW}, on the other hand, is the worst-case estimate of the number of 1/2-h aqueous washes it would have taken to wash out the residual drug found in the hair digest if the residue had been contained in the semiaccessible domain. It is the worst-case estimate because the slope of the wash kinetics operating during the third wash is assumed to remain constant — a highly unlikely situation under nonplateau conditions. R_{EW} values become very large numbers which, in the limit, approach infinity as perfect plateau conditions are attained.

R_{EW} and R_C ratios were used in conjunction with the truncated safety zone ratio, R_{TSZ}, to describe the results obtained with populations of known drug users.[11] The percentage distributions of values for these wash kinetic parameters are given in Tables 2, 3, and 4. Based on the data from these populations of known drug users and the wash kinetic values obtained with artificially contaminated hair, the following cutoff levels were set for the truncated wash procedure: $R_C = 1.3$, $R_{EW} = 10$, $R_{TSZ} = 0.33$. Samples are judged to be contaminated if any one of the three wash kinetic parameters falls below its cutoff value.

TABLE 2.

Percentage Distribution of Extended Wash Ratio Values[a] Using the Truncated Wash Procedure

Extended wash ratio	Percentage distribution of positive samples			
	Cocaine	Opiates	PCP	Methamphetamine
10–20	19.6	0	20	42.6
21–40	24.0	0	25	32.7
41–100	28.2	23.7	20	14.8
101–200	14.9	29.3	15	8.3
>200	13.3	47.0	20	1.7

[a] Extended wash ratio = $\dfrac{\text{amount of drug per 10 mg hair in digest}}{\text{amount of drug per 10 mg hair in last PO4 wash}}$

TABLE 3.

Percentage Distribution of Safety Zone Ratio Values[a] Using the Truncated Wash Procedure

Safety zone ratio	Percentage distribution of positive samples			
	Cocaine	Opiates	PCP	Methamphetamine
0.33–1.0	9.3	1.7	4.4	22.7
1.0–5.0	39.2	22.8	52.2	56.9
5.0–10.0	17.1	24.6	17.3	11.5
>10	34.4	50.9	26.1	8.9

[a] Safety zone ratio = $\dfrac{\text{amount of drug per 10 mg hair in digest}}{\text{amount of drug per 10 mg hair in all PO4 washes}}$

As with urinalysis cutoffs, the further the wash kinetic results are from their cutoff values, the more certain is a finding of drug use. In general, our data showed that the values of the wash kinetic parameters of porous hair samples tended to cluster in regions close to the cutoff levels whereas those of nonporous hair tend to be further

TABLE 4.

Percentage Distribution of Curvature Ratio Values[a] Using
the Truncated Wash Procedure

Curvature ratio	Percentage distribution of positive samples			
	Cocaine	Opiates	PCP	Methamphetamine
1.3–1.5	10.3	1.8	4.4	8.3
1.5–2.0	14.6	7.0	30.4	30.2
2.0–5.0	50.6	31.5	39.1	52.1
5.0–10.0	17.4	19.3	4.4	7.3
>10	7.1	40.4	21.7	2.1

[a] $\text{Curvature ratio} = \dfrac{\text{amount of drug per 10 mg hair in three PO4 washes}}{3 \times \text{amount of drug per 10 mg hair in last PO4 wash}}$

removed from the cutoff levels. Consequently, an evaluation of hair porosity by methylene blue staining enhances the certainty of the interpretation of a positive result.[11]

C. Measurement of Metabolites

Wash kinetic parameters are not the only criteria for differentiating between exogenous and endogenous drugs. In conjunction with wash kinetic criteria we also use metabolite criteria. In the case of benzoylecgonine and amphetamine, these are expressed as percentages relative to the parent drugs cocaine and methamphetamine. These percentages also have cutoff levels. With benzoylecgonine it is 4% above the hydrolysis control. The latter is included in all assay runs, since the digestion of hair at pH 6.4 causes a small degree of cocaine breakdown (approximately 6%) by hydrolysis reactions. In this particular example, therefore, the cutoff will be at 10%. The cutoff level of the amphetamine/methamphetamine percentage is 3%.

Other definitive metabolites which are used in our laboratory as indicators of drug use are cocaethylene, norcocaine, Δ9-carboxy-tetrahydrocannabinol (carboxy-THC) (a substance which is not found in smoke from a marijuana cigarette), and the heroin metabolites 6-MAM and morphine glucuronide. Drugs and metabolites are measured by ultrasensitive GC/MS/MS or chemical ionization ion-trap GC/MS techniques requiring 5 mg or less of sample.[10]

It is important to note that in some instances the validity of the measurement of metabolites depends critically on the efficacy of the wash procedures. This pertains to substances whose metabolite can be present as an impurity in the parent drug. This applies to benzoylecgonine, amphetamine, and 6-MAM, but not to cocaethylene, carboxy-THC, norcocaine, or morphine glucuronide. However, as indicated previously, by washing hair specimens to near plateau conditions, we measure only the metabolites present in the inaccessible domain, and these are not likely to have been deposited there through contact with exogenous drugs containing metabolite impurities.

D. Wash Kinetics for 6-Monoacetylmorphine

The importance of washing for evaluating the significance of metabolites is best illustrated with 6-MAM because 6-MAM is readily formed by the hydrolysis of heroin under conditions of even neutral pH. Thus, 6-MAM may be found in hair as a result of contamination by heroin rather than heroin use. In view of the important application of hair analysis for the investigation of positive urinalysis results which were overturned by medical review officers because of the absence of 6-MAM or clinical signs of heroin use, we will illustrate the combined use of our wash procedures and metabolite measurements with 6-MAM.

That 6-MAM-contaminated hair can be rapidly and completely decontaminated by phosphate buffer washes was shown in studies in which negative hair of low, intermediate, and high porosity was contaminated by exposing it at room temperature for 1 h to a 0.01-M phosphate buffer solution (pH 5.5) containing 6-MAM at a concentration of 1000 ng/mL (Figure 2). Although, as expected from earlier studies, increasing amounts of 6-MAM were taken up with increasingly porous hair, decontamination occurred rapidly and none of the hair digests contained any significant quantities of residual 6-MAM once washes attained plateau conditions.

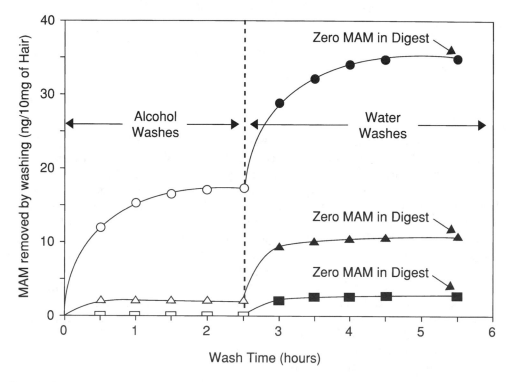

Figure 2
Wash kinetics of 6-monoacetylmorphine-contaminated hair. □ = nonporous hair; △ = semiporous hair; ○ = porous hair.

These results are in striking contrast to those obtained with the hair from heroin users (Figure 3). Essentially perfect plateau wash kinetics were obtained for 6-MAM, morphine, and codeine. Most important, the amounts of 6-MAM released from the inaccessible domain by digestion easily meets the R_{ESZ} criterion of heroin use. It should also be mentioned that had the alcohol washes been continued, then the alcohol wash kinetics would have been similar to curve AB in Figure 1.

E. Advantages of Hair Digestion Procedures for Clinical and Forensic Testing

For forensic testing, achieving positive results free from the ambiguities of false positives due to external contamination is the most critical element in the analytical process. For medical testing there are two additional critical criteria. One of these is that the hair has not been rendered porous by cosmetic treatment, since this can lead

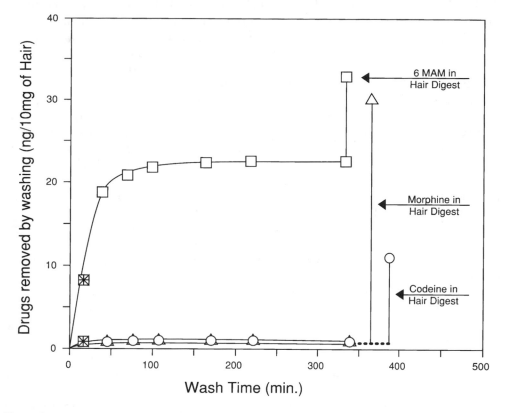

Figure 3
Wash kinetic/digestion profile of 6-monoacetylmorphine (□), morphine (△), and codeine (○) in the hair of a heroin user. All samples received an initial wash with dry ethanol (*). This was followed by washes with phosphate buffer (pH 5.5) and release of remaining drugs by enzymatic digestion.

to a partial loss of analyte. The porosity status of hair can be readily ascertained by methylene blue staining. The other criterion is more complicated and involves the assurance that the analyte has been extracted from the protein matrix of the hair with near 100% efficiency. Only when all three criteria are met can one expect the analytical result to provide valuable information on the severity and pattern of drug use.

The most effective and convenient way of meeting the required cleanup and extraction criteria is to wash hair specimens to near plateau conditions with phosphate buffer solutions and then to extract the analyte from the protein matrix by the enzymatic digestion of hair. The melanin pellet is separated from the protein digest by centrifugation. The removal of melanin ensures that the analytical result is not biased by hair color, i.e., by the preferential accumulation of drug in the melanin fraction, a fraction which constitutes approximately 5% of the hair mass. Although analytes are not removed from the melanin during digestion, such release can occur in the subsequently applied solvent-based cleanup procedures for mass spectrometric analysis.

We have developed two types of digestion procedures. One of these is suitable only for GC/MS confirmations,[11] and the other, a proprietary procedure,[12] can be used for both the initial RIA screen as well as GC/MS confirmations. The latter digestion procedure is more challenging than the first, for it requires the application of special measures to guarantee that the integrity of the antibody proteins used for the RIA analysis are maintained during their contact with the hair digest. In addition, sample matrix effects have to be kept to a minimum. The task of maintaining antibody integrity is made more difficult by the fact that hair is one of the most difficult protein structures to digest, requiring, in addition to enzymes, such chemicals as dithiothreitol and detergents, i.e., chemicals which can degrade antibodies, even in the absence of enzymes, by disulfide bond cleavage and denaturation.

The proprietary digestion procedure has the advantage of providing the most effective and therefore the forensically safest screening test results. In contrast to the RIA screening methods used by other laboratories which use solvent-based procedures that extract drugs mostly from the accessible and semiaccessible domains, our RIA of the hair digest measures drugs exclusively in the inaccessible domain. This, of course, is in the domain which the drug user cannot cleanse by extensive washing prior to submission of a hair specimen. Consequently, unlike some of the methods used by other laboratories, our procedure is not subject to false negative results. An additional advantage of the digest RIA method is that by measuring drugs in the inaccessible domain it provides accurate clinical information on the pattern and severity of drug use, i.e., information that can only be obtained with the other digest procedures by expensive MS.

Both hair digestion procedures have additional advantages over solvent-based analyte extractions. Drug "extraction" or, rather, solubilization via digestion, is universally applicable to all drugs, liberating these quickly (within 1–6 h) from the protein matrix with 100% efficiency under mild conditions of temperature (25–35°C) and neutral pH. In contrast to this, solvent-based procedures have several undesirable features: they extract analytes from both melanin and the protein matrix; they do this with variable and unknown efficiencies depending on whether hair is thick or thin, porous or nonporous; generally, elevated temperatures are required; extraction times tend to be long; different solvents generally have to be used for different analytes, thereby requiring more hair for analysis; and toxic solvents have to be evaporated (with possible loss of analyte). The result of all these difficulties is that wash kinetic criteria for the distinction between external contamination and drug use cannot be effectively applied with solvent-based procedures.

We also found that variations in matrix effects in different hair samples with the RIA-compatible digest are essentially nonexistent because of the constant protein and chemical composition of the digest. We have found that such variations are far smaller than those generally encountered with urine or blood RIA procedures. This, in turn, allows one to utilize the sensitivity of antibodies to the fullest extent through maximization of the signal-to-noise ratio for the assay. Excellent RIA detection of drug use is obtained with digests containing as little as 0.5 mg of hair. MS confirmations, which are necessary for forensic testing, are, of course, not required for clinical testing if noncross-reacting antibodies are used.

It is important to realize that ultrasensitive RIA and GC/MS analytical procedures are a necessary requirement for drug detection in human subjects. This is in contrast to animal studies. For example, with mice, it is quite easy to obtain approximately 0.1 g of hair from a 20-g mouse. Since the total hair of a mouse represents a significant fraction of its body weight, the amount of collected hair contains also a significant fraction of the administered drug dose. Hence assay sensitivity requirements are at a minimum. With human subjects, particularly with workplace testing, the RIA screens and GC/MS confirmations are performed on only 20–25 mg of hair. This is a very small and cosmetically insignificant fraction of the total amount of available head hair and an even smaller fraction of the total body weight. Consequently, this amount of hair contains also a very small fraction of the total ingested drug.

Another important difference between animal and human hair analysis studies relates to the problem of contamination. With drug-injected animals we are faced with the significant problem that the drugs found in hair may have been deposited there by exogenous contamination mechanisms involving contact with the endogenous drugs present in urine, feces, or saliva. The detection of drugs which were incorporated into hair by nonendogenous mechanisms can only be characterized as trivial. In our animal experiments[7] we made considerable efforts to develop appropriate wash procedures capable of removing drugs from hair that were deposited there through contact with urine. Unfortunately, many laboratories that have reported animal studies, particularly the early studies, do not appear to have taken such precautions.

III. RELIABILITY OF HAIR TESTING

A. General Considerations

Most of the controversy surrounding hair testing has revolved around what the degree of reliability of hair testing should be before its application to workplace testing can be justified. The reliability issue, unfortunately, has sometimes become entangled with ideological concerns regarding the appropriateness of workplace drug testing. A number of questions have been raised in this regard.

Concerning the appropriateness of the 3-month-wide detection window of hair analysis, we believe this not to be a significant issue. For one, it is not much wider than that of the urine marijuana test in the case of heavy marijuana users.[14,15] In addition there are valid clinical reasons for choosing a 3-month window. For example, patients who have completed a drug treatment program are tested on an individual basis for much longer periods by urinalysis as checks on possible recidivism. And scheduled hair analysis certainly does not suffer from the legal problems that unscheduled urinalysis incurs. Nevertheless, in a recent evaluation by the Society of Forensic Toxicologists (SOFT),[16] the opinion was expressed that hair analysis met the reliability requirements of forensic and medical testing, but not those for workplace testing. The reason cited was the "stand-alone nature of the test." What is meant by "stand alone" is that the positive result of the hair test and no other evidence is used in making employment decisions. With urinalysis the stand-alone issue is generally justified on the basis that the reliability and effectiveness of urine testing has been established by extensive field testing and external quality control programs.

We concur, of course, with the opinion that hair analysis should not be used for workplace testing until the reliability of the procedure equals at least that of urinalysis. However, we intend to show in the following discussion that hair analysis is more reliable than urinalysis because of its many probative advantages. And it is for this reason that we[11] and others[17] have suggested that a challenged urinalysis result from an employee should be investigated by hair analysis. On the other hand, with pre-employment testing, a check of a positive urinalysis result with hair analysis does not appear to be as critical, for it has been argued that job applicants frequently fail to gain employment on the basis of such subjective criteria as inappropriate appearance or demeanor. Hence, as in the criminal justice system, the question whether a urine positive result was caused by drug use or passive exposure by being around drugs or drug users is not as critical an issue as with the testing of employees.

B. Technical False Positives

The most critical issue of drug testing is the avoidance of false positive test results. One can distinguish between two fundamentally different types of false positives: technical false positives and interpretive false positives (Table 5). Technical false positives arise from faulty analytical procedures which erroneously indicate the presence of a drug in a negative specimen. Avoidance of technical false positive results requires strict adherence to GC/MS principles of identification and quantitation. Greater certainty (as well as greater sensitivity) can be achieved by the use of more advanced forms of MS, such as positive or negative chemical ionization procedures coupled with quadrupole or ion-trap MS/MS instrumentation. We have successfully applied these advanced ultrasensitive technologies to hair analysis under mass production conditions.[10]

TABLE 5.

Types of False Positive Results in Drug Testing

Types of false positives	Protection against false positive	
	Urine	Hair
A. Analytical false positives (Faulty mass spectrometry)	Easy	Easy
B. Interpretive false positives		
1. Subversive (spiked food or drink)	Impossible	Easy
2. Endogenous (passive ingestion or inhalation of drugs)	Main issue (problems)	Easy
3. Exogenous (contamination of specimens)	Difficult (but low probability)	Main issue (many solutions)

C. Interpretive False Positives

Technical false positives can be avoided relatively easily, for the analytical result is defined by the exact laws of physics and chemistry governing reliable MS operations. The same, however, does not hold true for interpretive false positives, since the interpretation of a positive result relies on the inexact laws of biology and physiology. For example, we will show that it is very difficult for urinalysis, but not for hair analysis, to establish whether a positive result was caused by passive exposure to drugs (e.g., ingestion of poppy seed), subversive activities (spiked food or drink), or by deliberate drug use. In the early days of urine testing, considerable confusion was caused by the fact that the certainty of urine testing was largely cast in terms of the avoidance of technical false positives rather than the avoidance of erroneous interpretations of positive results.

1. Subversive Interpretive False Positives

Examples of subversive false positives for urinalysis are positives caused by the ingestion of food or drink that has been spiked with a drug or by the willful contamination of urine specimens. It is essentially impossible to detect such subversive activities in the case of urinalysis; however, we will show that they are readily detected by hair analysis. Although subversive activities may be a rare occurrence, they cannot be ignored since they are frequently used as challenges to a positive urinalysis result (the "brownie" defense).

Obviously, one cannot distinguish between a one-time voluntary and a one-time involuntary drug use by hair analysis. However, in case of a one-time subversive act against an individual, the drug will be confined to a very small region of the hair, and this type of drug exposure or use can be readily identified by fine-grained segmental hair analysis. This is true for cocaine, opiates, phencyclidine (PCP), and methamphetamine, but not for marijuana, the least-sensitive hair assay.

The question has arisen whether or not a positive urine result is of necessity invalidated by a negative hair result. In defense of urine it has been argued that even fine-grained hair analysis will not always detect one-time drug use, particularly when only a small amount of drug is involved. We do not disagree that this is a possibility. However, we believe that it is a remote one, for field studies with cocaine users have shown (see below) that essentially all positive urine results are confirmed by hair analysis. Furthermore, the effectiveness of detecting infrequent drug use is up to 10 times greater by hair analysis than by unannounced urine testing (see below). On the basis of these considerations, we have adopted the position that although a negative hair analysis result does not necessarily disprove a positive urinalysis result, an individual presenting such a finding should be given the benefit of the doubt in light of the superior performance of hair analysis.

2. *Endogenous Interpretive False Positives*

Endogenous interpretive false positives are positives which result from the passive inhalation or ingestion of drugs present in food or in the environment. Such exposures generally involve quantities of drugs which are too low to produce psychotropic effects, e.g., the ingestion of morphine present in poppy seeds. Urinalysis and hair analysis have set endogenous cutoff levels as protection against false positives of this type. The endogenous cutoff levels used by hair analysis, however, are considerably more effective than those used by urinalysis. This has been demonstrated by theoretical considerations[4] and confirmed by experiments.[7]

The endogenous cutoff levels for urinalysis have no underlying theoretical basis and even the empirical procedures used in their definition are in most cases only weakly linked to passive exposure experiments. In contrast to this, the cutoffs chosen for hair analysis have a strictly defined theoretical basis, i.e., the observed correlation between drug levels in hair and ingested dose. Table 6 shows the results of blind studies demonstrating the existence of good correlations between ingested dose and drug concentrations in hair in the case of cocaine, morphine, methamphetamine, and PCP. With cocaine, this correlation was demonstrated in three independent studies with very good agreement in results.

TABLE 6.

Drug Intake Required to Produce Hair Drug Levels at the Cutoffs Set as Protection against Passive Endogenous Drug Exposure[a]

Drug	Study	N[b]	R[c]	Cutoff level (ng/10 mg hair)	Drug intake to reach cutoff level		
					Average	Range	Units
Cocaine	1	36	0.77	5	260	130–450	mg/month
	2	24	0.90	5	116	60–150	mg/month
	3	33	0.82	5	416	180–660	mg/month
PCP	4	23	0.80	3	0.44	0.26–0.75	Sherms/week
Methamphetamine	5	11	0.97	5	330	—	mg/week
Morphine	6	22	0.99	5	0.69	—	Bags/day

[a] Estimated drug intake is based on correlations between drug concentrations in hair and self-reports of drug use.
[b] Number of participants in study.
[c] Correlation coefficient.

The technically limiting factor in hair analysis is the minimum drug concentration that can be consistently detected in small quantities (5–10 mg) of hair by forensic-quality MS. These concentrations were also chosen as the endogenous cutoff levels. Their effectiveness was demonstrated by theoretical as well as by experimental evaluations. Theoretical validation simply involves calculating from dose/hair level correlation graphs the amount of drug which has to be ingested to produce drug levels in hair equaling the values set for the cutoff level. The magnitude of these calculated values and their ranges are listed in Table 6. It is evident from these values that passive exposure to drugs cannot result in endogenous interpretive false positive results.

The validity of these theoretically determined cutoff levels was evaluated by the various field studies described below. In some of these this involved comparing the results of hair analysis to those of urinalysis and self-reports. In the case of cocaine and morphine the validation of endogenous cutoff levels involved also the controlled administration of these substances.[7,18,19] With morphine this was done with poppy seed and involved a comparison of hair and urine results. These experiments clearly demonstrated the greater effectiveness of the endogenous cutoff levels used by hair analysis, for all attempts to exceed the opiate cutoff level of hair analysis by the massive ingestion of poppy seeds failed. In contrast to this, the cutoff level of urinalysis was exceeded by wide margins. That this particular problem of urinalysis is not unique to poppy seeds has been demonstrated experimentally by Baselt and Change[20] in the case of cocaine.

These considerations show clearly that endogenous interpretive false positives are the main problem confronting urinalysis. It is, however, a problem with which hair analysis can provide considerable assistance, since passive exposure as a cause for a positive urinalysis result can be excluded by a positive hair analysis result. As indicated previously, this support is particularly critical in the case of opiates, where considerations of public safety demand that the numerous overturned urinalysis results be further investigated.

a. Endogenous Cutoff Levels and Dormant Hair Effects

Before a hair analysis result is interpreted as a positive due to deliberate drug use it must meet all criteria of our diagnostic algorithm: three wash kinetic criteria, the metabolite criteria, and the drug level in hair must exceed the endogenous cutoff level for the particular drug. However, there is one important exception to this algorithm, and this occurs when one determines the pattern of an individual's drug use by segmental hair analysis. For this determination, an additional rule is required, and this to compensate for dormant hair effects.

Dormant hair effects occur because approximately 15% of head hair is at any one time in the nongrowing or dormant phase. Most dormant head hair is lost within approximately 2 months. Although hair dormancy has a negligible effect on dose correlation, it is of major significance when segmental analysis is used to demonstrate cessation of drug use. We demonstrated this aspect of segmental hair analysis in several studies with drug treatment organizations where patients entered a drug-free environment in which their drug-free status was guaranteed by appropriate monitoring procedures.[4]

A typical example of dormant hair effects in an individual who stopped drug use is shown in Figure 4. The 0- to 1-month and 1- to 2-month segments correspond to the period when the individual was in a drug-free environment and the 2- to 3-, 3- to 4-, and 4- to 5-month segments to the time when the subject was engaged in drug use. The important result of this particular experiment is the 26 ng of cocaine/10 mg hair that was found in the 1- to 2-month segment. The interpretation of this positive result is not that the patient persisted in a small amount of drug use, but rather that the positive result in this segment is an "echo" of his former drug use because of the influence of dormant hair. That is, at the time the individual entered the drug-free environment, a small fraction of his hair stopped growing and now is in phase with the hair that grew during the drug-free period. However, after 1 month, most of this dormant hair has fallen out, thereby giving rise to the insignificant amount of drug in the 0- to 1-month segment.

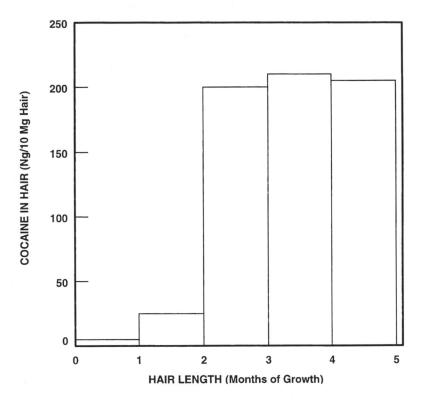

Figure 4
Segmental hair analysis illustrating the effects of dormant hair. The individual in this study ceased drug use at the 2-month mark. Drug residues in the 0- to 1-month and the 1- to 2-month hair segments are caused by dormant hair.

On the basis that approximately 15% of hair is in the dormant phase and that dormant hair is rapidly lost from the head, we have formulated the rule that the drug content of a hair segment must be greater than 15% of that of a preceding segment in order to indicate that some drug use has taken place. This rule also requires that both hair segments are of equal lengths.

It is important to note that dormant hair effects with marijuana, opiates, amphetamines, and PCP result in hair levels which fall below their endogenous cutoff levels if drug use has ceased 3 months ago. This is a consequence of the averaging of dormant hair effects over the 3-month detection window. In most cases this applies also for the ultrasensitive cocaine test. This is true because most cocaine-using job applicants (67%) fall into the light to intermediate cocaine user category (see below). Only the relatively small fraction of heavy cocaine-using applicants, those who pose higher risks for recidivism, are likely to exceed the endogenous cutoff level by dormant hair effects. However, any claims of cessation of cocaine use within the

3-month period can be readily checked by segmental analysis. If cessation of cocaine use has occurred 4 months ago or longer, then virtually all dormant hair will have fallen out and then even heavy cocaine users will present a negative test result for the most recent 3-month period. It must be emphasized, however, that the success of a recently completed drug treatment program is best ascertained by segmental hair analysis and not by the analysis of a single 1-month hair segment.

However, alternative, but more subtle, strategies utilizing single hair segments can also be used to establish cessation of drug use without interference from dormant-hair effects. We demonstrated this with patients who entered a guaranteed drug-free environment in which they were treated by aversion therapy. Hair specimens were obtained from these patients at the time of entry into the program and after 6–8 weeks of aversion therapy. A hair specimen corresponding to the patient's last 2 weeks of therapy was analyzed. The results in Table 7 show clearly that dormant-hair effects do not interfere with our ability to demonstrate cessation of drug use. An essentially 100% drop in the drug content of hair is evident during the last 2 weeks of therapy.

TABLE 7.

Dormant-Hair Effects in Hair Analysis for Determining Cocaine Use

1st sample[a] (ng cocaine/10 mg hair)	2nd sample[b] (ng cocaine/10 mg hair)	Time between 1st and 2nd samples (months)	Dormant-hair effect (%) (2nd/1st × 100)
32.5	0	1	0
264	2.5	2	0.95
56	0	2	0
57	0	2	0
88	0	2.5	0
317	3.3	2	1.0
101	0	2	0
113	0	1.5	0
115	0	2	0
193	2	2	1.0
41	0	2	0
286	8.3	1	2.9
168	3.8	2	2.3

[a] 1st sample was a 1.3-cm section.
[b] 2nd sample was a 0.6-cm section.

b. Endogenous Cutoff Levels: A Safeguard Against Possible Testing Biases

Several investigators have speculated that hair analysis may be biased with respect to hair color, race, sex, and age.[21] Ignoring the fact that these issues have not been settled for urinalysis, it should be mentioned that it is highly unlikely that such a bias can occur to any statistically significant extent, and this for biological as well as methodological reasons. The main reason for the former is the dominant influence of biochemical individuality on the efficiency of drug incorporation. That is, biochemical individuality already causes a large statistical scatter in the amount of drug that is incorporated into hair for a given administered dose. Consequently, it is unlikely that the existence of any additional factors such as gender or race will make any statistically significant contributions to drug incorporation. True, certain pharmaceutical agents, such as haloperidol (but not drugs of abuse), have been shown to accumulate preferentially in the melanin fraction of hair, thereby accumulating in greater quantities in black hair than in white hair. But the melanin issue is successfully circumvented by our hair digestion procedures in which we exclude the melanin fraction from the analysis. Of course, this is not the case for solvent-based extractions.

Some of the experiments cited in support of racial or sexual bias did not involve studies where drugs were ingested under controlled conditions, but rather they were studies where hair was exposed to drug-containing solutions.[22] Obviously, such experiments do not mimic *in vivo* drug incorporation mechanisms. At best they were models for exogenous contamination. In this respect it is important to recall that almost 10 times more 6-MAM is incorporated into porous than into nonporous hair (Figure 2). And since Asian hair is generally less porous than fine blond hair, or even African hair, it is not surprising that we and others have found that Asian hair is less readily contaminated than other types of hair. But this does not constitute a racial bias, for such contamination is readily removed by properly applied wash procedures.

Porosity effects are also the most likely cause for the recently reported increased uptake of exogenous drugs in female hair.[22] For one, female hair, because of its length, is more porous than that of males. Porosity is also likely to be greater in female hair than in males because of the application of a variety of cosmetic treatment regimens, e.g., curling, perming, etc. It is therefore important for data interpretation that exogenous contamination studies be guided by an evaluation of the porosity status of the hair. This is readily done by the methylene blue staining technique.[11] Once again, however, we do not consider contamination experiments to mimic *in vivo* processes.

In our laboratory, we have investigated the possibility of racial bias with 315 African-American and 846 Caucasian subjects, involving a variety of procedures, including a comparison of self-reports, urinalysis, and hair analysis data (see below). To date we have found no statistically significant evidence for the existence of a racial bias.

Finally, it is important to realize that the possibility of racial, sexual, or other bias does not pose a problem for hair analysis, since any biases can be readily corrected for by simply adjusting the endogenous cutoff levels for a particular subgroup. Thus if it were demonstrated by a statistically reliable study that Group A incorporates into hair twice the amount of drug per given dose than Group B, then one would compensate for this bias by simply doubling the endogenous cutoff levels. Such adjustments are possible for hair, but not for urine, since the cutoff levels of the latter are not correlated with dose. That is, the drug level in urine may be low because a little drug was ingested recently or because a large quantity of drug was ingested in the more distant past.

3. Exogenous Interpretive False Positives

The main issue in hair testing is the avoidance of exogenous interpretive false positives, i.e., positives caused by external contamination of hair by drugs present in the environment, e.g., smoke, powder. This type of false positive is not the major issue for urinalysis where endogenous interpretive false positives are the main concern. But, the effective avoidance by urinalysis of exogenous false positives due to specimen contamination in the laboratory depends critically on the exclusion of drug-using personnel, and this can best be achieved by evasion-proof hair analysis. However, when such false positives occur, or when urinalysis labs are unable to guarantee that they have taken effective measures to exclude such contamination, then very little can be done to remedy the problem. For, in contrast to hair, the collection of a new urine specimen identical to the first one is not possible.

In contrast to the single laboratory safeguard of urinalysis against its main problem, i.e., the endogenous cutoff level, hair analysis is favored by many protective measures against its main challenge — exogenous interpretive false positives. More-over, all of the measures used by hair analysis can be shown to be more effective than the endogenous cutoff level used by urinalysis.

When considering the relative safety of urinalysis and hair analysis, one is frequently misled by a false sense of security concerning the urine test. This misper-ception arises from the erroneous picture of a large human body and a small mass of hair exposed to the same drug-containing environment.

This picture is misleading on several counts. It is readily calculated that less than 1 mg of drug ingested in the relatively large mass of the human body, when con-centrated within 2–3 d in the much smaller mass of 1–2 L of urine, is sufficient to create a positive urinalysis result. For sake of comparison, one can make the initial assumption that the same amount of drug will accumulate in a tenfold smaller mass of head hair. But this does not mean that hair incurs a ten times higher risk of producing a false positive, and this for several reasons. For one, we have shown that hair is highly resistant to the penetration of drugs, particularly when these are applied in the form of smoke and powders. In contrast to this, the gastrointestinal tract and the lungs offer no resistance to penetration; actually these drugs cross the membranes by active transport mechanisms. And when it comes to the net transport of drugs into urine vs. deposition onto hair, then active breathing is a highly effective mechanism for increasing the drug concentration in urine vs. those for depositing passively onto hair.

So from a simple mass transport point of view, hair is less susceptible to con-tamination by exogenous mechanisms than urine by endogenous mechanisms. But then hair, but not urine, is cleansed by normal hygienic practices and by special laboratory wash procedures. The latter are further strengthened by special kinetic analysis of the wash data. And, finally, the deposition of exogenous drugs onto hair, unlike their accumulation in urine, is not associated with the formation of metabolites.

If one now adds to these safety features against exogenous false positives the ability to collect a second hair sample, safer endogenous cutoff levels, the trilevel reporting system (negative, contaminated, or positive), as well as the subversion-proof nature of the hair test, one can readily appreciate that hair analysis is a safer procedure than urinalysis. One also recognizes that hair analysis, unlike urinalysis, is not a stand-alone test.

D. Wash Kinetics and the Role of Contamination Models

Several recent papers by Blank and Kidwell[23-25] have questioned the reliability of our wash kinetic procedures. Their criticism, which in many instances is marred by the faulty application of our technology, has demonstrated the need for further clarification of the rationale upon which our procedures, particularly the wash kinetic procedures, are based.

1. Contamination and Decontamination Mechanisms

The most probable mechanism for the contamination of hair by exogenous drugs involves their deposition in the form of vapors or powders. Most workers in the field believe that such contaminants are readily removed by washing even by normal hygienic practices.[25] However, several investigators, including us, have taken the issue of external contamination one step further.[11,23,26] Some have postulated that drug contaminants on the surface of the hair may be dissolved by sweat and thereby carried into interior regions of the hair from which they are less readily removed by conventional laboratory wash procedures involving the use of nonpolar solvents. We agree with the suggestion that contamination deposited in aqueous form, because of penetration into the semiaccessible domain, requires more extensive decontamination procedures than vapors or powders of drugs deposited in the accessible domain. However, we disagree with the suggestion that such contamination cannot be effectively removed by aqueous wash procedures, or that it cannot at least be distinguished from endogenous drugs by the kinetic analysis of the wash data and by metabolite criteria. To this we should also add that we seriously doubt that the postulated sweat contamination mechanisms have any practical relevance or that the soaking of hair for 1 h at elevated temperatures (37°C) and in concentrated drug solutions is a realistic model for any exogenous contamination processes.[23,24]

It is not surprising, however, that contaminants deposited in hair from aqueous solutions are not as readily removed by normal hygienic practices as those deposited by vapors.[27] For one, normal shampooing and showering involves only short contact times with the aqueous wash medium, i.e., in the vicinity of 5 min. Such contact times are too short for the effective removal of drug contaminants deposited from aqueous solutions. It is precisely because of this situation that we use protracted contact times of at least 30 min in our laboratory wash procedures.

2. Complete Decontamination: An Unnecessary Requirement

Another point of confusion is the misleading notion that washing, particularly a limited number of washings, should remove all exogenous drug residues in order to effect a distinction between exogenous and endogenous drugs. For one, simple consideration of desorption kinetics should show that complete removal is a theoretically impossible notion. For, although the removal of contamination is not a simple (first-order) exponential process because of the involvement of different binding sites on hair, it is nevertheless a process which approaches (like other equilibrium processes) a state of zero contamination asymptotically. What is significant in decontamination is not that an absolute zero in contaminants is reached, but rather that drug levels fall either below the endogenous cutoff level and/or below the safety zone level.

Furthermore, this does not even have to be achieved operationally (i.e., with our three phosphate buffer washes), since washing can be extrapolated mathematically. With our methodology this is done by choosing a cutoff value of 10 for the extended wash ratio R_{EW}. What this does mathematically is to remove an amount of drug that 10 additional 30-min phosphate buffer washes would have removed if no further curvature in the wash kinetics were to take place. This, of course, is a highly unrealistic assumption. Hence, in actual fact, it would have taken many more real washes to remove the amount of drug projected by the extended wash ratio. In other words, in developing our wash kinetic approaches we have decided to err on the side of safety, for the sample is considered contaminated if 10 times the amount of drug found in the last phosphate buffer wash exceeds the amount found in the hair digest. To appreciate the additional safety that the extended wash ratio imparts, it must be recognized that three 30-min phosphate buffer washes are sufficient to reduce the number of positive samples obtained with unwashed hair by an average of 30%.

3. Importance of the Curvature Criterion

Another aspect of our diagnostic algorithm that is not sufficiently appreciated is that the validity of the extended wash ratio and the safety zone ratio depends on the wash kinetics showing a significant amount of curvature as measured by the curvature ratio. If the curvature falls below its empirically determined cutoff value, and there are several reasons for this (e.g., excessive contamination or excessive decontamination), then the wash procedures are invalidated and the sample is provisionally diagnosed as contaminated; i.e., the result will be solely evaluated by its metabolite content. It is therefore incorrect to maintain that the curvature ratio "has no diagnostic value."[25]

Finally, with respect to our wash kinetic approach, it must be understood that all three wash kinetic ratios have to be met simultaneously before a hair sample is judged positive due to drug use. If only one kinetic ratio fails, the sample is judged to be contaminated. This ruling, however, can be overturned by a highly definitive metabolite finding, e.g., by the presence of cocaethylene, norcocaine, or high benzoylecgonine ratios. In the ultimate analysis, however, it must be understood that an element of doubt will always remain in any scientific finding,[28] and this doubt in the case of hair analysis is adjudicated by the benefit of doubt considerations, outlined previously.

4. The Nature of the Inaccessible Domain

From the comments of Blank and Kidwell[23,24] it appears that the nature of the inaccessible domain also needs further clarification. Contrary to their comments, we do not define the inaccessible domain as regions in hair which cannot be penetrated by external contamination, but rather as the region which cannot be accessed with sufficient ease by water for the effective extraction of endogenous drugs. That there are regions in hair which are inaccessible to water is well known. As we have stated below, these have been shown by X-ray crystallography and other means to correspond to the microfibril structures of hair. And, of course, if water cannot penetrate these regions, neither can water-borne drug contaminants. But endogenous drugs entrapped in the microfibril structures are not necessarily the only ones which cannot be effectively extracted by water. We have long recognized that unrealistically severe contamination conditions may deposit exogenous drugs into hair which cannot be extracted/washed out with water. What we do maintain, however, is that such penetration is not likely to occur to a significant extent under realistic contamination conditions. "Significant" in this respect is defined by the endogenous cutoff levels and by the safety zone ratio.

E. Flawed Evaluations of Wash Kinetic Criteria

In a recent review, Blank and Kidwell[25] have questioned the adequacy of the wash procedures used in various hair analysis laboratories. In their review they also extended their earlier criticism[23,24] of our wash kinetic approach.

Their earlier criticism of our methodology was severely flawed in that their procedures differed in virtually every aspect from those used in our laboratory. This included the wash and digestion procedures and their substitution of isotopes in place of our direct analysis of drug contaminants; even the data manipulation was incorrectly performed. These obvious shortcomings in their evaluation of our technology were extensively criticized.[29,30] In spite of this exchange of ideas, little has changed in their recent extension of their criticism. The only response to our concerns has been to make mathematical corrections for the use of four instead of our three phosphate buffer washes; no attempt, however, was made to correct for the distorting effects that prior washing with moist alcohol produced. They also addressed some minor methodological issues relating to their tritium counting techniques.

One striking feature of their most recent evaluation is that their results differ drastically from those of their earlier evaluation, and this in spite of the claim that identical procedures were used. What is significant about this difference is that the results of their first study were very close to our results in spite of their marked deviation from our methodology. We showed that these slight differences between their first results and ours could be reconciled by small mathematical corrections to their results. These corrections related to the use of 4 instead of 3 wash procedures and to the inappropriate use of moist ethanol. However, such reconciling of results is not possible with their second evaluation of our procedures. Furthermore, they also provide no data on the curvature criterion. Since very little experimental detail is provided by them, it is difficult to identify the causes for the large discrepancies between their two studies. Nevertheless we will offer two suggestions (see below) which, although applicable to both of their studies, may have affected the second one more seriously than the first.

In light of the obvious flaws in their evaluation, we once again contend that their experiments did not constitute a valid evaluation of our wash kinetic procedures. Nor do we agree with them that their described isotope procedure is suitable for contamination/decontamination studies. As a matter of fact, it is likely that an inherent difficulty in their isotope procedures, which we shall describe shortly, may have created the observed difference between their first and second studies. It is also likely that insufficient attention was given to critical elements in their contamination/decontamination experiments. Since we have not described the latter in our previous publications, we will briefly summarize these now and then comment on a general problem with their isotope method.

1.   Decontamination Studies: Sample Preparation Techniques

In our experiments, hair (20 mg) is soaked in 2 mL of phosphate buffer (pH 5.5) containing cocaine hydrochloride (1,000–10,000 ng/mL) for 1 h at 37°C. To avoid interference from a variety of factors (see below) including possible adsorption of drugs on the walls of the polystyrene test tubes, the contaminated hair is subjected to the following treatments before commencing with the wash kinetic experiments in a new test tube.

Contamination due to loosely adhering drug in phosphate buffer is removed from hair by a 5- to 10-s rinse with fresh phosphate buffer solution. The hair sample is then immediately dried. This first involves an initial blotting with laboratory tissue or filter paper (a simulation of towel drying) followed by storage in a dessicator until the hair is completely dry. Complete dryness of the hair can be ascertained by the rapid loss of approximately 30% of the weight of a blotted specimen, the maximum water content of hair. Under the appropriate conditions of low humidity and initial blotting, drying occurs within 15–30 min at room temperature.

There are several reasons for drying the hair. For one, it is the natural condition of the hair as it is presented to the laboratory. Dry hair is essential for the proper application of the first wash with dry isopropanol (or dry ethanol). Lack of dry initial conditions created major distortions in Blank and Kidwells' first study. It is highly unlikely that appropriate procedures were used in their second study, for they indicate that some hair samples were dried and others not. The second reason for commencing the water washes with dry hair is that it may take 5–15 min before dry hair gains its maximum moisture content,[13] and this in turn has an effect on the kinetic profile of the water washes. This effect is probably also the main reason why drugs deposited in hair from aqueous solutions are less effectively removed by several 5-min shampooing/hair-drying cycles than drugs deposited in vapor form.[27]

2. Isotopic Contamination Studies: Potential Problems

Although our initial concerns about possible interferences due to chemilumines-
cence and quenching effects have been somewhat mitigated by recent disclosures of
Blank and Kidwell's tritium counting procedures, an entirely different problem exists
concerning their isotopic method, one that in our opinion casts serious doubt upon
the general suitability of their approach for doing hair contamination/decontamina-
tion studies.

The nature of our concern is best illustrated by a specific example. Blank and
Kidwell use a cocaine solution of 100,000 ng/mL for their contamination experiments, to which they add approximately 1 µCi of tritium-labeled cocaine, i.e., approximately one million counts per minute. Therefore, they have approximately a sensitivity of 10 cpm/ng of sample. Decontamination of hair means that residual drug
concentration must drop below the endogenous cutoff level of 5 ng/10 mg of hair,
i.e., to 50 cpm/10 mg hair. Now if the labeled cocaine has a radiochemical impurity
of as little as 0.1%, this corresponds to 1000 cpm or to 100 ng of residual cocaine
equivalents. Since self-irradiation of tritium-labeled material tends to form polymeric
impurities, and since these are likely to preferentially bind to hair,[13] one incurs a
major risk of concluding erroneously that the residual radioactivity represents residual cocaine contamination rather than contamination by polymeric degradation
products.

Our criticism, of course, is not directed at isotope techniques in general, but only
to contamination/decontamination studies of hair where only a small fraction of the
total isotope binds to hair and where residual drug levels in hair are exceedingly
low when compared to those in the contaminating solution. In light of these considerations it is our opinion that there is no excuse for substituting a risky technique,
no matter how convenient, for a reliable one involving the direct measurement of
analytes. This is particularly true when it is the objective to validate the latter
procedure.

3. Studies with Children of Drug Users

Finally, Smith and Kidwell[31,32] have attempted to evaluate our wash kinetic
algorithm under real-world conditions in a study involving the participation of
young children living with their cocaine-using parents. Since very young children
can be assumed not to be drug users, Smith and Kidwell[31,32] based their study on
the assumption that any cocaine found in the hair of children must have been the
result of external contamination. They erroneously ignored the possibility of positives due to passive ingestion of the drug. This, of course, is a serious flaw in their
study. For one, it ignores the high sensitivity of the cocaine hair test (Table 6), i.e.,
the relatively low endogenous cutoff value. The influence of the high sensitivity of
the cocaine assay is further aggravated by the small body weight of the young
children. For example, if their body weight was one tenth that of the adult population,
then, because of the correlation between hair levels and dose/kg of body weight,
the endogenous cutoff level should have been raised by a factor of 10. Also, the wash
procedure used in that study deviated considerably from ours; for example, methanol
was used instead of dry isopropanol, and only two of the three wash kinetic criteria
were applied.

Even though the majority of results were below the adult endogenous cutoff
level (not to mention the tenfold higher cutoff level for children) and in spite of the
presence of benzoylecgonine which indicated passive exposure, Smith and
Kidwell[31,32] drew the erroneous conclusion that our wash kinetic criteria failed to

detect that the hair of the children were externally contaminated. To us their data indicate exactly the opposite, namely, that the children had very low positive hair results which failed our wash kinetic contamination criteria because of passive ingestion of low and, hopefully, harmless concentrations of cocaine. This conclusion is supported by well-established clinical observations.

IV. FIELD STUDIES

A. Drug Treatment Programs

The need for effective diagnostic tools for monitoring drug use has long been recognized by drug treatment professionals. Handicapped by the limitations of self-reports of drug use, the lack of obvious intoxication which is a feature of some types of abuse, and the well-documented tendency of addicts to deceive themselves and others, therapists have searched for a more objective means of monitoring drug use. In a recent clinical study where urinalysis, clinical evaluations, and self-reports were compared to hair analysis, Brewer[33] concluded that a new diagnostic dimension was opened up with the advent of hair analysis. Although the unprecedented diagnostic power of hair analysis had become evident in our earlier clinical studies,[34-37] these were handicapped by lack of contact with the patients and therapists. Because of this deficiency, we will focus in this review mainly on the experience of Brewer[33] who has applied hair analysis in a variety of clinical settings. The use of hair analysis in determining prenatal drug exposure, in therapeutic drug monitoring, and treatment in the criminal justice system will be discussed separately in their respective sections below.

In the diagnostic application of hair analysis, all studies to date have found hair analysis to be more effective than urinalysis at identifying drug use. In the case of light drug users the detection rate of hair analysis can be ten times that of urinalysis.[38] Such low drug use is important information for the therapist in view of Brewer's[33] observation that relapse is frequently preceded by dabbling in drug use. Consequently, it was found that hair analysis was a particularly sensitive method for detecting relapse.

With highly motivated and reliable patients, Brewer[33] also noticed that there was a good correlation between drug levels in hair and self-reports of the amount of drug used. Thus properly performed hair analysis (which must include porosity evaluations) promises to be a useful tool for the initial evaluation of the severity of drug use, thereby facilitating referral to appropriate treatment programs, e.g., inpatient vs. outpatient programs.

Of course, it may turn out that with some drugs one will obtain better correlation between drug levels in hair and the clinical condition of patients, i.e., with a toxicological or addiction index. This will occur in situations where the absorption, metabolism, and excretion of the drug vary greatly from individual to individual, thereby negating the simple correlation among drug levels in hair, the ingested dose, and clinical condition. In situations such as this, hair analysis may provide an index for the bioavailability of the drug. This possibility, which will be discussed in greater detail in the section dealing with the role of hair analysis in therapeutic drug monitoring, is currently being investigated by us in a study comparing the results of hair analysis with extensive psychiatric evaluations and positron emission tomography measurements of drug-induced neurobiological changes.

Brewer[33] observed that hair analysis is not only a diagnostic tool, but also an important component of the treatment process. Several features of hair analysis were found to be critical in this respect. The most important one is its unprecedented effectiveness at identifying drug use. This is not only a property of the sensitivity of the test, resulting from the wide window of detection, but also a consequence of the evasion-proof nature of the test. Even shaving one's head is not effective for evasion, since hair analysis procedures can be applied with equal effectiveness to fingernail clippings or fingernail shavings. We prefer the use of fingernail shavings to clippings or body hair, since shavings access essentially the same time frame as head hair.[4]

Brewer[33] noted that the evasion-proof nature of the hair test was found to result in a "therapeutically useful change in the balance of power between patient and clinical staff," especially in programs where adverse consequences accompanied the resumption of drug use. In contrast to hair analysis, urinalysis was found to be readily evaded in clinical settings, either by temporary abstention from drug use, tampering with the sample, or substitution of the sample by the use of synthetic external bladders simulating natural urination. Even such extreme measures as catheterization of the bladder have been observed. To thwart some of these evasive measures, urine had to be collected under close observation — a highly embarrassing and humiliating experience for both therapist and patient. This is further aggravated by the fact that some patients find it genuinely difficult to pass urine under observation, even when they have no illicit drug use to conceal.

In light of all these difficulties, it is not surprising that urinalysis is generally viewed as having an adverse effect on the patient-therapist relationship. In contrast to this, Brewer[33] found that hair analysis was acceptable to both parties and resulted in improved client-therapist relationship, frequently manifesting itself in more candid self-reporting of drug use prior to a scheduled hair test. The patients' estimates of drug use were often in good agreement with those obtained by hair analysis. And in case of disagreement both the patient and therapist are comforted by the fact that the first result can be checked by the collection of a new hair specimen.

Finally, the ability to monitor progress in recovery by segmental hair analysis was found to be the most valuable feature in virtually all clinical hair analysis studies. It has even been suggested that segmental hair analysis could have a major impact on the philosophy of addiction treatment. Thus Brewer[33] states, "the difficulty of quantifying drug use has reinforced the existing tendency in some treatment programmes and philosophies to regard any drug use as a treatment failure, because total abstinence is the acceptable goal. However, in most conditions, patients and their doctors are generally pleased if there is significant improvement in the problem being treated, even if the relief is less than total."

Another field that is likely to be affected in a positive way by hair analysis is the diagnosis of PCP-induced toxic psychosis.[39] The differentiation between toxic and nontoxic psychosis is difficult by conventional psychiatric evaluations; however, it can be readily made by analyzing hair for PCP. This was not possible with urine, presumably because toxic psychosis results from distant rather than recent PCP use.

In our study with the California State Hospital in Norwalk, patients at the time of admission were given a hair and a urine test for PCP. Of admitted patients, 19% were identified as former PCP users by hair analysis; none were positive by urinalysis, even though the latter was performed by GC at a detection limit of 1 ng/mL. The positive hair analysis results were found to be of considerable value in establishing a diagnosis of toxic psychosis. Because of this diagnostic advantage and the much lower cost of treating toxic psychosis than nontoxic psychosis, the Psychopharmacology Committee of the State of California recommended that hair analysis be used for routine clinical testing.

B. Prenatal Drug Exposure

The first investigation of prenatal drug exposure by hair analysis was reported by us in 1987.[7] The investigation was initiated in response to the concerns of a family member who became alarmed by the abnormal behavior pattern of a baby. The submitted hair samples from the baby and the mother were found to contain PCP. The drug levels in hair indicated low PCP use and segmental analysis showed that PCP use had occurred during the entire pregnancy. It was disconcerting to see that PCP-related abnormalities in infants are so striking as to be readily discernable by medically untrained individuals. Although the devastating effects of prenatal PCP exposure are well recognized by the medical profession, we were somewhat surprised at the medical professions' focus on prenatal cocaine exposure to the virtual exclusion of other drugs, at least as far as hair analysis studies were concerned.

Our first hair analysis research of prenatal drug exposure in a clinical setting was performed in collaboration with Dr. Lance Parton at the Los Angeles Childrens Hospital in Los Angeles.[40] The study involved the analysis of hair and urine from 15 babies aged 0 d to 3 months who were clinically suspected of being exposed to cocaine (4/15) or whose mothers had a positive history of cocaine use some time during their pregnancy (11/15). Seven positive hair analysis results were obtained. These were in excellent agreement with five positive urinalysis results and self-reports of cocaine use. Clinical suspicion was poorly correlated with hair analysis, urinalysis, and self-reports, except for one case where the baby died of congenital heart disease.

The second collaborative prenatal drug exposure study which was initiated by our laboratory was performed with Chasnoff's group.[4] Because clinical studies are characterized by a close relationship between researcher and volunteer subject and by the ability to apply frequent urine testing under ideal, i.e., evasion-free conditions, it was the purpose of this particular prenatal study not so much to investigate everyday clinical problems, but rather to obtain epidemiological data and to validate hair analysis by the best possible self-reports and urinalysis data. The study involved the collection of hair specimens from 24 mother/baby pairs 1 week after delivery. Specimens were analyzed for cocaine, PCP, and opiates. All subjects admitted to drug use sometime during their pregnancies. The mother's hair was cut into nine segments, each representing approximately 1 month of growth. Urine toxicological evaluations were performed weekly throughout the pregnancy. The results in Table 8 show that in spite of frequently applied urine testing, hair analysis identified more mothers using drugs during their pregnancies than did urinalysis. The number of babies yielding positive results with hair samples was twice that of urinalysis.

TABLE 8.

Comparison of Hair and Urine Positive Results
in Mother/Baby Pairs

Test	Subjects[a]	Number of positive patients		
		Cocaine	PCP	Heroin
Hair	Mother	23	2	6
	Baby	19	2	4
Urine	Mother	17	2	2
	Baby	11	0	1

[a] Total number of mother/baby pairs = 24.

Figure 5 is a typical example of the power of segmental hair analysis for providing clinically significant data on both relative and absolute amounts of drug use. The severity and relative changes in drug use during the three trimesters of pregnancy are likely to be important clinical parameters for assessing clinical consequences to the fetus or the mother. More important, evidence of decreased drug use, as indicated in this particular example during the last 3 months of pregnancy, may provide the clinician with valuable feedback information for facilitating further improvements in the patient's drug habit. With urinalysis the only information that the clinician receives is that drug use of some undefined intensity is continuing — that is, no insight can be gained from urinalysis whether or not the client shows improvement in her drug habit.

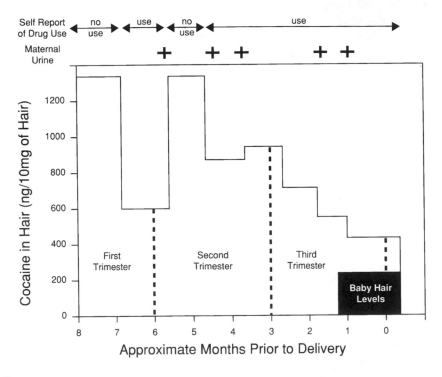

Figure 5
Cocaine use during pregnancy determined by segmental hair analysis on the mother's hair, self-reports, and urinalysis.

The following prenatal drug studies were initiated and designed by the independent investigators, with our laboratory providing the analytical services on a blind basis. In one of these studies, the drug use detection efficiency of maternal hair analysis was compared to that of meconium.[41] The specimens were analyzed for cocaine, opiates and marijuana. The results were compared to self-reports. Meconium was analyzed by an independent laboratory. There were 26 pregnant women enrolled in the study who delivered at the participating hospital and who were known to have used alcohol and illicit drugs during the pregnancy. The results in Table 9 showed an essentially identical positive rate for hair analysis and meconium for all three drugs and a generally good agreement with self-reports.

Although hair and meconium exhibited essentially identical detection rates, we believe that the clinical significance of the data are not identical. For one, meconium analysis allows the detection of drug use only after birth, whereas analysis of the mother's hair during pregnancy facilitates intervention and reduction of risk to the

TABLE 9.

Comparison of Hair Analysis with Meconium
Analysis and Self-Reports

	Percent positive rate		
Drug	Meconium	Hair	Self-report
Cocaine	61.5	73.3	73.1
Opiate	34.6	26.7	7.7
Cannabinoid	26.9	21.4	19.2

fetus and the mother. Also, meconium analysis cannot match the 9-month retrospective information of hair analysis on the severity and pattern of drug use by the mother.

In a blind study with the same research group, but in an everyday clinical prenatal setting, the detection efficiency of hair analysis was compared to that of urinalysis.[42] The results of this study showed that the detection of maternal cocaine use by urinalysis, applied at the first prenatal visit, was only 26% that of hair analysis.

The correlation between cocaine levels in the hair of mothers and their babies was investigated with 60 mother/baby pairs.[43] All mothers had used crack cocaine during pregnancy. Hair that grew during the last trimester was used for this correlation. All mothers who reported cocaine use were found to have cocaine in their hair. For optimum correlation (r = 0.67) between drug levels in the hair of the mothers vs. that in the hair of their babies, it was necessary to exclude hair that was cosmetically treated from the comparison.

In a study with Callahan et al.,[44] the effectiveness of hair, urine, and meconium analyses for identifying cocaine use was compared and validated by self-reports. The correlation between drug levels in the hair of mothers and babies was also investigated, but this was done under conditions where treated hair was not excluded from the study. The study involved 59 mother-infant pairs. The women had given birth at a teaching hospital and were recruited for a prospective study comparing neurobehavioral outcome of cocaine-exposed infants with that of control infants.

The "gold standard" used for defining fetal cocaine exposure was the presence of cocaine in the proximal segment of maternal hair obtained during the third trimester. All 40 women who gave a history of consistent cocaine use during pregnancy gave a positive hair sample; 78% of their babies yielded positive hair results. For meconium, when the analysis was performed by GC/MS, the positive rate relative to hair was 74%; when this was measured by fluorescent polarization immunoassay, the positive rate relative to hair decreased to 52%. The urine positive rate was 38% relative to hair. The correlation coefficient between maternal and infant hair was 0.77.

In another study with the same group,[45] post-partum women (N = 405) with sufficiently long hair to measure cocaine use during all three trimesters were recruited for a prospective study for an evaluation of the neurodevelopmental outcome of *in utero* cocaine exposure of their infants. A potentially significant feature of the study was that the women were primarily white (86%), with African-American women constituting 14% of the test subjects. Hair analysis results were compared to a detailed, structured and confidential interview concerning the time and amount of cocaine use during pregnancy.

While 41% of women were identified as cocaine users by hair analysis, self-reports identified 36.5% as users. Cocaine levels in the hair of the mothers were correlated with self-reported amounts of cocaine and clearly distinguished between heavy, intermediate, and light cocaine users. Self-reports confirmed 87% of positive hair analysis results. Of the women who self-reported cocaine use, 18% were missed by hair analysis. These women, however, indicated that their cocaine use had been very low and mostly during the early stages of their pregnancies. It is likely, therefore, that they were missed by hair analysis because of variable hair growth rates (assumed average = 1.3 cm/month) or as a result of the porosity of old or cosmetically treated hair.

The greatest discrepancy between self-reports and hair analysis results was found with African-American women. In light of the observation of Chasnoff et al.[46] that African-American women were nearly ten times more likely than white women to be reported to public health authorities although toxicological studies revealed similar rates of drug use during pregnancies, the authors attribute the above observed discrepancy to fear concerning biased attitudes and reporting rather than to any bias in the hair testing procedure. This conclusion is supported by the lack of racial bias in hair testing (see below).

C. Therapeutic Drug Monitoring

It has been shown in several studies[7,9,18,48] that a doubling of drug dose in the same individual results in a doubling of the drug content of the hair. This phenomenon illustrates a property which is unique to hair and quite unattainable by urine or blood, and that is the capacity of hair to function as a chronometrically operating relativistic dosimeter. The relativistic dosimeter function arises from the fact that hair acts as an irreversible terminal excretory pool whose drug content is permanently recorded in tape-recorder fashion as the hair grows out from the scalp. Thus hair is admirably suited for accurately monitoring relative changes in drug intake in the same individual. Studies of this compliance monitoring function of hair analysis have to date focused on illicit drug use.

Blood cannot provide the chronometric function of hair. Consequently, any attempt at monitoring compliance requires frequent blood sampling. And even then the results cannot match the accuracy of hair analysis, since a doubling of the dose does not result in the same accurate relative changes in blood drug levels as in the case of hair. The lower accuracy results from interfering "peak" and "trough" kinetic effects of blood analysis and from the fact that even pseudo-steady-state kinetic conditions in blood cannot approach the accuracy of the dosimetric function of hair. The situation with urinalysis is even less certain.

Although most hair analysis research to date has focused on compliance with respect to illicit drug use, we have used segmental hair analysis also to measure compliance of methadone intake. Patients who received constant methadone doses under close supervision showed considerable constancy in their methadone levels in hair (Table 10). In situations where the inaccessible domain of a drug is not well defined by wash kinetics, or where there is a change in its properties over time (i.e., changes in porosity due to age, but not due to treatment), then the accuracy of drug monitoring by hair analysis can be improved by confining it to the same hair segment, e.g., to 2–3 months of the most recent hair growth.

Another desirable goal of therapeutic drug monitoring is the correlation among ingested dose, drug levels in blood or hair, and therapeutic effectiveness. Pseudo-steady-state drug levels in blood are considered to be a measure of the bioavailability of the drug and consequently they sometimes bear a relationship to either therapeutic effectiveness or toxic side effects. The therapeutic/toxicological correlations with blood levels are purely empirically determined relationships and, in light of the more accurate dosimetric function of hair, such correlations should also be investigated with hair. However, in the absence of such separately determined correlations for hair, one can investigate correlations between drug levels in hair and blood.

TABLE 10.

Constant Methadone Levels in Hair in Methadone
Maintenance Therapy

Subjects	Methadone in hair segments (ng/10 mg)		
	0–1 month[a]	1–4 months	4–7 months
1	223	218	—
2	200	192	—
3	211	211	243
4	179	184	208
5	221	208	—
6	178	167	—
7	765	722	578
8	178	199	205
9	156	187	178

[a] Root segment.

In our early studies we investigated such a correlation with digoxin.[7] However, we obtained a correlation coefficient of only 0.43 between blood and hair digoxin levels. It is not clear whether this poor correlation was caused by the analytical challenges of measuring picogram quantities of digoxin in hair with the less refined techniques that we used at that time and/or by the interfering effects of peak/trough pharmacokinetics of digoxin in blood. Furthermore, highly significant correlations were also precluded by the small therapeutic range of digoxin levels in blood (0.5–1.8 ng/mL). Highly significant correlations between blood and hair levels were obtained, however, in the case of haloperidol by Uematsu et al.[9]

Bioavailability, and hence toxicological or therapeutic effectiveness, bears a relationship to administered dose/kG of body weight if the efficiency of absorption, metabolism, and excretion of the drug do not vary too widely from individual to individual. In such situations a simple correlation between administered dose and concentrations in hair and, to a lesser extent, concentrations in blood are observed. Good correlations were observed with abused drugs (Table 6).

With therapeutic agents, we investigated the possibility of a correlation between methadone dose and hair levels. The study was performed with 76 patients who were given methadone under close supervision. Hair specimens were analyzed according to the truncated wash kinetic procedure. The average dose of methadone administered over a 12-week period was calculated and compared to the corresponding hair segment on the assumption of an average growth rate of 1.3 cm/month.

The results showed no correlation between hair levels and administered methadone dose (r = 0.14). We consider this lack of correlation with methadone dose not surprising, since it is known that the gastrointestinal absorption of methadone can vary greatly among individuals. This variability in absorption and uncertainty concerning pharmacokinetic parameters has no doubt contributed to the controversy about what constitutes hazardous overdosing or therapeutically ineffective underdosing with methadone.[49,50] We believe that hair analysis can potentially make important contributions to the resolution of this controversy.

D. Criminal Justice System Applications

1. Validation and Operational Evaluation of Hair Analysis

The first study undertaken to evaluate the validity, usefulness, and effectiveness of hair analysis in the criminal justice system was conducted in collaboration with

Baer et al.[38] during 1986 to 1988. Urine was screened by methods in effect in the federal probation system: THC metabolite and opiates were screened by EMIT® and confirmed by high performance liquid chromatography (HPLC) at cutoffs of 100 and 500 ng/mL, respectively. Cocaine metabolites and PCP were screened by thin-layer chromatography (TLC) at cutoffs of 2000 and 500 ng/mL, respectively, and confirmed by GC. Hair testing was by RIA. The RIA procedures, validated by GC/MS, showed no false positives due to cross-reactivity effects.

a. Validation Study

The study had two general objectives: the validation of hair analysis by urinalysis and the evaluation of the usefulness of hair analysis in the criminal justice system. The validation of hair analysis by urinalysis was based on the assumption that frequently applied (approximately once per week), unannounced, and observed urine collection would identify essentially all drug users in the population under study. Thus all positive hair analysis results were expected to be validated by positive urine results. Two types of populations were investigated: in one, clients were already in a urine testing program and thus had a record of many urinalysis results (Group 1, Table 11) and in the other, clients were just entering the program and their prior urinalysis history was limited to fewer results (Group 2, Table 11).

TABLE 11.

Hair and Urine Analysis Results in Probationer/Parolee Population

Group[a]	Drug	N[b]	H+U+[c]	H–U–	H–U+	H+U–
1	Cocaine	86	17	48	1	20
	Morphine	81	6	70	1	4
	PCP	83	1	80	0	2
2	Cocaine	77	6	40	0	31
	Morphine	90	6	78	1	5
	PCP	98	2	95	0	1

[a] Group 1 consisted of clients already in a drug testing program and Group 2 consisted of clients beginning hair and urine testing at the same time.
[b] Number of clients.
[c] H and U signify hair and urine specimens, respectively, and + and – indicate drug-positive or drug-negative.

As the data in Table 11 show, the expectation that all drug users would be identified by urinalysis was not realized. This result was most likely caused by the fact that urinalysis was not applied in a research setting, but under conditions where a positive urinalysis result had adverse consequences to the client, thereby encouraging the use of evasive maneuvers. Thus hair analysis, against which evasive maneuvers are not effective, identified considerably more drug users than urinalysis. For example, in the case of cocaine, with clients from Group 1, urinalysis identified 18 cocaine users, 17 of which were confirmed by hair analysis. However, hair analysis identified an additional 20 cocaine users which urinalysis missed. That these additional positives were not false positives was established from the fact that 13 out of the 20 cocaine users which urinalysis missed admitted to drug use in their confidential self-reports. That confidential self-reports are not a perfect means for confirming positive results was established from the observation that only 15 of the 18 positive urinalysis results were supported by positive self-reports.

That the efficacy of the urine test for identifying drug users is also a function of the frequency of the applied test is shown by the results of Group 2 in Table 11. Here urine identified only six cocaine users (all confirmed by hair analysis). In contrast to this, hair analysis identified an additional 31 cocaine users. The results obtained with opiates and PCP follow the same pattern and can be explained in a similar manner.

b.   Relative Detection Efficiencies of Hair and Urine Analyses

The effectiveness of hair analysis and urinalysis at identifying drug-using individuals during a particular 3-month period are compared under identical conditions by the experiments summarized in Table 12. This comparison was performed in a prospective study, i.e., where the same population (subjects under probation/parole conditions) was tested for up to 1 year by urinalysis approximately once per week on an unannounced basis and by hair analysis once every 3 months. Urine specimens were collected under observed conditions and checked for flushing (increased water intake) by clinical refractometry.

TABLE 12.

Relative Effectiveness of Hair and Urine Tests in Identifying Drug Use

	Urine			Hair			Effectiveness of hair test relative to urine test based on:	
Drug	n^a	$U+^b$	% U+	n^a	$U+^c$	% U+	Positive individuals[d]	Positive tests[e]
Cocaine	1697	11	0.7	176	41	23.3	33	24
Morphine	1783	11	0.6	187	15	8.0	13	7
PCP	1515	2	0.1	158	8	5.1	51	38
Marijuana	295	16	5.4	295	12	4.1	0.8	0.8

[a] n = number of tests performed.
[b] U+ = number of individuals with positive urine results within a given 3-month period.
[c] H+ = number of individuals with positive hair results within a given 3-month period.
[d] [% H+ (individuals)]/[% U+ (individuals)].
[e] [% H+ (tests)]/[% U+ (tests)]; % U+ tests not shown in this table.

The study showed that hair analysis identified approximately four times the number of cocaine users as urinalysis in each 3-month time frame and this with approximately one tenth the number of tests. This shows hair analysis to be 33 times more effective than urinalysis on a per-test basis. Considering the cost of urine testing, test scheduling, and observed collection, this translates into a considerable savings per identified drug user. If the relative efficiency of urinalysis is calculated on the basis of positive cocaine test results (since cocaine users in a 3-month interval were occasionally identified more than once by urinalysis), then hair analysis is 24 times more effective than urinalysis. Except for marijuana, the test results with the other drugs follow a similar pattern.

Only in the case of marijuana is unannounced urinalysis slightly more effective than hair analysis, i.e., by a factor of 1.3. This is attributed to the wider detection window of marijuana urinalysis relative to the other urine drug tests, and to the lower sensitivity of the marijuana hair test used in this study as compared to the hair assays for the other drugs. Since that time the sensitivity of the hair marijuana assay has been improved to a point where the positive rate of hair analysis is approximately twice that of urinalysis in arrestee populations.[51] However, incomplete concordance between the two tests shows that both urinalysis and hair analysis still miss some marijuana users, most likely the light users.

c.   Determining Severity of Drug Use by Hair Analysis

The study evaluated also the possibility of obtaining from hair analysis results a measure of the severity of drug use. This was done by investigating the correlation between cocaine levels in hair (the drug use severity index) and self-reports of the intensity of cocaine use (grams of cocaine used per month). The results showed a good correlation between the two parameters ($r = 0.82$). On the basis of this correlation, it was possible to differentiate with considerable accuracy between at least three categories of cocaine use: light use (5–30 ng of cocaine/10 mg hair); intermediate use (31–100 ng/10 mg hair), and heavy use (greater than 100 ng/10 mg hair).

The study also found that the efficacy of urinalysis for confirming positive hair analysis results increases with increasing drug levels in hair, i.e., with an increase in the hair severity index of drug use. Since the probability of identifying light cocaine use by urinalysis is expected to be low, it was not surprising to find that urinalysis identified only 4% of the light users. Those showing hair levels corresponding to intermediate users were identified by urinalysis with 33% effectiveness; the positive rate of urinalysis increased to 50% in the case of the heavy user. This theoretically expected correlation provides further independent evidence for the validity of the drug use severity index of hair analysis.

The fact that hair analysis can provide a measure of the severity of drug use is of considerable practical importance for criminal justice applications. For one, hair analysis can assist in making appropriate treatment and custody referrals on the basis of the severity of drug use. The results of hair analysis were also found to serve as a predictor of relapse into drug use. Thus 25% of former drug users who provided a negative hair result for the period 2 months prior to entering a drug diversion program relapsed into drug use as compared to 53% of those who provided a positive hair analysis result. The small sample size of positive results precluded the investigation in this study of a possible correlation between severity of drug use and relapse probability.

d.   Operational Advantages of Hair Analysis

The study[38] showed that hair analysis has many advantages for application in the criminal justice system. Because of its wide window of detection and evasion-proof application, hair analysis identified, with the exception of marijuana, up to four times as many drug users with one tenth the number of tests. Thus, by all indications, one expects hair testing to be more cost effective than urinalysis. In addition, hair analysis provides both a measure of the severity and pattern of drug use. Hair analysis promises, therefore, to be particularly useful as clients enter the criminal justice system, for its retrospective overview of drug use allows one to identify clients who are likely to be at risk of resuming or continuing drug use. In particular, the results of retrospective hair analysis can assist in making appropriate treatment or custody referrals. The study also showed that retrospective hair analysis is useful for identifying drug use in individuals who temporarily absconded from supervision.

2.   Validation of Hair Analysis in a Research Setting

In 1988, Magura et al.[52] initiated a second study for the validation of hair analysis in the criminal justice system. Attempts were made to improve the conditions under which this comparison was performed. The most important difference consisted in operationally removing the study from the adverse consequences of the criminal justice system, i.e., the participants clearly understood that the results of urinalysis, self-reports, and hair analysis could have no adverse consequences. The sensitivities of the urine tests were also improved: opiate and cocaine results were compared at urinalysis cutoffs of 300 ng/mL and 1000 ng/mL, respectively.

In spite of these improvements in validation procedures, essentially similar results were obtained as in the study of Baer et al.,[38] i.e., hair analysis identified more drug users than unannounced urinalysis. Thus, urine and hair agreed on 26 cocaine negatives and 83 positives, but hair detected 24 additional positives that urine did not detect and failed to detect one urine positive. Of the 24 additional positive hair analysis results, 10 were confirmed by positive self-reports. However, the fact that 15 positive urines which were confirmed by positive hair analysis results were not confirmed by positive self-reports indicates that self-reports (even under the nonpunitive conditions of the present study) cannot be taken as an absolute indicator of drug use. Hence the 14 self-report-unconfirmed hair analysis results cannot be construed as false positives. Also, such an interpretation was excluded by extensive washing of the hair sample and kinetic analysis of the wash data.

Concerning the opiate results, urine and hair agreed on 65 negative opiates and 28 positives, but hair detected 34 additional positives which urinalysis failed to detect, and urine detected 7 positives which hair failed to detect. Of the 34 additional hair positives, 22 were confirmed by self-reports of drug use; and of the 7 urine positives, 2 were confirmed by self-reports.

3. Cocaine Use Among Criminally Involved Youth

In collaboration with Magura et al.,[53] we examined the utility of hair analysis for determining the prevalence of cocaine use among criminally involved youth. Personal interviews and scalp hair specimens were obtained from 121 male youths (median age = 19) who had been in jail in New York City and were followed up in the community after their release. Of these hair specimens, 67% were positive for cocaine at a mean concentration of 68 ng/10 mg of hair. Only 23% of the youths reported any use of cocaine or crack during the previous 3 months, and 36% reported any lifetime use. Associations were found between cocaine in hair and several behavioral variables: prior number of arrests, rearrest after release from jail, not continuing education, and no legal employment.

4. Drug Treatment in the Criminal Justice System

Mieczkowski et al.[54] describe the essential elements of an initial experience of a drug treatment program in the criminal justice system which uses both hair and urine analysis for monitoring treatment outcome. The Orleans Parish District Attorney's Office was funded for this program in January 1993 by the NIJ to implement a diversionary program for first-time, nonviolent offenders with substance abuse problems. The question has arisen regarding the efficacy of this process: Does compelled treatment work? Does it work as effectively as treatment which was entered into on a voluntary basis? While opinion on this topic is not unanimous, a significant body of evaluation research has found support for the position that offender populations which are legally compelled into drug treatment do as well as volunteer populations, at least as determined by a variety of measures of successful treatment completion and rates of regression to drug use upon discharge.[55]

The objective of the diversionary program is to reduce recidivism rates as well as to reduce overburdened criminal court dockets and correctional probation caseloads. Intended as a cost-effective alternative to prosecution, it is premised on the findings of earlier research which has shown that without effective drug testing of clients, supervision is very reduced in its ability to effect positive changes in client behavior. Furthermore, it recognizes that without meaningful, reliable, and objective drug testing of clients it is highly unlikely that the courts, prosecutorial agencies, correctional agencies, and the police would be supportive of such a program. The program is entirely voluntary; the incentive for the offender is that upon successful completion of program requirements after a prescribed period of time (a minimum of 6 months for felonies, 3 months for misdemeanors), the charges are dismissed and the person does not appear in court.

Hair samples are taken at intake and at approximately 2-month intervals during the program. Additionally, random urine testing is also employed in the program. Participants call a recorded message daily for notification about whether or not they must report that day for a urine test. Because of the efficacy of hair testing, the need for frequent random urine testing is reduced, but urine testing is still an important instrument in the treatment program.

Violations of program conditions result in termination from the program. This would occur if a participant continues to have positive drug screens (despite an intensified drug treatment regimen) or other serious noncompliance to recommendations. Additionally, if the participant violates any other laws, except minor traffic violations, during the program period, this would lead to program dismissal.

The initial drug use of 91 clients who entered this program on the basis of hair analysis and urinalysis is shown in Table 13.

TABLE 13.

Hair and Urine Analysis Results on 91 Clients at Entry into a Diversionary Substance Abuse Program

Drug	Number of positive results	
	Hair analysis	Urinalysis
Cocaine	50	12
Marijuana	35	24
Opiates	3	1
PCP	1	0
Amphetamines	0	0

E. Epidemiological Studies

In collaboration with Mieczkowski and associates,[56-58] we have performed extensive epidemiological studies of drug use in arrestee populations in Pinellas County, Florida. The studies, by comparing hair, urine, and self-reports, also provided independent validation of our hair analysis technology. In addition, these studies provided further evidence for the absence of racial bias in hair testing, a correlation between frequency of cocaine use and cocaine levels in hair, and information on the relative effectiveness of hair and urine analysis for identifying drug use.

The specimens and survey data were collected from volunteers at the Pinellas County Jail. Twice a year 250–300 male and female arrestees were confidentially interviewed about their drug habits, and hair and urine specimens were collected at the booking stage of arrest. Data on 1245 arrestees have been published to date.[56] Hair and urine samples were analyzed for cocaine, cannabinoids, opiates, amphetamines, and PCP. The sample is 71.5% Caucasian, 26.9% African-American, and 1.6% Hispanic or of other ethnic origin.

In this population 41.2% were identified as cocaine users by hair analysis; urinalysis identified 18.2% and self-reports 12.5%. With respect to marijuana, hair identified 36.7% as users, urine 40%, and self-reports 46.3%. The incidence of use of the other drugs was less than 1%.

The study also provided further evidence for the validity of the three diagnostic levels of hair analysis: light cocaine use, 2–30 ng/10 mg of hair; intermediate cocaine use, 31–100 ng/10 mg of hair; and heavy cocaine use, greater than 100 ng/10 mg of hair. The light cocaine users were identified by urinalysis with 14.7% efficiency as compared to hair analysis, the intermediate-users group with 40.7% efficiency, and the heavy-users group with 67.4% efficiency. These findings confirm the results of the earlier study[38] where this trend was first demonstrated with a less sensitive urine test. NIDA cutoff levels were used in the present study.

Whether hair analysis was racially biased or not was also investigated in collaboration with Mieczkowski and Newel's group.[57] In this study, which involved 315 African-American and 846 Caucasian arrestees, the frequency of cocaine use was determined from self-reports and positive urinalysis and hair analysis data. The results showed that all three measures of cocaine use indicate that the drug use in the African-American population was almost twice that of the Caucasian population: the African/Caucasian ratio of drug use was found to be 1.71 by hair analysis, 2.17 by urinalysis, and 1.70 by self-reports. Thus, on the assumption that neither urinalysis nor self-reporting are racially biased, hair analysis, by providing the same relative positive rates for the African-American and Caucasian populations as urinalysis and self-reports, is also not racially biased.

Another interesting feature of the study was the unreliability of the absolute number of individuals who self-reported cocaine use and the underestimation of cocaine use by urinalysis: there is an approximately threefold under-reporting of cocaine use as compared to the positive rate by hair analysis, whereas the positive rate by unannounced urine testing was approximately half that of hair analysis.

An epidemiological study conducted by the Task Force on Violent Crime was undertaken as an extension of the Drug Use Forecasting (DUF) urinalysis project led by Feucht et al.[59] and funded by the NIJ. The drug testing results were linked to arrest records and provided a means of characterizing the drug use patterns of juvenile arrestees. The results of the DUF studies are used in determining the funding levels for drug enforcement and treatment. The study included interviews, urinalysis by a NIDA-certified laboratory at NIDA cutoffs, and hair analysis by our laboratory. Hair results revealed that 50 of the 88 test subjects (57%) had used cocaine; in contrast, urinalysis identified 7 (8%) as cocaine users, and self-reports, 7.4%. Urinalysis and hair analysis were in concordance for subjects who were heavy users of cocaine and who had used cocaine within the last 30 days (as determined by segmental analysis). The researchers concluded that self-reports of drug use and urinalysis data severely underestimate cocaine use in this population.

A study was performed[60] to determine the prevalence of drug use among driving-while-impaired (DWI) arrestees, as measured by self-report, urine test, and hair test, and to discover differences in behavior and attitudes between drunk drivers and drugged drivers (those who drive under the influence of drugs or drugs and alcohol). In this study, 96 first-time DWI arrestees referred to a county DWI education program participated. The results of the various measures of drug use are shown in Table 14. As expected, hair analysis identified more drug users than urinalysis.

TABLE 14.

Hair Analysis, Urinalysis, and Self-Reports
of 96 First-Time DWI Arrestees

Drug	Number of positive results		
	Hair analysis	Urinalysis	Self-reports
Marijuana	19	12	16
Cocaine	20	4	5
PCP	1	2	0
Opiates	6	1	0

The authors conclude that drug use among DWI offenders may be much more widespread than is indicated by self-report. Drunk drivers and drugged drivers also reported significant differences in attitudes about driving under the influence of alcohol and other drugs.

In a NIDA-funded workplace study,[61] a random sample of 1200 employees of a steel-manufacturing plant were randomly assigned to four different self-report methods of assessing illicit drug use: (1) individual interview in the workplace, (2) group-administered questionnaire in the workplace, (3) telephone interview, and (4) individual interview off the worksite. Urine specimens were collected and analyzed on 928 subjects participating in the study, and our laboratory performed hair analysis on 307 of the subjects. Although self-reports produced the highest drug use prevalence rate, analyses combining the results of the three assessment methods showed that the actual prevalence rate was approximately 50% higher than the estimate produced by self-reports. The authors concluded that the findings cast doubt on the validity of self-reports as a means of estimating drug use prevalence, and suggest the need for multiple assessment methods.

This experience illustrates the desirability of implementing the recent recommendations of the U.S. General Accounting Office[62] that hair analysis be tested as an objective measure of drug prevalence and validation of self-reports.

F.   Workplace Drug Testing

Our laboratory has had extensive experience with workplace drug testing. Over half a million hair samples have been analyzed to date for cocaine, marijuana, opiates, methamphetamine, and PCP. All testing was done according to the regulations promulgated by the NIDA for urine testing.

The greatest advantage of hair testing in the workplace is in its application to pre-employment testing where urinalysis is readily evaded by temporary abstention. This was demonstrated in a side-by-side study of urinalysis and hair analysis with a midwest manufacturing organization. The study comprised 774 applicants. The results, shown in Table 15, clearly illustrate the advantages of hair analysis in pre-employment testing situations. For cocaine, the hair test produced 16.8 times the number of positives compared to urinalysis, and for marijuana, 7 times the number of positives. Over 90% of individuals who tested positive by hair analysis admitted their drug use in a follow-up interview.

TABLE 15.

Comparison of the Positive Rates of Urine and Hair Analysis in a Side-by-Side Pre-Employment Study[a]

Drug	Percentage positive	
	Urine	Hair
Cocaine	0.5%	8.4%
Marijuana	0.5%	3.5%
Other drugs	1.7%	6.1%

[a]   Number of participants: 774.

Positive cocaine results obtained from workplace testing were grouped into light, intermediate, and heavy user classes. The percentage distribution among these use categories from pre-employment testing was compared to those obtained in the criminal justice system and drug treatment programs. The results show that the majority of applicants, 66.8%, fall into the light user class with the balance being approximately equally distributed between the intermediate and heavy user categories, 17.2% and 16.0%, respectively. Exactly the opposite is true for individuals in a drug treatment program; there the majority of positives, 51.5%, fall into the heavy user class, with 20% in the intermediate category and 28.5% in the light category. Arrestees exhibit an intermediate distribution, with 47.8% in the light, 18.9% in the intermediate, and 33.3 in the heavy category.

G. Forensic Applications

We have used hair analysis in over 300 forensic cases, approximately half of this number in collaboration with Siegel,[37] who utilized hair analysis as an adjunct to his forensic psychopharmacological evaluations.

The majority of cases in which our laboratory was directly involved were child custody cases where parental drug use was suspected. Other cases dealt with probation, criminal, and employment violations, including challenges to positive urinalysis results in various employment settings. The latter included several military cases, some prior to initiation of court-martial proceedings. The previously outlined probative advantages of hair analysis played a critical role in the presentation and defense of the evidence. In cases where a positive urinalysis result was challenged by a negative hair analysis result, the previously outlined benefit-of-the-doubt argument was successfully applied in most instances. This included one U.S. Navy court-martial.

The Society of Forensic Toxicology (SOFT) indicated in their consensus evaluation of the scientific status of hair analysis that they considered it a useful forensic tool for the investigation of drug use.[16] At least 12 forensic laboratories in Europe and in the U.S., including those of the FBI, have used hair analysis in hundreds of forensic investigations. There have been 12 challenges to hair analysis to date in the U.S. courts. Hair analysis was successfully defended in all instances. An extensive review of these cases has been published.[63]

H. Environmental Toxin Exposure

One essentially unexplored area for hair analysis is its application to the investigation of environmental toxin exposure. We received a research grant during 1979 from the National Institute of Occupational Safety and Health to explore the possibility of using hair analysis for monitoring exposure to polychlorinated biphenyl compounds (PCBs). Although the analytical chemical problems of this project were successfully solved, we were unable to mount successful field studies with human subjects.

Our negative experience in this field suggests that it may be more effective to measure environmental toxins which, in contrast to PCBs, produce measurable metabolites in hair. Since exposure to environmental toxins involves the passive endogenous and exogenous deposition of chemicals in hair, and since the latter is the predominant process, the ratio of endogenous signal/exogenous noise will be maximized more effectively if metabolites rather than the parent toxins are measured.

This has been our experience with the challenging task of differentiating between the hair of marijuana users and contaminated hair from nonusers who were only exposed to marijuana smoke. Here only the measurement of the metabolite, carboxy-THC, but none of the wash procedures or wash kinetic approaches that we have successfully applied to the other drugs have led to satisfactory results. Our ability to measure digoxin in small hair samples, even though this was taken in doses as small as 0.1 mg/d, suggests that analytical sensitivity should not pose problems for the measurement of endogenously incorporated metabolites of environmental toxins. On the basis of these considerations, the measurement of pesticide metabolites may be an appropriate starting point for the exploration of occupational toxin exposure.

In light of the incidence of health problems and mortality associated with cigarette smoking, it is of considerable interest to health professionals and the life insurance industry to obtain quantitative measures of the extent of an individual's smoking habit. Because of this interest, we have made considerable efforts to develop immunoassay procedures for measuring the severity of cigarette smoking. To date these efforts have failed to provide a satisfactory margin between smokers and nonsmokers, and this apparently because of the ubiquitous presence of nicotine. It has been suggested that our problem may have been caused by the presence of nicotine in certain vegetables rather than by passive exposure to smoke. Alternatively, in view of the publications by other investigators,[64] the problem may have been due to antibody cross-reactivity effects. We expect to resolve this issue by the MS determination of nicotine and cotinine levels in hair.

I. Historical and Anthropological Hair Analysis

Our laboratory has been involved in the analysis of several historical hair samples. The first sample analyzed[7] was that of the English poet John Keats, who in his capacity as pharmacological chemist and physician had prescribed for himself the morphine-containing pain killer laudanum while dying of tuberculosis. The hair sample, stemming from this period of the poet's life, was found to contain morphine 167 years after his death. This result is a striking demonstration of the unique properties of hair as a storage medium for incorporated drugs.

Subsequently, our laboratory was requested, as part of an ongoing anthropological investigation,[65] to analyze hair specimens taken from 500-year-old Peruvian mummies. The analysis established that the hair came from an individual who either used cocaine or was in a cocaine-containing environment. However, virtually all of the cocaine had broken down to benzoylecgonine.

One noteworthy and common feature of these historic hair samples was the abnormally high porosity of the specimens. That virtually all of the inaccessible domain had been degraded was demonstrated by methylene blue staining procedures and by the complete extraction of the drugs by the aqueous wash solutions.

This situation poses potential problems for historical hair studies, because of the difficulty of distinguishing by conventional wash procedures between drugs which were deposited in the distant past by endogenous processes and drugs which may have been introduced into or onto the hair sample by more recent environmental contamination. In light of several recent controversial reports concerning the presence of cocaine and nicotine in the hairs of Egyptian mummies,[66] it appears desirable to develop special wash procedures for historical hair samples.

One possible approach that we would like to suggest is to perform contamination and decontamination studies with negative historical hair samples where the drug in question is deposited on hair as powder or vapor. Attempts should then be made to remove such contamination by the application of nonhair-swelling solvents, starting initially with nonpolar solvents and concluding with dry or slightly moist isopropanol or ethanol. The final extraction should be performed with aqueous solvents at the appropriate pH and with enzymatic digests. Procedures such as these may prove useful in resolving this latest controversy in the field of hair analysis.

ACKNOWLEDGMENT

The authors gratefully acknowledge the technical support of Akram Gorgi. This publication was supported in part by the Veterans Administration Medical research funds.

REFERENCES

1. Saitoh, M., Uzuka, M., Sakamoto, M., and Kobori, T., In *Advances in Biology of Skin and Hair Growth*, Montagna, W. and Dobson, R. L., Eds., Pergamon Press, Oxford, 1969.
2. Goldberger, B. A., Caplan, Y. H., Maguire, T., and Cone, E. J., Testing human hair for drugs of abuse. III. Identification of heroin and 6-acetylmorphine as indicators of heroin use, *J. Anal. Toxicol.*, 15, 226, 1991.
3. Moeller, M. R., Fey, P., and Sachs, H., Hair analysis as evidence in forensic cases, *Forensic Sci. Int.*, 63, 43, 1993.
4. Baumgartner, W. A. and Hill, V. A., Hair analysis for drugs of abuse: clinical, epidemiological and criminal justice field studies, in *Proceedings of the 2nd International Meeting on Clinical and Forensic Aspects of Hair Analysis*, Genoa, Italy, 1994.
5. Buskirk, H., SAMHSA studies changing how opiate positives are evaluated, *Drug Detection Report*, July 20, 1993, p. 1.
6. Willims, R. J., *Biochemical Individuality*, University of Texas Press, Houston, 1974.
7. Baumgartner, W. A., Hill, V. A., and Blahd, W. H., Hair analysis for drugs of abuse, *J. Forensic Sci.*, 34, 1433, 1989.
8. Sramek, J., Baumgartner, W. A., Ahrens, T. N., Hill, V. A., and Cutler, N. R., Detection of benzodiazepines in human hair by radioimmunoassay, *Ann. Pharmacother.*, 26, 469, 1992.
9. Uematsu, T., Sato, R., Suzuki, K., Yamaguchi, S., and Nakashima, M., Human scalp hair as evidence of individual dosage history of haloperidol: method and retrospective study, *Eur. J. Clin. Pharmacol.*, 37, 239, 1989.
10. Baumgartner, W. A., Cheng, C. C., Donahue, T. D., Hayes, G. F., Hill, V. A., and Scholtz, H., Forensic drug testing by mass spectrometric analysis of hair, *Forensic Applications of Mass Spectrometry*, CRC Press, Boca Raton, FL, 1995, 61.
11. Baumgartner, W. A. and Hill, V. A., Sample preparation techniques, *Forensic Sci. Int.*, 63, 121, 1993.
12. Baumgartner, W. A., Ligand assays of enzymatic hair digests. U.S. Patent No. 5,324,642, June 28, 1994.
13. Robbins, C. R., *Chemical and Physical Behavior of Human Hair*, 2nd ed., Springer-Verlag, New York, 1988.
14. Ellis, G. M., Jr., Mann, M. A., Judson, B. A., Schramm, N. T., and Tashchian, A., Excretion patterns of cannabinoid metabolites after last use in a group of chronic users, *Clin. Pharmacol. Ther.*, 38, 572, 1985.
15. Dackis, C. A., Pottash, A. L. C., Annitto, W., and Gold, M. S., Persistence of urinary marijuana levels after supervised abstinence, *Am. J. Psychiatry*, 139, 1196, 1982.
16. Hearn, W. L., Chairman, Advisory Committee on Hair Analysis for Drugs of Abuse. Report to the Annual Meeting of the Society of Forensic Toxicologists, October 13–17, 1992, Cromwell, CT.
17. Harkey, M. R. and Henderson, G. L., Hair analysis for drugs of abuse, in *Advances in Analytical Toxicology*, Vol. 2, Baselt, R. C., Ed., Yearbook Medical Publishers, Chicago, 1989, 298.
18. Henderson, G., The kinetics of the uptake of drugs into hair, in *Proceedings of the 2nd International Meeting on Clinical and Forensic Aspects of Hair Analysis*, Genoa, Italy, 1994. To be published by the National Institute on Drug Abuse.
19. Sachs, H., Schmidl, D., Hages, G., and Schwilk, B., Morphine concentration in hair after consumption of poppy seed and oral administration of MST, *Int. J. Leg. Med.*, in press.
20. Baselt, R. S. and Change, R., Urinary excretion of cocaine and benzoylecgonine following oral ingestion in a single subject, *J. Anal. Toxicol.*, 11, 81, 1987.
21. Kidwell, D. A. and Blank, D. L., Hair analysis: techniques and potential problems, in *Recent Developments in Therapeutic Drug Monitoring and Clinical Toxicology*, Sunshine, I., Ed., Marcel Dekker, New York, 1994.
22. Cone, E. J., Pharmacology and pharmacokinetics of cocaine in hair, presented at the Society of Forensic Toxicology Conference on Drug Testing in Hair, Tampa, FL, Oct. 29–30, 1994. To be published by National Institute on Drug Abuse.

23. Blank, D. L. and Kidwell, D. A., External contamination of hair by cocaine: an issue in forensic interpretation, *Forensic Sci. Int.*, 63, 145, 1993.
24. Kidwell, D. A. and Blank, D. L., Comments on the paper by W. A. Baumgartner and V. A. Hill: sample preparation techniques, *Forensic Sci. Int.*, 63, 137, 1993.
25. Blank, D. L. and Kidwell, D. A., Decontamination procedures for drugs of abuse in hair. Are they sufficient?, *Forensic Sci. Int.*, 70, 13, 1995.
26. Henderson, G. L., Mechanisms of drug incorporation into hair, *Forensic Sci. Int.*, 63, 19, 1993.
27. Cone, E. J., Yousefnejad, D., Darwin, W. D., and Maguire, T., Testing human hair for drugs of abuse. II. Identification of unique cocaine metabolites in hair of drug abusers and evaluation of decontamination procedures, *J. Anal. Toxicol.*, 15, 1991.
28. Popper, K., *Logic of Scientific Discovery*, Hutchinson, London, 1974.
29. Baumgartner, W. A. and Hill, V. A., Comments on the paper by D. L. Blank and D. A. Kidwell: External contamination of hair by cocaine: an issue in forensic contamination, *Forensic Sci. Int.*, 63, 157, 1993.
30. DuPont, R. L. and Baumgartner, W. A., Hair analysis: complementary features and scientific issues, *Forensic Sci. Int.*, 70, 63, 1995.
31. Smith, F. P. and Kidwell, D. A., Cocaine in children's hair when they live with drug-dependent adults. Pharmacology and Pharmacokinetics of Cocaine in Hair, presented at the Society of Forensic Toxicology Conference on Drug Testing in Hair, Tampa, FL, Oct. 29–30, 1994. To be published by National Institute on Drug Abuse.
32. Smith, F. P. and Kidwell, D. A., Children of cocaine users show passive drug incorporation in their hair, in *Proceedings of the 2nd International Meeting on Clinical and Forensic Aspects of Hair Analysis*, Genoa, Italy, 1994. To be published by the National Institute on Drug Abuse.
33. Brewer, C., Hair analysis as a tool for monitoring and managing patients on methadone maintenance: a discussion, *Forensic Sci. Int.*, 63, 277, 1993.
34. Baumgartner, A. M. and Jones, P. F., (Aerospace Corporation) and Baumgartner, W. A., and Black, C. T. (Wadsworth V. A. Medical Center), Radioimmunoassay of hair for determining opiate abuse histories, *J. Nucl. Med.*, 20, 749, 1979.
35. Baumgartner, A. M., Jones, P. F., and Black, C. T., Detection of phencyclidine in hair, *J. Forensic Sci.*, 26, 576, 1981.
36. Baumgartner, W. A., Jones, P. F., Black, C. T., and Blahd, W. H., Radioimmunoassay of cocaine in hair, *J. Nucl. Med.*, 23, 790, 1982.
37. Siegel, R. K., Repeating cycles of cocaine use and abuse, in *Treating Drug Problems*, Vol. 2, Gerstein, D. R. and Harwood, H. J., Eds., National Academy Press, Washington, D.C., 1992, 289.
38. Baer, J. D., Baumgartner, W. A., Hill, V. A., and Blahd, W. H., Hair analysis for the detection of drug use in pretrial, probation and parole populations, *Fed. Probation*, 55, 1, 1991.
39. Sramek, J. J., Baumgartner, W. A., Tallos, J., Ahrens, T. N., Meiser, J. F., and Blahd, W. H., Hair analysis for detection of phencyclidine in newly admitted psychiatric patients, *Am. J. Psychiatry*, 142, 950, 1985.
40. Parton, L., Baumgartner, W. A., and Hill, V., Quantitation of fetal cocaine exposure by radioimmunoassay of hair, *Pediatr. Res.*, 21, 1987, A372.
41. Ostrea, E. M., Comparison of hair analysis and meconium in detecting drug use during pregnancy, *Substance Abuse*, 11, 214, 1990.
42. Welch, R. A., Martier, S. S., Ager, J. W., Ostrea, E. M., and Sokol, R. J., Radioimmunoassay of hair; a valid technique of determining maternal cocaine abuse, *Substance Abuse*, 11, 214, 1990.
43. Marques, P. R., Tippetts, A. S., and Branch, D. G., Cocaine in the hair of mother-infant pairs: quantitative analysis and correlations with urine measures and self report, *Am. J. Drug Alcohol Abuse*, 19, 159, 1993.
44. Callahan, C. M., Grant, T. M., Phipps, P., Clark, G., Novacek, A. H., Streissguth, A. P., and Raisys, V. A., Measurement of gestational cocaine exposure: sensitivity of infants' hair, meconium, and urine, *J. Pediatr.*, 120, 763, 1992.
45. Grant, T., Brown, Z., Callahan, C., Barr, H., and Streissguth, A. P., Cocaine exposure during pregnancy: improving assessment with radioimmunoassay of maternal hair, *Obstet. Gynecol.*, 83, 524, 1994.
46. Chasnoff, I. J., Landress, H. J., and Barrett, M. E., The prevalence of illicit drug or alcohol use during pregnancy and discrepancies in mandatory reporting in Pinellas County, Florida, *N. Engl. J. Med.*, 322, 1202, 1990.
47. Baumgartner, W. A. and Hill, V. A., Hair analysis for drugs of abuse, in *Recent Developments in Therapeutic Drug Monitoring in Therapeutic Drug Monitoring and Clinical Toxicology*, Sunshine, I., Ed., Marcel Dekker, New York, 1992, 577.

48. Cone, E. J., Testing human hair for drugs of abuse. I. Individual dose and time profiles of morphine and codeine in plasma, saliva, urine, and beard compared to drug-induced effects on pupils and behavior, *J. Anal. Toxicol.*, 14, 1, 1990.

49. Greenhouse, C. M., Study finds methadone treatment practices vary widely in effectiveness, *NIDA Notes*, July/Aug. 1992, 1.

50. Marks, J., Deaths from methadone and heroin, *Lancet*, 343, 47, 1994.

51. Mieczkowski, T., A research note: the outcome of GC/MS/MS confirmation of hair analysis on 93 cannabinoids positive cases, *Forensic Sci. Int.*, 70, 83, 1995.

52. Magura, S., Freeman, R. C., Siddiqi, Q., and Lipton, D. S., The validity of hair analysis for detecting cocaine and heroin use among addicts, *Int. J. Addict.*, 27, 51, 1992.

53. Magura, S., Kang, S. Y., and Shapiro, J. L., Measuring cocaine use by hair analysis among criminally-involved youth, *J. Drug Issues*, 25, 683, 1995.

54. Mieczkowski, T., Mumm, R., and Connick, H. F., The use of hair analysis in a pretrial diverson program in New Orleans, *Int. J. Offender Ther.*, in press.

55. Anglin, D., Brecht, M., and Maddshian, M., Pretreatment characteristics and treatment performance of legally coerced versus voluntatry methadone maintenance admissions, *Criminology*, 27, 537, 1989.

56. Mieczkowski, T. and Newel, R., Comparing hair and urine assays for cocaine and marijuana, *Fed. Probation*, 57, 59, 1993.

57. Mieczkowski, T. and Newel, R., An evaluation of racial bias in hair assays for cocaine: black and white arrestees compared, *Forensic Sci. Int.*, 63, 85, 1993.

58. Mieczkowski, T., Barzelay, D., Gropper, B., and Wish, E., Concordance of three measures of cocaine use in an arrestee population: hair, urine and self report, *J. Psychoactive Drugs*, 23, 241, 1991.

59. Feucht, T., Stephens, R. C., and Walker, M. L., Drug use among juvenile arrestees: a comparison of self-report, urinalysis and hair assay, *J. Drug Issues*, 24, 099, 1994.

60. Saylor, K. E., Wish, E., DuPont, R. L., Shiraki, S., Levitsky, R., Gray, T., Milgraum, M., Ramos, M., and Rivlin, R., Drug use prevalence, beliefs and attitudes in a DWI sample, *JAMA Brief Rep.*, December 7, 1994.

61. Cook, R. F., Bernstein, A., Arrington, T. L., Andrews, C. M., and Marshall, G. A., Methods for assessing drug use prevalence in the workplace: a comparison of self-report, urinalysis, and hair analysis, *Int. J. Addict.*, in press.

62. General Accounting Office. Drug use measurement: strengths, limitations, and recommendations for improvement. U.S. General Acounting Office Report (GAO/PEMD-93-18), 58–60 and 68–69, 1993.

63. McBay, A., The legal aspects of hair drug testing, in *Proceedings, Second Intern. Meeting on Clinical and Forensic Aspects of Hair Analysis*, Genoa, 1994. To be published by National Institute on Drug Abuse (in press).

64. Kintz, P., Evaluation of nicotine and cotinine in human hair, *J. Forensic Sci.*, 37, 72, 1992.

65. Cartmell, L. W., Aufderhide, A., and Weems, C., Cocaine metabolism in pre-Columbian mummy hair, *J. Okla. State Med. Assoc.*, 84, 11, 1991.

66. Balabanova, S., Parsche, F., and Pirsig, W., First identification of drugs in Egyptian mummies, *Naturwissenschaften*, 79, 358, 1992.

Chapter **11**

Clinical Applications of Hair Analysis

Pascal Kintz

CONTENTS

0-8493-8112-6/96/$0.00+$.50
© 1996 by CRC Press, Inc.

I. INTRODUCTION

Morphological, serological, and chemical examination of human hair for medical purposes was initiated some years ago. In the 1960s and 1970s, hair analysis was used to evaluate exposure to toxic heavy metals. At this time, examination of hair for organic substances, especially drugs, was not possible because analytical methods were not sensitive enough. Since the early 1980s, the development of highly sensitive and specific assay methods such as radioimmunoassay (RIA) or gas chromatography/mass spectrometry (GC/MS) has permitted the analysis of organic substances trapped in hair. This, theoretically, offered the possibility of revealing an individual's recent history of drug exposure beginning at sampling day and dating back over a period of weeks or months.

Until now, most studies have focused on forensic considerations. However, another exciting application of hair anlysis may be clinical investigations. The determination of plasma or urine concentrations of the monitored drugs is commonly used for that purpose; however, such analyses may reflect only the exposure within a few days prior to obtaining the specimen. Hair analysis may represent an appropriate alternative, by providing information on the degree of exposure over a long time scale — weeks or months.

The present review aims to summarize and discuss the various clinical applications that have been published.

II. HAIR AS A SCREENING PROCEDURE OF PSYCHIATRIC PATIENTS

By providing information on exposure to drugs over time, hair analysis may be useful in verifying self-reported histories of drug use in any situation in which a history of past rather than recent drug use is desired. Hair analysis may be especially useful when a history of drug use is difficult or impossible to obtain, such as from psychiatric patients.

Sramek et al.[1] implemented a procedure for analyzing phencyclidine content in hair of newly admitted psychiatric patients. Hair analysis identified 11 patients who had used phencyclidine, whereas blood and urine analyses did not identify any, among the sample population (47 patients hospitalized with acute psychiatric illness). Depending on the length of hair, analysis of hair was used to confirm previous phencyclidine exposure and clarify the chronicity of use. This latter evaluation was achieved by sectional hair analysis. The authors concluded that hair analysis may be particularly useful in understanding atypical patients who do not easily fulfill the criteria of any standard mental disorder.

In another study conducted by Tracqui et al.,[2] hair analysis was presented as a drug-exposure screening test for the antidepressant amitriptyline. Sixty subjects were included in the study, and patients were admitted under the diagnosis of acute or chronic depression. Subjects were placed into two groups, one treated by amitriptyline during part of or all of the past 2 months, and the second was a control group, not having received amitriptyline during the past months.

After determination of the amitriptyline concentrations by GC/MS, the sensitivity of the exposure screening test was 93.3% and the specificity 100%. Numerous drugs may be screened in the same manner. Such screening tests offer a large range of applications of considerable interest, including the study of past history of patients admitted to the intensive care unit (ICU) or psychiatry units who are unable (or unwilling) to answer questions.

III. HAIR AND EPILEPTIC MANAGEMENT

Generally, therapeutic drug monitoring during antiepileptic therapy is based on clinic observation, but also on blood measurement. However, this concentration may reflect only dosage over several hours prior to the sampling of the specimen and does not necessarily indicate treatment compliance. To assess compliance over longer periods it would be an advantage to sample a readily accessible tissue which provides a more permanent marker of drug intake.

The first report on the detection of phenobarbital in hair from an individual on long-term therapy was published by Smith and Pomposini.[3] The presence of phenobarbital in hair (8.84 ng) was proposed as an additional discriminating factor in forensic sciences. More recently, Goulle et al.[4] investigated phenobarbital in 40 vertex hair samples obtained from subjects under antiepileptic treatment. The range of concentrations was 1.5 to 194.0 ng/mg, with a mean value of 36.4 ng/mg. It was possible to establish a significant correlation between phenobarbital in hair and posology ($r = 0.692$, $p < 0.001$). However, only a group correlation was observed, but not an individual correlation due to individual intervariation. Three applications were described where hair analysis was particularly helpful in the management of epileptics: error of prescription (10 mg instead of 100 mg), irregular treatment, and bad observance of the treatment. The authors concluded that determination of phenobarbital in human hair appears to complement blood analysis during trade drug monitoring.

Carbamazepine and its metabolites were investigated in hair samples of 30 epileptics by Rothe et al.[5] using GC/MS. The concentrations ranged from 7.6 to 205 ng/mg for carbamazepine and from 1.1 to 23 ng/mg for carbamazepine-10,11-epoxide. No significant correlation was found between these hair concentrations and the daily dosage of carbamazepine, leading the authors to consider the investigation of hair samples not suitable for compliance analysis.

Kintz et al.[6] evaluated the potential usefulness of hair analysis in carbamazepine monitoring, i.e., index of dosing history and compliance in 30 epileptic patients under the same medical treatment for at least 6 months. Carbamazepine concentrations in the first proximal 3-cm hair portion ranged from 1.2 to 57.4 ng/mg. The concentration in hair was significantly ($p < 0.0001$) correlated with the daily dose of the drug ($r = 0.793$).

However, the variance of the obtained data is a parameter at least as important as the correlation that they exhibit together, and what is true for a population does not necessarily suit one individual. In fact, while reporting significant correlation, all available statistics show enormous interindividual variations in the hair concentrations of subjects presenting similar amount of drug ingestion. For example, 10 patients had a daily dose of 1200 mg carbamazepine for at least 6 months. Hair concentrations of these subjects ranged from 11.2 to 41.8 ng/mg.

By cutting strands of hair into sections (for example, 1-month intervals), it is possible to obtain information on the pattern of an individual's drug use, that is, whether use is constant (good observance of the treatment) or with variation. For example, a measure of compliance may be obtained by comparing the drug levels in hair corresponding to the time a patient received medication under the controlled conditions of a hospital setting to those noted after discharge. In these circumstances, the patient has become his own control.

The results of sectional hair analysis in two patients are shown in Table 1. Hair strands were cut into small sections, each corresponding to approximately 1 month. In both cases, the prescribed daily dose was 1200 mg carbamazepine for 12 months. Case A probably represents good compliance with carbamazepine therapy, while Case B the converse. In that case, hair analysis revealed that the patient has omitted numerous doses. This was first denied, and then admitted by the subject. Thus, relative changes in an individual's carbamazepine use can be established with a high degree of certainty.

TABLE 1.

Pattern of Carbamazepine Use. All the Concentrations are in ng/mg.

Section	Case A	Case B
1 (root)	24.3	31.2
2	26.0	21.7
3	21.4	42.1
4	20.8	32.8
5	20.4	33.7
6	18.7 (end)	20.7
7	—	39.7
8	—	28.7 (end)

IV. HAIR AS EVIDENCE OF GESTATIONAL DRUG EXPOSURE

Failure to identify maternal drug exposure is extensive owing to the limitations of the four methods currently used to verify drug use. Maternal self-reported drug history, the first method, has been shown to be unreliable: many women who deny use during pregnancy exhibit drug metabolites in their urine.[7] Systematic urinalysis, the second method, is hampered by the short elimination half-life of the drugs. This test is not suitable for validation of survey data, since the quantification of drugs in urine only reflects exposure during the preceding 1–3 days and does not indicate the frequency in subjects who might deliberately abstain for several days before biomedical screening. Evaluation of drug concentrations in the amniotic fluid measured during pregnancy or at delivery, the third method, cannot provide information on the duration and degree of fetal exposure.[8] The same disadvantages are noted with the analysis of meconium, the last method, which is only a qualitative test at the moment of delivery.[9]

Because of immediate and long-term problems, newborns born to women exposed to drugs during pregnancy should be identified soon after birth so that appropriate intervention and followup can be done. Methods to detect substance abuse in a pregnant woman should ideally address not only the types of drug abused, but also the amount, frequency, and duration of drug exposure.

Hair analysis may remedy the disadvantages of currently used methods with a wide window of detection ranging from weeks to months and may provide information concerning the severity and pattern of an individual's drug use, when a maternal drug history is not available or is in doubt.

Since 1987, and the first report of fetal cocaine exposure by Parton et al.,[10] 10 papers have presented data on gestational exposure revealed by hair analysis. The most important results[10-20] are presented in Table 2. All the reports have clearly demonstrated that the hair analysis technique is better able to detect previous drug use than is the standard urinalysis. However, hair analysis cannot completely replace urine testing, since drug use occurring only in the recent few days would not be detectable by hair analysis, yet would be detectable through urinalysis.

In three cases, it was possible to establish a significant correlation between the drug concentrations in the hair of the neonates and of their corresponding mothers: $r = 0.72$ for cocaine,[15] $r = 0.83$ ($p < 0.001$) for nicotine,[16] and $r = 0.78$ ($p < 0.01$) for

TABLE 2.

Results of the Analysis of Hair Obtained from Neonates

Compound	Nb of cases	Concentrations (ng/mg)	Ref.
Cocaine	15	?	10
Cocaine	7	0.2–27.5	11
Haloperidol	3	?	13
Cocaine	?	?	14
Cocaine	25	?	15
Nicotine	40	0.15–11.80	16
Nicotine	10	6.0 ± 9.2 (mean)	17
Diazepam	8	3.36–17.55	18
Oxazepam	3	0.78–31.83	18
Cocaine	2	0.71–2.47	18
Amphetamine	1	1.21	18
Nicotine	40	0.15–11.80	19
Morphine	9	0.61–3.47	19

nicotine.[17] These correlations clearly indicate a dose-dependent transfer of maternal drug to her baby.

V. HAIR NICOTINE AS A MARKER OF PASSIVE EXPOSURE TO TOBACCO

Noninvasive validation of tobacco smoking behavior is necessary for large population health studies. Moreover, a main problem in the risk assessment of passive smoking is the lack of a suitable methodology for the quantification of exposure. Measurements of nicotine in hair could prove to be a reliable marker for passive exposure. Several reports have presented data on nicotine in hair.[16,17,20-25] Some have included the monitoring of cotinine, the major metabolite of nicotine.

The following nicotine concentrations have been published:

TABLE 3.

Concentrations (ng/mg) of Nicotine Reported

Smokers	Nb of cases	Nonsmokers	Nb of cases	Ref.
0.37–63.50	40	—	—	16
21 ± 18	10	0.9 ± 0.8	11	17
3.0–38.7	10	0.3–11.3	10	21
39 ± 8	24	1.4 ± 0.2	24	22
0.91–33.89	42	0.06–1.82	22	23

When evaluated, the nicotine concentration in hair was approximately proportional to the number of cigarettes consumed daily,[16,21-25] excepted in one report.[17]

It was demonstrated that cotinine is not concentrated within hair shafts like nicotine, and the concentration of nicotine was about 30 times more than that of cotinine.[21,23] Haley and Hoffmann[21] were not able to establish a nicotine cutoff value for distinguishing smokers from nonsmokers. Zahlsen and Nilsen[22] demonstrated that no overlap existed between smokers and nonsmokers in a population of 48 subjects. Although it was difficult to determine an absolute cutoff level, Kintz et al.[23] proposed that an amount greater than 2 ng/mg could be used to differentiate smokers from nonsmokers. Moreover, in the nonsmoker population, these authors were able to distinguish passive smokers from other nonsmokers. The nicotine content was over 0.5 ng/mg in the group of passive smokers and lower in the nonexposed nonsmokers. This was recently confirmed by Goulle et al.,[26] with a reported cutoff of 0.5 ng/mg to evidence passive exposure.

All the authors concluded that measurement of nicotine in hair can prove to be a reliable marker for passive exposure. Nicotine is specific to environmental tobacco smoke, and hair nicotine gives retrospectively a time-weighted average measurement of exposure for months.

VI. CLINICAL SURVEY OF HEROIN ADDICT

Regular urine testing is a feature of most detoxification programs. Attempted evasion is a persistent problem. Methadone treatment in France or in Germany is linked to very strict conditions. One of them is the surveillance of the subjects to exclude/detect the additional consumption of other drugs. This is done by urine tests in irregular time intervals. However, if a time difference of 2–3 days between two tests is exceeded, the concentration of the drugs to be checked is normally under the cutoff of the screening procedures. This is well known to the addicts. So, even a permanent, but well-planned consumption of additional drugs cannot be excluded by urine analysis.

Hair testing avoids most of the problems seen with urinalysis. As important as the ability to quantify drug use and to measure changes in intake from one to the next is the virtual impossibility of evasion. The collection of hair samples involves minimal time and minimizes the risk of infection.

Major benefits of such investigations are observed in the relation between drug addict and medical personnel. Patients often spontaneously talk about their illicit drug use as they imagine no limit on the investigation by hair. They usually assume that any drug or change can be identified. If a positive result is disputed, it is usually possible to collect a second sample from the same growing period and repeat the test.[27]

Hair analysis provides a unique means of obtaining an overall assessment of the subject's drug-taking behavior. In samples obtained from subjects of a methadone treatment program, Moeller et al.[28] reported 95% positive segments for methadone. In the same time 69 and 43% were positive for opiates and cocaine, respectively. In another study, Goldberger et al.[29] reported, in a population of 20 patients, 18 positives for methadone and 14 and 4 positives for cocaine and opiates, respectively. Both authors concluded that testing hair for methadone and illicit drugs of abuse may be useful to drug treatment specialists as a means of verifying drug use history, monitoring compliance, and providing a broad measure of drug exposure. The same observations were made by Kintz et al.,[30] who investigated buprenorphine in hair of 14 subjects admitted in a detoxification center. The authors mentioned a good observance of the buprenorphine treatment, revealed by hair analysis.

Hair testing can also provide information on the behavior of drug addicts. After examination of individual hair samples, it is sometimes obvious that some heroin consumers have switched to another drug, easier to find and of lower cost. Sachs et al.[31] reported the use of dihydrocodeine, while Kintz et al.[32] mentioned ethylmorphine.

VII. EVALUATION OF PHARMACEUTICAL EXPOSURE

Testing hair specimens for drugs offers a much broader window for detection than most biological specimens. In fact, hair analysis may provide a good complement to plasma and urine. Hair is easily collected and can be stored without damage.

Overall, hair analysis provides convincing evidence of past exposure to a drug. Viala et al.[33] identified by GC/MS chloroquine and its major metabolite in hair samples of patients who received the antimalaria drug for several months.

In a study conducted on 13 patients with chronic psychiatric conditions who had received therapeutic dosage of benzodiazepines for several months, Sramek et al.[34] were able to detect diazepam (0.2–1.6 ng/mg), but alprazolam and lorazepam were not detected. Inspection of diazepam results were not suggestive of a correlation between dosage and hair concentration.

The utility of hair in the detection of chronic beta-blocker administration was examined by Kintz and Mangin[35] in 8 hypertensive patients. Betaxolol (3 cases, 1.2–2.7 ng/mg), sotalol (2 cases, 4.4–5.3 ng/mg), atenolol (1 case, 0.9 ng/mg), and propanolol (2 cases, 1.6–2.4 ng/mg) were identified. Relative changes in observance of treatment was revealed by sectional hair analysis in a case of betaxolol treatment.

More recently, head hair samples obtained from surgery patients who received fentanyl were analyzed by RIA. Wang et al.[36] detected fentanyl equivalents in eight subjects. Concentrations ranged from 0.013 to 0.048 ng/mg. No correlation between hair fentanyl concentration and administered dose was found for the 13 fentanyl subjects.

Finally, meprobamate was detected by Kintz and Mangin[37] in 2 patients receiving a daily dose of 200 mg/d for 6 months. Concentrations were 3.32 and 4.21 ng/mg of hair.

VIII. HAIR ANALYSIS AS A TOOL OF CLINICAL DIAGNOSIS

Kintz[38] reported a case of amineptine abuse in a 45-year-old man admitted to a dermatology unit with extremely severe acne lesions. As it was not possible to evaluate the degree of severity of the intoxication by blood analysis (decided 10 d after admission), a strand of hair (6 cm) was collected in the vertex region. The concentrations of amineptine in the hair segments were 12.23 and 8.06 ng/mg. After the presentation of the results to the subject, he admitted a daily consumption of 60 tablets of 100 mg amineptine. This observation clearly demonstrates the usefulness of hair testing when classic biological samples (plasma, urine) are not available or without interest.

In another study, Kintz et al.[39] identified dextromoramide in the hair of a subject who denied drug abuse. In three segments, dextromoramide concentrations ranged from 1.09 to 1.93 ng/mg. During the first medical questionnaire, the subject denied the use of an illicit compound. Urinalysis did not reveal any drug of abuse by EMIT® investigations. After the patient was informed of the presence of dextromoramide in his hair, he recognized its use. It appears that the evasive measures are ineffective against hair analysis. Urine-based techniques are essentially limited to testing for the presence or the absence of drugs over a short retrospective period. This problem can be solved by using hair instead of urine.

IX. HAIR AS A TOOL FOR MONITORING NEUROLEPTICS

In a series of papers published between 1989 and 1991, Uematsu and Nakashima specially focused on the possibility of using hair scalp to monitor the dosage history of neuroleptics, by testing according to various methodologies the dose-related hair incorporation of haloperidol (HL),[40-42] then chlorpromazine.[43] These important experiments may be summarized as follows:

A retrospective study[40] was conducted in 40 psychiatric patients who had received fixed daily doses (3–10 mg/d) of HL for more than 4 months each. RIA-determined hair concentrations of HL were found to correlate both with the drug concentration in plasma at steady state ($r = 0.772$, $p < 10^{-3}$, $n = 39$), and with the daily dose ($r = 0.555$, $p < 10^{-3}$, $n = 40$).

In eight patients[41] to whom HL had been administered at fixed doses (2–10 mg/d) for more than 1 month, and in whom either therapy had just been discontinued or doses halved, a cross-sectional analysis of hair collected at the time of dose changing and at fixed times (1–3 months) thereafter revealed variations of hair concentrations that were time related (assuming a hair growth rate of 1.0–1.5 cm/month) to the changes in drug intake.

In 59 patients[42] treated with HL at fixed daily doses (3–10 mg/d) over at least 4 months prior to sampling, HPLC determinations of both HL and reduced metabolite (RHL) content in hair showed significant correlations with the daily dose of HL ($r = 0.682$, $p < 10^{-3}$, $n = 59$ for HL, and $r = 0.813$, $p < 10^{-3}$, $n = 59$ for RHL). A cross-sectional analysis, carried out in five patients whose HL dose regimen had been decreased or discontinued a few months before sampling, showed in all cases an abrupt drop of both HL and RHL hair levels at the time at which dosage changed.

Using an HPLC procedure with coulometric detection, chlorpromazine[43] was assayed in hair samples of 23 subjects who had been taking the drug in fixed daily doses. Chlorpromazine concentrations ranged from 1.6 to 27.5 ng/mg and was significantly correlated with the daily dose ($r = 0.788$, $p < 10^{-3}$). With the assumption of a hair growth rate of 1 cm per month, the individual history of chlopromazine doses in all patients could be deduced from the distribution of chlorpromazine along the hair shaft. The authors concluded that these results indicate that hair could serve as an indicator of individual exposure to neuroleptics and could yield retrospective information.

X. HAIR ANALYSIS FOR COMPLIANCE MONITORING

The determination of plasma or urine concentrations of the monitored drugs is commonly used for compliance monitoring; however, such analyses may reflect only the dosage regimen within a few days prior to obtaining the specimen. Hair analysis may represent an appropriate alternative by providing information on the quantity of drug administered over a longer timescale — weeks or months. The main requirement for such an application would be the existence of a strong enough dose-response relationship between the amount of drug taken and its hair levels. To date, the relatively few results that have been published show major controversy between pros and cons. Reported relationships between hair concentration and daily dose are presented in Table 4.

Two recent reviews have concluded in an opposite point of view. Uematsu[46] claimed that human scalp hair is a useful tissue for therapeutic drug monitoring because it retains information on the amount and duration of ingested drug along the hairshaft over a period of months to years.

At the opposite, Tracqui et al.[47] demonstrated that the idea of using hair analysis to ascertain whether or not a patient has taken his treatment exactly as prescribed appears inapplicable due to large interindividual variations of drug incorporation.

Although a number of studies dealing with various pharmaceuticals have undoubtedly ascertained the occurrence of these correlations, other experiments, however, failed to bring such evidence. In fact, these contradictory results are not surprising, considering the many intervening events that could screen, weaken, or

TABLE 4.

Reported Relationship in the Literature

Compound	Nb of subjects	Correlation coefficient	Ref.
Amitriptyline	30	0.563, $p < 0.002$	2
Phenobarbital	40	0.692, $p < 0.001$	4
Carbamazepine	30	No correlation	5
Carbamazepine	30	0.793, $p < 0.0001$	6
Nicotine	10	No correlation	17
Nicotine	36	0.685, $p < 0.001$	24
Methadone	5	0.63	28
Methadone	20	No correlation	29
Diazepam	8	No correlation	34
Fentanyl	13	No correlation	36
Haloperidol	40	0.555, $p < 0.001$	40
Haloperidol	59	0.682, $p < 0.001$	42
Chlorpromazine	23	0.788, $p < 0.001$	43
Morphine	13	No correlation	44
Ofloxacin	12	0.868, $p < 0.001$	45

eliminate those relationships. Among these factors, the methological biases (especially those related to the collection of data on drug intake), of course, stay at first place. In such respects, it is highly significant that the best correlations are to be observed in prospective studies with controlled drug administration and/or large subject populations; on the other hand, retrospective studies where the amount of drug taken is estimated on the basis of a self-questionnaire (even the subjects who are assumed to be of "good compliance") obviously do not represent an appropriate tool for demonstrating these relationships — in particular when the investigated compounds are under law prohibition measures.

The uncertainties regarding the stage of hair growth also represent a potential flawing factor. It is well known that each hair bulb exhibits its own growth cycle of 2 to 8 years or more, including 2 to 8 years of growing stage (anagen, the only one where drugs are supposed to undergo incorporation), some weeks of intermediate stage (catagen), and some months of resting stage (telogen). This phenomenon may contribute to an important intraindividual variability, especially when analyzing only one or a few strands of hair.

A number of other parameters may intervene more or less as biasing factors in the measurement of those dose-concentration relationships: the coloration of hair may have some effect, since it has been shown that the concentration of HL excreted in hair was significantly related to the amount of color pigment;[42] cosmetic hair treatments have been incriminated for possible removal of sequestered drug; finally, the nature and extent of drug metabolism may probably exert a certain influence which may be of prime importance for drugs exhibiting great interindividual variations in their metabolism.

In conclusion, when considering the hugeness of the above described interindividual variability of data, the announced future application of hair analysis to compliance monitoring becomes hardly conceivable. The results of these investigations might be somewhat helpful by bringing information about a *yes-or-no* taking of the prescribed medication within the period tested, since the reliability of such *qualitative* interpretations appears to be excellent. However, unless more is known about the factors that may influence the incorporation of drugs into hair and the manner to reduce the observed variability, the idea of using quantitative drug measurements in hair to determine the quantity of drug consumed will remain unapplicable.

REFERENCES

1. Sramek, J. J., Baumgartner, W. A., Tallos, J. A., Ahrens, T. N., Heiser, J. F., and Blahd, W. H., Hair analysis for detection of phencyclidine in newly admitted psychiatric patients, *Am. J. Psychiatry*, 142, 950, 1985.

2. Tracqui, A., Kressig, P., Kintz, P., Pouliquen, A., and Mangin, P., Determination of amitriptyline in the hair of psychiatric patients, *Hum. Exp. Toxicol.*, 11, 363, 1992.

3. Smith, F. P. and Pomposini, D. A., Detection of phenobarbital in bloodstains, semen, seminal stains, saliva, saliva stains, perspiration stains, and hair, *J. Forensic Sci.*, 26, 582, 1981.

4. Goulle, J. P., Noyon, J., Layet, A., Rapoport, N. F., Vaschalde, Y., Pignier, Y., Bouige, D., and Jouen, F., Phenobarbital in hair and drug monitoring, *Forensic Sci. Int.*, 70, 191, 1995.

5. Rothe, M., Pragst, F., Hunger, J., and Thor, S., Determination of carbamazepine and its metabolites in hair of epileptics, presented at the TIAFT-SOFT meeting, Tampa, November 1 to 4, 1994.

6. Kintz, P., Marescaux, C., and Mangin, P., Testing human hair for carbamazepine in epileptic patients: is hair investigation suitable for drug monitoring, *Hum. Exp. Toxicol.*, 14, 812, 1995.

7. Ostrea, E. M. and Chavez, C. S., Perinatal problems (excluding neonatal withdrawal) in maternal drug addiction: a study of 830 cases, *J. Pediatr.*, 94, 292, 1979.

8. Ripple, M. G., Goldberger, B. A., Caplan, Y. H., Blitzer, M. G., and Schwartz, S., Detection of cocaine and its metabolites in human amniotic fluid, *J. Anal. Toxicol.*, 16, 328, 1992.

9. Ostrea, E. M., Brady, M., Gause, S., Raymundo, A. L., and Stevens, M., Drug screening in newborns by meconium analysis: a large-scale, prospective, epidemiologic study, *J. Pediatrics*, 89, 107, 1992.

10. Parton, L., Warburton, D., Hill, V., and Baumgartner, W., Quantification of fetal cocaine exposure by radioimmunoassay of hair, *Pediatr. Res.*, 372A, 1987.

11. Graham, K., Koren, G., Klein, J., Schneiderman, J., and Greenwald, M., Determination of gestational cocaine exposure by hair analysis, *JAMA*, 262, 3328, 1989.

12. Ostrea, E. M. and Welch, R. A., Detection of prenatal drug exposure in the pregnant woman and her newborn infant, *Clin. Perinatol.*, 18, 629, 1991.

13. Uematsu, T., Yamada, K., Matsuno, H., and Nakashima, M., The measurement of haloperidol and reduced haloperidol in neonatal hair as an index of placental transfer of maternal haloperidol, *Ther. Drug Monit.*, 13, 183, 1991.

14. Forman, R., Schneiderman, J., Klein, J., Graham, K., Greenwald, M., and Koren, G., Accumulation of cocaine in maternal and fetal hair; the dose response curve, *Life Sci.*, 50, 1333, 1992.

15. Callahan, C. M., Grant, T. M., Phipps, P., Clark, G., Novack, A. H., Streissguth, A. P., and Raisys, V. A., Measurement of gestational cocaine exposure: sensitivity of infants' hair, meconium, and urine, *J. Pediatr.*, 120, 763, 1992.

16. Kintz, P., Kieffer, I., Messer, J., and Mangin, P., Nicotine analysis in neonates' hair for measuring gestational exposure to tobacco, *J. Forensic Sci.*, 38, 119, 1993.

17. Klein, J., Chitayat, D., and Koren, G., Hair analysis as a marker for fetal exposure to maternal smoking, *New Engl. J. Med.*, 328, 66, 1993.

18. Kintz, P. and Mangin, P., Determination of gestational opiate, nicotine, benzodiazepine, cocaine and amphetamine exposure by hair analysis, *J. Forensic Sci. Soc.*, 33, 139, 1993.

19. Kintz, P. and Mangin, P., Evidence of gestational heroin or nicotine exposure by analysis of fetal hair, *Forensic Sci. Int.*, 63, 99, 1993.

20. Koren, G., Measurement of drugs in neonatal hair; a window to fetal exposure, *Forensic Sci. Int.*, 70, 77, 1995.

21. Haley, N. J. and Hoffmann, D., Analysis for nicotine and cotinine in hair to determine cigarette smoker status, *Clin. Chem.*, 31, 1598, 1985.

22. Zahlsen, K. and Nilsen, O. G., Gas chromatographic analysis of nicotine in hair, *Environ. Technol.*, 11, 353, 1990.

23. Kintz, P., Ludes, B., and Mangin, P., Evaluation of nicotine and cotinine in human hair, *J. Forensic Sci.*, 37, 72, 1992.

24. Mizumo, A., Uematsu, T., Oshima, A., Nakamura, M., and Nakashima, M., Analysis of nicotine content of hair for assessing individual cigarette-smoking behavior, *Ther. Drug Monit.*, 15, 99, 1993.

25. Uematsu, T., Utilization of hair analysis for therapeutic drug monitoring with a special reference to ofloxacin and to nicotine, *Forensic Sci. Int.*, 63, 261, 1993.

26. Goulle, J. P., Noyon, J., and Leroux, P., Nicotine in hair and passive exposure in pediatric pathology, presented at the 2nd international meeting on clinical and forensic aspects of hair analysis, Genova, June 6 to 8, 1994.

27. Bewer, C., Hair analysis as a tool for monitoring and managing patients on methadone maintenance. A discussion, *Forensic Sci. Int.*, 63, 277, 1993.

28. Moeller, M. R., Fey, P., and Wennig, R., Simultaneous determination of drugs of abuse (opiates, cocaine and amphetamine) in human hair by GC/MS and its application to a methadone treatment program, *Forensic Sci. Int.*, 63, 185, 1993.

29. Goldberger, B. A., Darraj, A. G., Caplan, Y. H., and Cone, E. J., Detection of methadone, methadone metabolites, and other illicit drugs of abuse in hair of methadone treatment subjects, presented at the 2nd international meeting on clinical and forensic aspects of hair analysis, Genova, June 6 to 8, 1994.

30. Kintz, P., Cirimele, V., Edel, Y., Jamey, C., and Mangin, P., Hair analysis for buprenorphine and its dealkylated metabolite by RIA and confirmation by LC/ECD, *J. Forensic Sci.*, 39, 1497, 1994.

31. Sachs, H., Denk, R., and Raff, I., Determination of dihydrocodeine in hair of opiate addicts by GC/MS, *Int. J. Leg. Med.*, 105, 247, 1993.

32. Kintz, P., Jamey, C., and Mangin, P., Ethylmorphine concentrations in human samples in an over-dose case, *Arch. Toxicol.*, 68, 210, 1994.

33. Viala, A., Deturmeny, E., Aubert, C., Estadieu, M., Durand, A., Cano, J. P., and Delmont, J., Determination of chloroquine and monodesmethyl chloroquine in hair, *J. Forensic Sci.*, 28, 922, 1983.

34. Sramek, J. J., Baumgartner, W. A., Ahrens, T. N., Hill, V. A., and Cutter, N. R., Detection of benzodiazepines in human hair by radioimmunoassay, *Ann. Pharmacother.*, 26, 469, 1992.

35. Kintz, P. and Mangin, P., Hair analysis for detection of betablockers in hypertensive patients, *Eur. J. Clin. Pharmacol.*, 42, 351, 1992.

36. Wang, W. L., Cone, E. J., and Zacny, J., Immunoassay evidence for fentanyl in hair of surgery patients, *Forensic Sci. Int.*, 61, 65, 1993.

37. Kintz, P. and Mangin, P., Determination of meprobamate in human plasma, urine and hair by gas chromatography and electron impact mass spectrometry, *J. Anal. Toxicol.*, 17, 408, 1993.

38. Kintz, P., Amineptine abuse detected by hair analysis, *Toxichem + Krimtech*, 61, 70, 1994.

39. Kintz, P., Cirimele, V., Edel, Y., Tracqui, A., and Mangin, P., Characterization of dextromoramide (Palfium) abuse by hair analysis in a denied case, *Int. J. Leg. Med.*, 107, 269, 1995.

40. Uematsu, T., Sato, R., Suzuki, K., Yamaguchi, S., and Nakashima, M., Human scalp hair evidence of individual dosage history of haloperidol: method and retrospective study, *Eur. J. Clin. Pharmacol.*, 37, 239, 1989.

41. Sato, R., Uematsu, T., Yamaguchi, S., and Nakashima, M., Human scalp hair as evidence of individual dosage history of haloperidol: prospective study, *Ther. Drug Monit.*, 11, 686, 1989.

42. Matsuno, H., Uematsu, T., and Nakashima, M., The measurement of haloperidol and reduced haloperidol in hair as an index of dosage history, *Br. J. Clin. Pharmacol.*, 29, 187, 1990.

43. Sato, H., Uematsu, T., Yamada, K., and Nakashima, M., Chlopromazine in human scalp hair as an index of dosage history: comparison with simultaneously measured haloperidol, *Eur. J. Clin. Pharmacol.*, 44, 439, 1993.

44. Püschel, K., Thomasch, P., and Arnold, W., Opiate levels in hair, *Forensic Sci. Int.*, 21, 181, 1983.

45. Miyazawa, N., Uematsu, T., Mizuno, A., Nagashima, S., and Nakashima, M., Ofloxacin in human hair determined by high performance liquid chromatography, *Forensic Sci. Int.*, 51, 65, 1991.

46. Uematsu, T., Therapeutic drug monitoring in hair samples, *Clin. Pharmacokinet.*, 25, 83, 1993.

47. Tracqui, A., Kintz, P., and Mangin, P., Hair analysis: a worthless tool for therapeutic compliance monitoring, *Forensic Sci. Int.*, 70, 183, 1995.

Chapter 12

DRUG ANALYSES IN NONHEAD HAIR

Patrice Mangin

CONTENTS

I. INTRODUCTION

The incorporation of drugs into hair was primarily ascribed to the transportation by passive diffusion of the substances present in blood into the growing cells located at the base of the follicle of the hair where the drugs, as the cells die and fuse to form hair strands, become trapped and tightly bound in the keratin matrices during subsequent keratogenesis. According to this model, drug incorporation into hair is dependent on the drug concentration in blood which, in turn, is dependent on the dose of drug ingested.

However, in the light of multiple experimental findings based upon more accurate and precise techniques, this model seems to be oversimplified since it does not take into account the potential transfer of drugs from sweat, sebaceous and apocrine gland secretions, nor the external contamination even via deep compartments located in the skin surrounding the hair follicle.

Thus, in this more complex model, incorporation of drugs into hair appears to be additionally dependent on local factors which, in turn, are dependent on the anatomical regions involved. In this respect, nonhead hair investigations are of the greatest interest to point out these possible interferences due to anatomically related factors as well as to carry out experimentally controlled studies on drug incorporation into hair or to allow the drawing of some useful conclusions in terms of forensic practice.

II. HAIR SPECIFICITIES ACCORDING TO THE ANATOMICAL AREA

Most data on nonhead hair analysis have been collected from beard, axillary, and pubic hair samples. Therefore, the differences in the biology of each of these types of hair should be considered before interpreting the results.[1]

Scalp hair has the greatest variability of growth rates, but also the highest one. Growth rates range from 0.2 to 1.12 mm/d. The growth/rest cycle for scalp hair is short (30 months/3–2.5 months on average), especially in the vertex region where 85% of the follicles are in the anagen phase. Scalp hair is exposed to sebaceous and sweat secretions as well as contaminants in air, water, or dust and may be modified in its chemistry and physiology by cosmetic treatments. Finally, scalp hair is influenced in its major part by androgen impregnation (male sexual hair follicles).

Beard hairs have the slowest growth rate (approximately 0.27 mm/d) and a long growth/rest cycle duration of 14–22 months/9–12 months. Beard hairs are exposed to sweat and sebum secretions, the latter being, on the contrary to scalp hairs, excreted through a duct that opens directly onto the surface of the skin. Like scalp hair, beard hair may be subject to environmental contamination and/or cosmetic treatments. In addition, when beard hair samples are obtained by shaving, they are likely to be contaminated by pieces of epidermis. Like scalp hair, beard hair follicles are of the male sexual type.

Axillary and pubic hairs are quite similar in terms of growth rate (approximately 0.3 mm/d) and growth/rest cycle durations (11–18 months/12–17 months). They are both exposed to sweat and sebum secretions in addition to the secretions of the apocrine glands which are present only in the axilla and pubic area and discharge directly into the hair follicle rather than onto the surface of the surrounding skin. Axillary hair is less exposed to environmental contamination, but may be subjected to cosmetic treatments. Pubic hair may be contaminated by urine. Both types of hair are composed of ambosexual follicles.

Scalp hair is easier to collect than nonhead hair, but presents the disadvantage of its great variation in growth rates in various regions of the scalp, making the choice of the sampling area particularly important. Beard hair can be collected on a daily basis, allowing kinetic studies on drug incorporation. Axillary hair is not always available, and pubic hair may be difficult to collect for practical reasons.

III. SCALP HAIR VS. AXILLARY VS. PUBIC HAIR

Axillary and pubic hair are the nonhead hair specimens which have been the more investigated in comparison with scalp hair in drug analyses. Studies dealt mainly with opiates and subsidiarily with cocaine, cannabis, and other drugs.

A. Opiates

All the studies demonstrated clearly that opiates could be detected in extracts from both head and axillary or pubic samples of drug abusers (Table 1). Concentrations measured in the different kinds of hair are similar in magnitude whatever the opiate analyzed. With one exception (Balabanova and Wolf[2]), the highest concentrations were found in pubic hair, followed by scalp hair and axillary hair. This has been confirmed in our laboratory by a compilation of results obtained recently in 16 fatal heroin overdoses (Table 2). Besides, we found a very significant correlation between head hair and axillary or pubic hair concentrations for any of the opiates tested. On the other hand, even in patients receiving daily methadone treatment (Balabanova and Wolf[3]), no correlation has been reported between blood and head or nonhead opiate concentrations. This is not surprising since blood concentration represents a measure at the present moment and hair concentration represents a chronic accumulation.

TABLE 1.

Opiate Levels in Hair of Head, Axillary, and Pubic Regions

	Head		Axillae		Pubis			
	Range	Mean	Range	Mean	Range	Mean	No.	Ref.
Morphine (GC/MS)	0.62–27.10	6.71	0.40–24.20	4.73	0.80–41.34	10.47	20	4
Codeine (GC/MS)	0.15–1.87	0.76	0.12–1.56	0.53	0.22–2.34	0.95	20	4
6-Monoacetylmorphine (GC/MS)	5.31–36.85	18.1	4.71–31.07	13.9	7.98–80.51	33.30	20	4
Morphine (RIA)	0.23–5.30	1.3	0.18–1.60	0.63	0.18–31.70	5.95	14	5
Morphine (RIA)	0.9–8.2	3.12	2.7–23.7	8.92	2.0–11.7	5.25	4	2
Ethylmorphine (GC/MS)	—	0.12	—	0.08	—	0.18	1	6
Methadone (RIA)	0.5–2.7	—	1.3–8.0	—	1.0–4.0	—	11	3

Note: All values are in ng/mg; No. = number of cases.

TABLE 2.

Opiate Concentrations in Hair of the Head, Axillary, and Pubic Regions in 16 Fatal Heroin Overdoses

	6-Monoacetylmorphine		Morphine		Codeine	
	Range	Mean	Range	Mean	Range	Mean
Head	0.40–37.15	8.44	0.09–21.13	4.20	0.00–3.45	1.10
Axillae	0.00–31.07	6.79	0.00–14.56	3.22	0.00–2.88	0.71
Pubis	0.14–80.51	14.23	0.13–24.58	5.30	0.00–6.37	2.12

Note: All values are in ng/mg.

Like head hair, axillary or pubic hair can be easily tested with gas chromatography/mass spectrometry (GC/MS) for detection and quantification of the opiates after decontamination by washing and acidic hydrolysis according to previous reports.[4,6] In heroin fatalities confirmed by the presence of 6-acetylmorphine in urine, scalp hair as well as axillary and pubic hair investigations could demonstrate the simultaneous presence of morphine, codeine, and 6-acetylmorphine, allowing the medical examiner to consider the cause of death as a possible opiate overdose occurring in a chronic drug-abuse situation.

In some cases, when the codeine-to-morphine ratio is higher than one, codeine abuse shall be considered as highly probable in association with heroin addiction whether 6-acetylmorphine has been detected or not. Finally, in all cases of chronic heroin abuse, 6-acetylmorphine, despite its very short plasma life, predominated over morphine and codeine in head and nonhead hair samples as well. The 6-acetylmorphine concentrations were approximately two- to threefold greater than metabolite morphine whatever the anatomical origin of the hair tested.

B. Cocaine

To our knowledge, only two studies dealing with cocaine concentrations in head vs. axillary vs. pubic hair have been reported in the literature. Analyses in both studies were performed after washing decontamination by radioimmunoassay in hair samples of chronic addicts. In both studies (Table 3), data were reported in terms of cocaine equivalents, since antibodies reacted with cocaine and its metabolites. Cocaine was found in all types of hair tested, with higher concentration in pubic hair followed by scalp and axillary hair according to Offidani et al.[5] and in axillary, then pubic and head hair according to Balabanova and Wolf.[2] However, in the last study, owing to the small number of cases and the wide range of values, the difference between head and nonhead drug concentrations is obviously not statistically significant. In addition, in both studies, no correlation was found between hair and blood, urine, or bile drug concentrations.

TABLE 3.

Cocaine Concentrations (ng/mg) in Hair of the Head, Axillary, and Pubic Regions

	Head		Axillae		Pubis			
	Range	Mean	Range	Mean	Range	Mean	No.	Ref.
Cocaine (RIA)	1.7–70.0	17.6	1.1–60.0	11.7	1.1–166.7	20.8	13	5
Cocaine (RIA)	0.0–6.8	3.14	0.6–8.1	4.38	0.6–7.9	4.02	5	2

Note: No. = number of cases.

C. Cannabis

Although numerous extraction procedures have been published for testing cannabinoids in blood and urine, only some review papers have presented screening methods for testing human hair for cannabis. This is due to the very low reported concentrations of the cannabinoids. Therefore, studies concerning cannabinoid concentrations in nonhead hair are scarce (Table 4). The oldest one, conducted by Balabanova et al.[7] dealt with the determination by radioimmunoassay of total tetrahydrocannabinols and metabolites in pubic, axillary, and head hair of six hashish smokers. For the same reasons as those previously mentioned, interpretation of the data is limited by the lack of specificity of the analytical methods and the absence of significant difference between the concentrations measured in the different kinds of hair. Despite the small number of cases, a more recent study carried out by Cirimele et al.[8] offers the greatest interest in measuring separately with GC/MS the parent drug Δ9-tetrahydrocannabinol (THC) and one of its metabolites 11-nor-Δ9-THC carboxylic acid (THC-COOH). In seven subjects exhibiting positive head hair samples for THC and THC-COOH, it was possible to detect both cannabinoids in all pubic hair except one for THC-COOH, probably because of the very low concentrations usually observed for this metabolite. The highest cannabinoid concentrations were found in pubic hair with the same predominance of the parent drug (THC) over the metabolite (THC-COOH) already mentioned for opiates in hair of heroin abusers.

TABLE 4.

Cannabinoids Concentrations (ng/mg) in Hair of the Head, Axillary, and Pubic Regions

	Head		Axillae		Pubis			
	Range	Mean	Range	Mean	Range	Mean	No.	Ref.
THC (GC/MS)	0.28–2.17	0.84	—	—	0.34–3.91	1.35	7	8
THC-COOH (GC/MS)	0.00–0.18	0.06	—	—	0.07–0.83	0.28	7	8
Cannabinoids (RIA)	0.0–3.1	1.93	0.4–1.9	1.05	0.5–3.8	1.75	6	7

Note: No. = number of cases.

D. Miscellaneous

Other drugs, including nicotine, have been investigated in nonhead hair samples. However, the data (Table 5) are of less significance since all the analyses but one lack specificity due to the analytical methods used. The highest concentrations were observed sometimes in axillary hair (phenobarbital, zipeprol, nicotine), sometimes in pubic hair for benzodiazepines. No correlation between concentrations in hair samples and blood or urine have been reported. Nevertheless, these data have the merit to demonstrate that nonhead hair specimens could represent an alternative to head hair when the latter is not available for the detection of drugs.

TABLE 5.

Miscellaneous Drugs in Hair of the Head, Axillary, and Pubic Regions (Literature Data)

	Head	Axillae	Pubis	No.	Ref.
Zipeprol (GC/MS)	7.34	16.73	13.86	1	20
Phenobarbital (RIA)	2.37±2.7	4.92 ± 3.58	3.68 ± 2.47	4	2
Benzodiazepines (RIA)	0.5–1.13	0.8–1.9	0.36–3.00	9	5
Nicotine (RIA)	4.3 ± 3.5	5.3 ± 5.4	5.0 ± 5.2	19	9,10

Note: All values are in ng/mg; No. = number of cases.

IV. HEAD HAIR VS. ARM HAIR

There is a paucity of literature on the comparison of concentrations of drug in head hair vs. arm hair. Cone et al.[11] reported a study dealing with 20 paired head- and arm-hair samples collected from known heroin or cocaine abusers in which cocaine, heroin, and metabolites were analyzed simultaneously with GC/MS. Cocaine and 6-acetylmorphine were the major analytes present in both head (0.4–76 ng/mg and 0.0–0.8 ng/mg) and arm (0.0–109 ng/mg and 0.0–3.1 ng/mg) hair samples. Cocaine and benzoylecgonine concentrations tended to be higher in arm hair than in head hair samples, although the differences were not significant.

Benzoylecgonine concentrations in both arm and head hair were lower than cocaine concentrations in all cases. Heroin was detected in only two head hair samples, while 6-acetylmorphine was detected in extracts of 14 hair samples and in 6 arm hair samples. If present, 6-acetylmorphine concentrations also tended to be higher in arm hair than in head hair with no significant difference. Morphine was detected only in extracts of 3 head hair samples at a concentration lower than the respective 6-acetylmorphine one.

V. BEARD HAIR

Since it can easily be collected on a daily basis and despite the fact that it grows at a lower rate than scalp hair, beard hair has been proposed as a suitable alternate to head hair in order to determine the time course of drug appearance and disappearance in hair after administration and the presence of a dose-response relationship between dose and hair drug concentration. In this respect the literature data are summarized in Table 6. The interpretation of the results is not easy, owing to individual differences in growth rates and the lack of standardized procedures and a common detection limit or cutoff level. According to the literature data, the time lag between the administration of the drug and its appearance in beard hairs varies from 1 to 7 d depending on the drug and the analytical methods used. Considering the data from only controlled studies, where a drug is taken for the first time or after a free-drug window of sufficient duration, it could be stated that the time lag between the administration of the drug and its detection with GC/MS in beard hairs is on average 5–7 d. This delayed time could be interpreted as the required growth time for the hair shaft to emerge from the bulb area in the follicle to a height above the skin surface sufficient for razor collection. In the same way after simple dose administration, the time lag for disappearance of the drug from the beard hairs appears to be, on average, 2 weeks after having peaked between day 7 and 12 after intake.

According to Nakahara et al.,[17] after repeated doses for successive 7 d of methoxyphenamine (bronchodilatator), the duration of detection after the last dosage was 2–3 d longer than after a single dose, which could mean that long-term use would result in some accumulation of drug in the body longer than the simple use with a subsequent delayed release and incorporation into hair. On the other hand, Kintz et al.[16] clearly demonstrated that the pattern of meprobamate presence in beard hair was quite similar whatever the dosage for a period of 8–10 d after administration.

As for the relationship between dose and concentration, the same controlled studies, where drug is administered under close supervision, provide results which are strongly suggestive that such a dose-response relationship exists between drug levels in beard hairs and the administered dose. This is not the case in chronic abusers or treated patients, where daily doses vary significantly from day to day so that the establishment of a dose-response relationship would require a large amount of data to attenuate the effects of any bias due to individual differences.

VI. CONCLUSIONS

Although the limited number of studies dealing with drug analyses in nonhead hair preclude generalization, the data collected with the reservation of further confirmation allows the following conclusions.

Drugs can be detected in extracts from hair of any part of the body, i.e., scalp, beard, arm, axillae, and pubis. The overall pattern of drug and metabolites is similar

TABLE 6.

Drugs in Beard Hair

	Mode of administration	Time lag (min) Appearance	Peak	Disappearance	Dose-conc. relationship	No.	Ref.
Codeine	Single oral dose	1	3	6–8	n.i.	2	12
PCP	Chronic abuser	n.i.	n.i.	n.i.	n.i.	1	13
Morphine, codeine	2 single doses i.m. at 1-week interval	7–8	9–12	—	Yes	2	14
Pholcodine	3 oral doses on 1 d	5	n.i.	20	n.i.	1	15
Meprobamate	Single oral dose	4–5	7–9	12–14	Yes	12	16
Methoxyphenamine and metabolites	Single oral dose	1	3	10–12	Yes	6	17
Methoxyphenamine and metabolites	Single dose daily for 1 week	1	7–8	16–18	Yes	3	17
Methamphetamine	Chronic abusers	n.i.	2–3	15	n.i.	17	17
Amphetamine	Chronic abusers	n.i.	2–3 (increase beyond day 7)	16	n.i.	n.i.	17

Note: No. = number of cases; n.i. = not indicated.

in head and nonhead hair. Besides, whatever the kind of hair tested, the concentration of the parent drug is always higher than that of its primary metabolite.

Length for length, drug and metabolite concentrations vary according to the anatomical origin of the hair tested, but are significantly correlated with each other. With some exceptions, the highest drug and metabolite concentrations are observed in pubic hair, while the lowest are found in axillary hair.

The difference in concentration of drug and metabolite between head and nonhead hair remains to be explained. This is likely due to the differences in growth rate and the lack of equivalent matching of growth periods between hair from one to another area. However, the constant high parent drug to metabolite ratio in hair, whatever its origin, suggests also that incorporation of drugs and metabolites into hair may be connected more or less with sweat, sebum, and apocrine gland secretions, depending on the anatomical area involved. In the same way, the very high THC to THC-COOH ratio found in pubic hair is not consistent with the hypothesis of a persistent urine contamination to explain the highest drug levels measured in that hair compared with others, since THC is almost entirely metabolized by liver enzymes with less than 1% excreted unchanged in urine.[18] On the other hand, possible absorption of drugs secreted in sweat, sebum, or apocrine fluid onto clothing[19] could interfere mainly in axillae to explain the lowest concentrations observed in axillary hair.

While no clear correlations between chronic drug intake and drug concentration in hair have been demonstrated, results provided by controlled studies on drug incorporation into beard hair from samples collected on a daily basis suggest the existence of a dose-concentration relationship. In the same studies, when a drug is taken for the first time, the time lag between administration and detection of the drug in the beard hair varies from 5 to 8 d depending on the drug, the detection limit of the analytical method, and individual factors.

Finally, due to the highly significant correlation between scalp, axillary, and pubic hair concentrations, nonhead hair may be a useful alternative in forensic investigations when head hair is not available.

REFERENCES

1. M.R. Harkey. Anatomy and physiology of hair. *Forensic Sci. Int.* 63, 9–18, 1993.
2. S. Balabanova and H.U. Wolf. Bestimmungen von Cocain, Morphin, Phenobarbital und Methadon im Kopf-, Achsel- und Schamhaar. *Lab. Med.* 13, 46–47, 1989.
3. S. Balabanova and H.U. Wolf. Methadone concentrations in human hair of the head, axillary and pubic hair. *Z. Rechtsmed.* 102, 293–296, 1989.
4. P. Mangin and P. Kintz. Variability of opiates concentrations in human hair according to their anatomical origin: head, axillary and pubic regions. *Forensic Sci. Int.* 63, 77–83, 1993.
5. C. Offidani, S. Strano Rossi and M. Chiarotti. Drug distribution in the head, axillary and pubic hair of chronic addicts. *Forensic Sci. Int.* 63, 105–108, 1993.
6. P. Kintz, C. Jamey and P. Mangin. Ethylmorphine concentrations in human samples in an overdose case. *Arch. Toxicol.* 68, 210–211, 1994.
7. S. Balabanova, P.J. Arnold, V. Luckow, H. Brunner and H.U. Wolf. Tetrahydrocannabinole im Haar von Haschischrauchern. *Z. Rechtsmed.* 102, 503–508, 1989.
8. V. Cirimele, P. Kintz and P. Mangin. Testing human hair for cannabis. *Forensic Sci. Int.*, 70, 175-182, 1995.
9. S. Balabanova and E. Schneider. Nachweis von Nikotin im Kopf-, Achsel- und Schamhaar. In *Medizinrecht-Psychologie-Rechtsmedizin*, H. Schutz, H.J. Kaatsch und H. Thomsen, Eds., Springer-Verlag, Berlin, 1991, 334–337.

10. S. Balabanova, E. Schneider and G. Buhler. Nachweis von Nikotin im Haaren. *Dtsch. Apoth. Ztg.* 40, 2200–2201, 1990.
11. E.J. Cone, W.D. Darwin and W.L. Wang. The occurrence of cocaine, heroin and metabolites in hair of drug abusers. *Forensic Sci. Int.* 63, 55–68, 1993.
12. K. Puschel, P. Thomasch and W. Arnold. Opiate levels in hair. *Forensic Sci. Int.* 21, 181–186, 1983.
13. J.J. Sramek, W.A. Baumgartner, J.A. Tallos, T.N. Ahrens, J.F. Heiser and W.H. Blahd. Hair analysis for detection of phencyclidine in newly admitted psychiatric patients. *Am. J. Psychiatry* 142, 950–953, 1985.
14. E.J. Cone. Testing human hair for drugs of abuse. I. Individual dose and time profiles of morphine and codeine in plasma, saliva, urine, and beard compared to drug-induced effects on pupils and behavior. *J. Anal. Toxicol.* 14, 1–7, 1990.
15. H.H. Maurer and C.F. Fritz. Toxicological detection of pholcodine and its metabolites in urine and hair using radio immunoassay, fluorescence polarisation immunoassay, enzyme immunoassay and gas chromatography-mass spectrometry. *Int. J. Leg. Med.* 104, 43–46, 1990.
16. P. Kintz, A. Tracqui and P. Mangin. Pharmacological studies on meprobamate incorporation in human beard hair. *Int. J. Leg. Med.* 105, 283–287, 1993.
17. Y. Nakahara, K. Takahashi and K. Konuma. Hair analysis for drugs of abuse. VI. The excretion of methoxyphenamine and methamphetamine into beards of human subjects. *Forensic Sci. Int.* 63, 109–119, 1993.
18. L. Lemberger, N.R. Tamarkin and J. Axelrod. Δ9-Tetrahydrocannabinol: metabolism and disposition in long term marijuana smokers. *Science* 178, 72–74, 1971.
19. A. Tracqui, P. Kintz, B. Ludes, C. Jamey and P. Mangin. The detection of opiate drugs in non-traditional specimens (clothing). Report of ten cases. *J. Forensic Sci.* 40, 263-265, 1995.
20. P. Kintz, V. Cirimele, A. Tracqui and P. Mangin. Fatal zipeprol intoxication. *J. Leg. Med.,* 107, 267–268, 1995.

INDEX

289